NATURAL PHILOSOPHY

Praise for *Natural Philosophy*

"With the appearance of *Natural Philosophy*, Paul Thagard, one of the foremost proponents of philosophical naturalism in our time, establishes how the social, cognitive, and brain sciences, and Chris Eliasmith's Semantic Pointer Architecture, in particular, provide resources for a rigorous, scientifically-informed, and *systematic* approach to the entire range of classical philosophical problems. Thagard's *Natural Philosophy* is not a program of reduction but rather one of integration, which examines what are, in a scientific age, the inevitable interconnections and interdependence of these sciences and the perennial projects of philosophy—including metaphysics and mind, epistemology and ethics, and political philosophy and the philosophy of art. With the characteristic clarity, economy, and insight that have distinguished all of his work for more than four decades, Thagard demonstrates the strengths of a naturalistic philosophical program that attends to the relevant sciences, compared to its classical and contemporary competitors."

Robert N. McCauley, William Rand Kenan Jr. University Professor of Philosophy at the Center for Mind, Brain, and Culture, Emory University and author of *Why Religion Is Natural and Science Is Not* (OUP)

"Drawing on the many original positions he has developed throughout his distinguished career in philosophy and cognitive science, Paul Thagard provides a synoptic overview of natural philosophy in his flowing, easy to read style. He makes use of the now widely accepted view, that he helped to develop, of interactions between mechanisms at multiple levels—the molecular, neuronal, mental, and social. The work admirably shows that philosophy can be, as he puts it 'extraverted, directing its attention to real world problems.'"

Lindley Darden, Professor of Philosophy, University of Maryland, College Park

OXFORD SERIES ON COGNITIVE MODELS AND ARCHITECTURES

Series Editor
Frank E. Ritter

Series Board
Rich Carlson
Gary Cottrell
Robert L. Goldstone
Eva Hudlicka
William G. Kennedy
Pat Langley
Robert St. Amant

Integrated Models of Cognitive Systems
Edited by Wayne D. Gray

In Order to Learn: How the Sequence of Topics Influences Learning
Edited by Frank E. Ritter, Joseph Nerb, Erno Lehtinen, and Timothy O'Shea

How Can the Human Mind Occur in the Physical Universe?
By John R. Anderson

Principles of Synthetic Intelligence PSI: An Architecture of Motivated Cognition
By Joscha Bach

The Multitasking Mind
By David D. Salvucci and Niels A. Taatgen

How to Build a Brain: A Neural Architecture for Biological Cognition
By Chris Eliasmith

Minding Norms: Mechanisms and Dynamics of Social Order in Agent Societies
Edited by Rosaria Conte, Giulia Andrighetto, and Marco Campennì

Social Emotions in Nature and Artifact
Edited by Jonathan Gratch and Stacy Marsella

Anatomy of the Mind: Exploring Psychological Mechanisms and Processes with the Clarion Cognitive Architecture
By Ron Sun

Exploring Robotic Minds: Actions, Symbols, and Consciousness as Self-Organizing Dynamic Phenomena
By Jun Tani

Brain-Mind: From Neurons to Consciousness and Creativity
By Paul Thagard

Mind-Society: From Brains to Social Sciences and Professions
By Paul Thagard

Natural Philosophy: From Social Brains to Knowledge, Reality, Morality, and Beauty
By Paul Thagard

Natural Philosophy

FROM SOCIAL BRAINS TO KNOWLEDGE, REALITY, MORALITY, AND BEAUTY

Paul Thagard

OXFORD
UNIVERSITY PRESS

Oxford University Press is a department of the University of Oxford. It furthers
the University's objective of excellence in research, scholarship, and education
by publishing worldwide. Oxford is a registered trade mark of Oxford University
Press in the UK and certain other countries.

Published in the United States of America by Oxford University Press
198 Madison Avenue, New York, NY 10016, United States of America.

© Oxford University Press 2019

First issued as an Oxford University Press paperback, 2021

All rights reserved. No part of this publication may be reproduced, stored in
a retrieval system, or transmitted, in any form or by any means, without the
prior permission in writing of Oxford University Press, or as expressly permitted
by law, by license, or under terms agreed with the appropriate reproduction
rights organization. Inquiries concerning reproduction outside the scope of the
above should be sent to the Rights Department, Oxford University Press, at the
address above.

You must not circulate this work in any other form
and you must impose this same condition on any acquirer.

Library of Congress Cataloging-in-Publication Data
Names: Thagard, Paul, author.
Title: Natural philosophy : from social brains to knowledge, reality,
morality, and beauty / Paul Thagard.
Description: New York : Oxford University Press, 2019. |
Includes bibliographical references and index. Contents: —Volume 3. Natural
philosophy: from social brains to knowledge, reality, morality, and beauty.
Identifiers: LCCN 2018033277 | ISBN 9780190678739 (hardback) |
ISBN 9780197619681 (paperback)
Subjects: LCSH: Philosophy of mind. | Cognitive science.
Classification: LCC BD418.3.T39 2019 | DDC 128—dc23
LC record available at https://lccn.loc.gov/2018033277

9 8 7 6 5 4 3 2 1

Paperback printed by Marquis, Canada

To Laurette, naturally philosophical.

Contents

List of Illustrations xi
Foreword xiii
Preface xvii
Acknowledgments xix

1. *Philosophy Matters* 1
    Why Philosophy? 1
    What Is Philosophy? 3
    Issues and Alternatives: Ways of Philosophizing 5
    Elements of Natural Philosophy 12
    Overview of This Book 19
    Summary and Discussion 21
    Notes 23
    Project 24

2. *Mind* 25
    Mental Processes 25
    Issues and Alternatives 27
    Neural Mechanisms 29
    Semantic Pointers 31
    Inference to the Best Explanation to Multilevel Materialism 40
    Philosophical Objections 42
    Consciousness 47
    Summary and Discussion 54
    Notes 56
    Project 57

3. *Knowledge* 58
- Minds and Knowledge 58
- Issues and Alternatives 59
- What Is Knowledge? 61
- The Growth of Knowledge 65
- Justification 70
- Probability 78
- Knowledge Is Social 82
- Conceptual Change and the Brain Revolution 85
- Summary and Discussion 88
- Notes 89
- Project 91

4. *Reality* 92
- Make Reality Great Again 92
- Issues and Alternatives 93
- Existence 94
- Truth 104
- Space and Time 106
- Groups and Society 110
- Summary and Discussion 114
- Notes 116
- Project 117

5. *Explanation* 118
- Knowledge Meets Reality 118
- Issues and Alternatives 119
- Styles of Explanation 120
- Emotional and Social Aspects of Explanation 130
- Causality 132
- Reduction and Emergence 140
- Summary and Discussion 144
- Notes 145
- Project 146

6. *Morality* 147
- Right and Wrong 147
- Issues and Alternatives 148
- Values 150

Moral Emotions 154
Objective Values and Rational Emotions 156
Needs 157
The Needs of Others 163
Empathy 168
Conflicting Needs and Ethical Coherence 171
Why Is There Evil? 174
Summary and Discussion 176
Notes 179
Project 181

7. *Justice* 182

From Morality to Justice 182
Issues and Alternatives 184
Just Societies: Needs Sufficiency 186
Just Governments 192
Just Social Change 197
Basic Income 200
Summary and Discussion 202
Notes 205
Project 206

8. *Meaning* 207

Life and Language 207
Issues and Alternatives 209
Language and Mental Representation 210
The Meanings of Life 213
The Meaning of Death 222
Summary and Discussion 224
Notes 226
Project 227

9. *Beauty and Beyond* 228

Aesthetics 228
Issues and Alternatives 230
Beauty in Painting 233
Other Emotions in Painting 238
Creativity in Painting 242
Beauty in Music 245

Other Emotions in Music 247
   Creativity in Music 252
   Empathy in Literature and Film 253
   Summary and Discussion 254
   Notes 258
   Project 260

10. *Future Philosophy* 261
    Looking Backwards and Forwards 261
    Free Will 264
    Mathematical Knowledge and Reality 272
    Nonhumans: Animals and Machines 286
    Summary and Discussion 289
    Conclusion: 12 Rules for Philosophical Life 291
    Notes 292
    Project 293

REFERENCES 295
NAME INDEX 311
SUBJECT INDEX 319

List of Illustrations

1.1 How descriptive information can be relevant to normative conclusions 14
1.2 A vacation decision as a coherence problem 16
3.1 An explanation choice as a coherence problem 74
6.1 Value map for Donald Trump supporters 153
6.2 Value map for Hillary Clinton supporters 154
6.3 Maslow's hierarchy of needs as a pyramid 160
10.1 Natural philosophy as a coherent system 262
10.2 Natural philosophy and alternatives 263
10.3 Visual representation of a right triangle 279

Foreword

Frank E. Ritter

THREE DECADES AGO, Newell, Anderson, and Simon shared a desire for a unified theory of how cognition arises and what a mechanistic explanation would look like. Today, much still remains to be done to pursue that desire.

Allen Newell talked about narrow and deep theories, and broad and shallow theories, and that theories could differ in these ways. Many psychology theories are deep, explaining a few phenomena in great detail, but not explaining many phenomena nor how they interact and mutually constrain each other.

In this trio, Paul Thagard creates a much broader and accessible explanation than we have seen before of what a mechanistic explanation of mind and human behavior would look like. These books explain the cognitive science approach to cognition, learning, thinking, emotion, and social interaction—nearly all of what it means to be human—and what this means for a wide variety of sciences and philosophy. These books provide a good overview of cognitive science and its implications. Different readers will be drawn to the treatise in different ways. It does not matter where they start.

In his book, *Mind–Brain*, Thagard explains how the semantic pointer architecture (SPA) by Chris Eliasmith can be used to explain the mind, cognition, and related concepts. The SPA architecture is a very useful dynamic theory that can do multiple tasks in the same model, and it is explained in journal articles and by Eliasmith's (2013) book in the Oxford Series on Cognitive Models and

Architectures. Most of the implications based on SPA are also supported by and have lessons for other computational models of cognition, so these books can be useful to users of other cognitive architectures.

In his book, *Mind–Society*, Thagard examines what this approach means for social science and related professional fields and how the mechanisms account for successes and failures of major professional activities. This book provides a very broad, singular framework for explaining the breadth of human behavior.

In this book, *Natural Philosophy*, Thagard examines what this approach means for philosophy, including important topics of philosophy of mind and of beauty. It provides a useful and engaging overview of philosophy, particularly for those interested in cognitive science or working in cognitive science. In this book, he connects philosophy with current theories in psychology and neuroscience and with the best current computational theories of mind based on cognitive architectures, using these computational theories as the core of its own theory.

Is this three-book treatise useful? It is very much so. It starts to address some problems that I have seen in various fields by using multilevel analyses, with a cognitive architecture at its middle level.

This treatise also provides a detailed theoretical explanation of how the SPA provides explanations naturally for many phenomena directly and that many similar cognitive architectures may also provide. While this treatise does not note the linkages for other cognitive architectures, many architectures can be seen to provide most (but not all) of the support for this framework to explain how minds work in society.

These books introduce several useful theories and methods about how to do science as well. Beyond allowing and using explanations via multilevel mechanisms, particularly valuable are Thagard's introduction and use of three-analysis for definitions and coherence. The three-analysis definitions are a way to explain concepts without using simple definitions. They define a concept using *exemplars*, *typical features*, and *explanations*. This approach resolves several problems with simple dictionary definitions.

The books also champion coherence as a useful concept for reasoning. Coherence is used in this book as a way to describe the quality of theories—that theories are not just good when they predict a single result but also how they cohere with multiple sources of data and with other theories. Coherence is hard to quantify itself, but it is clearly a useful concept. But the use of coherence is not just normative—we should use it—it is also descriptive in that scientists and laypersons appear to already use it, at least implicitly. Making the use of coherence explicit will help us to apply, teach, and improve the process.

I particularly enjoyed the liberal education available through carefully reading Chapter 9 on beauty, checking the visual and audio references from online sources, and then reflecting on how the SPA cognitive architecture explains the different experiences.

These books will be useful to cognitive scientists and those interested in cognitive science. They will also be useful to those who simply want to learn more about the world and cognition. They offer one of the best and clearest explanations we have for cognition. Thus, it will be useful for humanists and social scientists interested in knowing how cognitive science works. There are, sprinkled throughout, pieces of liberal education because this book draws examples and support from a wide range of material.

These books contain powerful ideas by one of the most highly cited living philosophers. They can change the way you think about the world, including brains and mind, and how you might think that the mind works and interacts with the world. Thagard calls these trio of books a treatise, and I found them so compelling that I've decided to use them in a course this next semester.

REFERENCE

Eliasmith, C. (2013). *How to build a brain: A neural architecture for biological cognition.* New York, NY: Oxford University Press.

Preface

THIS BOOK IS part of a trio (Treatise on Mind and Society) that can be read independently:

Brain–Mind: From Neurons to Consciousness and Creativity
Mind–Society: From Brains to Social Sciences and Professions
Natural Philosophy: From Social Brains to Knowledge, Reality, Morality, and Beauty

Brain–Mind shows the relevance of Chris Eliasmith's Semantic Pointer Architecture to explaining a full range of mental phenomena concerning perception, imagery, concepts, rules, analogies, emotions, consciousness, intention, action, language, creativity, and the self. *Mind–Society* systematically connects neural and psychological explanations of mind with social phenomena, covering major social sciences (social psychology, sociology, politics, economics, anthropology, and history) and professions (medicine, law, education, engineering, and business).

This book, *Natural Philosophy*, extends the integrated mental–social approach to apply to the humanities, primarily philosophy but also the arts, particularly painting and music. A unified account can be given of all branches of philosophy—epistemology, metaphysics, ethics, and aesthetics—by applying the intellectual tools developed in *Brain–Mind* and *Mind–Society*. *Natural Philosophy* can be read independently of the other two books because chapters 1 and 2 provide succinct summaries of the key ideas about mental, neural, and social processes. My goal is to harmonize the cognitive sciences, social sciences, professions, and humanities as a coherent system, not to reduce one to the other.

The development of science-connected philosophy requires new methods that have broader application. The method of three-analysis characterizes concepts by identifying examples, features, and explanations rather than definitions. Value maps (cognitive-affective maps) provide a concise depiction of the emotional coherence of concepts. An exact account of explanatory coherence applies to many problems concerning knowledge and reality. Overall, such methods provide a philosophical procedure more powerful than traditional techniques of thought experiments, introspection, and pure reason.

Most of my papers can be found via paulthagard.com, which also contains live links for the Web addresses in this book.

Acknowledgments

MOST OF THIS book was newly written in 2016–2018, but I have incorporated some extracts from other works, as indicated in the notes. I have also used excerpts from my *Psychology Today* blog, *Hot Thought*, for which I hold the copyright.

I am grateful to Laurette Larocque for insightful suggestions, and to Frank Ritter and anonymous reviewers for useful comments. I thank Joan Bossert for editorial advice, Phil Velinov and Shanmuga Priya for organizing production, Alisa Larson for skilled copyediting, and Kevin Broccoli for professional indexing. CBC Radio 2 and Apple Music provided the accompaniment.

1

Philosophy Matters

WHY PHILOSOPHY?

Philosophy is the attempt to answer general questions about the nature of knowledge, reality, and values. Why should you care about philosophy? If you are interested in your health, your personal relationships, and your career, then you should be concerned about how to make these aspects of your life work better for you, which requires understanding of life's values. For a deeper understanding, you should be concerned not only with *what* is known about life, but also with *how* it is known, which raises fundamental questions about the nature of knowledge.

Questions about knowledge blend swiftly into philosophical questions about reality, for example concerning the existence of God and an afterlife. Questions about values are also inescapable if you care about how to be a good person and how to help others be good. If your interests extend to what kind of society people should live in, then you are already enmeshed in philosophical theories about the nature of justice. If you have ever worried about what makes life worth living, then you have run into philosophical questions about the meaning of life. Finally, if you are involved with artistic pursuits such as literature, painting, and music and have wondered about what makes them valuable, then you are engaged with philosophical questions about beauty and other values.

Yet according to the famous physicist Stephen Hawking, philosophy is dead, because it has been superseded by science. This judgment is naïve and premature, because philosophy deals with problems concerning the nature of knowledge, reality, and morality that are much more general than those pursued in particular sciences. Philosophy only sometimes smells dead because of the putrid obscurity and verbal triviality of some contemporary approaches. Philosophy has much to gain from sciences such as physics and psychology but also has much to contribute toward a deeper understanding of how knowledge develops, how reality can be grasped, and how judgments about right and wrong can be justified.

Philosophy has been dismissed as providing unintelligible answers to insoluble problems, but this charge is doubly wrong. First, the writings of many philosophers are highly comprehensible, for example Plato, Descartes, John Stuart Mill, Bertrand Russell, and Daniel Dennett. Second, progress can be made on solving philosophical problems when appropriate methods are applied. Whereas standard philosophical tools such as introspection and pure reason are of limited effectiveness, this book shows that progress can be made in answering classical philosophical questions using ideas and methods drawn from the cognitive and social sciences.

The following are some central questions to which I propose novel answers:

1. What is philosophy?
2. What is knowledge and how can people acquire it?
3. What is reality and how can we explain its operations?
4. What is the difference between right actions and wrong actions?
5. What is the moral basis of a just society?
6. What is the meaning of life?
7. What makes some works of art greater than others?

Answering these questions covers all the major areas of philosophy, including epistemology (about knowledge), metaphysics (reality), ethics (morality), and aesthetics (art).

I call my approach "natural philosophy" because it draws heavily on the sciences and finds no room for supernatural entities such as souls, gods, and possible worlds. My approach differs from other naturalistic work in philosophy because I employ resources developed extensively in the companion books, *Brain–Mind* and *Mind–Society*. *Brain–Mind* shows how new ideas in theoretical neuroscience provide comprehensive explanations of a wide range of psychological phenomena, from perception to the self. *Mind–Society* applies these ideas to important questions across six social sciences and five professions.

Natural Philosophy does not attempt the impossible task of reducing philosophy and the humanities to the cognitive and social sciences but shows their interconnections and interdependence. Philosophy differs from the sciences in two main respects: generality and normativity. The questions that philosophy asks are much more general than those posed by scientists, for example concerning knowledge as a whole rather than just knowledge about some particular aspect of nature such as gravitational force. Moreover, philosophy is more normative than the sciences, concerned with how the world ought to be rather than just with how it is.

Brain–Mind argues that Chris Eliasmith's new Semantic Pointer Architecture for understanding how the brain works serves to explain the full range of human thinking, including consciousness and creativity. *Mind–Society* extends this account of individual minds into a general theory of communication of cognitions and emotions that explains human interactions and social change. These new theories of brains and social interaction are highly relevant to philosophy. An empirically supported theory of mind is invaluable for figuring out how knowledge grows in individuals and what aspects of reality they are capable of apprehending. Minds do not achieve knowledge on their own, so a robust theory of social interaction is also required for appreciating how groups can acquire more knowledge than isolated individuals.

Emotions are often derided as impediments to rational thought, but contemporary neuroscience and psychology appreciate that successful human cognition employs emotions to provide focus, evaluation, and motivation. Emotions also connect thinking with important philosophical questions concerning morality, meaning, and beauty. I will show how natural philosophy can be developed by building on theories of brains, cognition, emotion, and interactive communication.

WHAT IS PHILOSOPHY?

A naïve view of concepts would suggest that a discussion of philosophy should begin with a definition that provides necessary and sufficient conditions for something being philosophical. Candidates might include the Greek etymology in which philosophy is the love of wisdom, or the simple characterization I gave earlier of philosophy as the attempt to answer fundamental questions about knowledge, reality, and values.

Fortunately, the much more comprehensive theory of concepts explained in chapter 2 suggests a richer way of characterizing concepts that I call "three-analysis." This method characterizes a concept by specifying exemplars, typical features, and explanatory roles. Table 1.1 presents a three-analysis of the concept of philosophy. Exemplars are standard examples, which here include recognized philosophers such as Plato, respected books such as Descartes' *Meditations*, and questions such as what is knowledge. The typical features that

TABLE 1.1

Three-Analysis of the Concept *Philosophy*

Exemplars	Philosophers such as Plato, Aristotle, Descartes, Hume, Kant. Books such as Plato's *Republic* and Descartes' *Meditations*. Questions such as what is knowledge, what is reality, and what is ethical
Typical features	Questions that are highly general and abstract
	Concern with normative questions about what ought to be
	High amounts of disagreement
Explanations	Explains: why some questions are contentious and difficult to answer, yet persist for millennia
	Explained by: the human need to understand life and the universe more deeply than science can do alone

set philosophy off from science and other enterprises are generality, abstractness, normativity, and lack of agreement. The concept of philosophy contributes to explanations of why some questions are so difficult but perennial. The persistence of philosophy is explained by people's needs for deep explanations of values as well as facts.

According to Wilfred Sellars and Daniel Dennett, philosophy aims to understand how things hang together in the broadest sense. But they never specify what things are supposed to hang together or what hanging together means. In contrast, I will use a rigorous theory of coherence to show how beliefs and emotions can hang together in knowledge about reality and in normative theories of morality, justice, and art. Coherence then connects how things are and how they ought to be. Many philosophers have assumed that such objectivity must be achieved by severing links between philosophy and science, but I use ideas about brains, emotions, and coherence to display surprising connections among the sciences and the humanities.

My approach to all the areas of philosophy in the following chapters uses the same *philosophical procedure*:

1. Identify the most important philosophical issues as questions.
2. Consider a range of available answers to these questions.
3. Evaluate these answers based on coherence with scientific knowledge and other defensible philosophical doctrines.

4. Reach philosophical conclusions by accepting some answers and rejecting others based on explanatory coherence with evidence and on emotional coherence with human goals.

This approach does not yield answers that reign with unchallengeable certainty but rather produces interconnected, revisable hypotheses relevant to everyday life and scientific understanding. Philosophy can avoid linguistic puzzles, historical trivia, and terminal obscurity, by instead striving to provide insights into how the world is and how it ought to be. Philosophy can aim both to interpret the world and to change it for the better, with accomplishments that afflict the comfortable and comfort the afflicted.

ISSUES AND ALTERNATIVES: WAYS OF PHILOSOPHIZING

I prefer naturalistic answers to questions about how to philosophize. But step 2 of my philosophical procedure requires considering alternative ways of answering general and normative questions. Hawking's remark and other skeptical criticisms of philosophy suggest that it simply be abandoned, but I will show the fecundity of natural philosophy. The following is a quick review of other approaches to philosophy that have been influential, including religious, historical, pure-reason, analytic, phenomenological, and postmodernist approaches.

Religious

Religion is still a major force in modern life, with 5 billion of the 7 billion people in the world professing adherence to some religion, most commonly Christianity and Islam. The role of philosophy might be to develop and deepen the insights of religion, adding reasons for beliefs and practices usually accepted merely on the basis of faith. But philosophy cannot simply be an adjunct to religion, because it demands probing investigation of fundamental religious assumptions about the nature of knowledge, reality, and morality. Philosophy might end up supporting religion after due deliberation, but the method of starting with religion founders on the problem of choosing which religion to assume. Hence religious philosophizing is not a plausible approach.

Historical

Scientists rarely read works of historical figures such as Newton, because it is much more productive to keep up with current journals. But philosophy still shows a keen interest in greats such as Plato, Aristotle, and Descartes, justified by the fact that the problems that they identified have yet to be resolved. This interest in history can be extreme when philosophy is viewed primarily as the study of its own history, an ongoing conversation with the past. Emphasis on historical rather than constructive philosophy has even been justified by the slogan "Nonsense is nonsense, but history of nonsense is scholarship."

This purely historical approach to philosophy ignores the point that what made the great philosophers important was their sincere and strenuous attempts to answer the difficult questions that they identified. Similarly, contemporary philosophy should strive to develop new answers to the huge old questions, drawing on knowledge about mind and society that was not available to earlier thinkers. Hence philosophy can continue to show interest in its historical figures while recognizing that there is much more to philosophy than its history.

Pure Reason

Many philosophers, from Plato and Kant to contemporary writers in analytic philosophy and phenomenology, have wanted philosophy to achieve a solid foundation through truths arrived at by reason alone. Where science can only aspire to ephemeral conclusions subject to replacement by future theories, philosophy can establish the bedrock on which knowledge and morality can be built. Disappointingly, such foundations have been impossible to find, for reasons presented in chapter 3 on knowledge. Ideally, pure reason should provide truths that are a priori (independent of experience) and necessary (true in all possible worlds, not just this one). The reasons for concluding that there are no a priori, necessary truths are the same as the reasons for believing that there are no unicorns or leprechauns: people have been looking for them for centuries and failed to find any.

The failure to find such absolute truths leads some to conclude that philosophy is impossible, but a more reasonable conclusion is that philosophy should abandon the quest for certainty. Instead, it can follow science in looking for truths that are reasonable given all that is currently known, while remaining open to possibilities that these may be overturned as knowledge develops. Natural philosophy embraces fallibility and moderate uncertainty as virtues rather than vices. The attempt to base philosophical conclusions on pure reason always degenerates into skepticism about the possibility of philosophy.

The pure-reason approach to philosophy is based on assumptions that chapters 2 and 3 provide ample reasons to distrust. Numerous philosophers have hoped that intuition and imagination can provide insights into the nature of reality, but understanding how the brain intuits and imagines makes it clear that these are unreliable guides to truth. Imagination is constrained by what one already believes, and many of these beliefs may be false. Intuitions are worthwhile only when they were based on extensive experience and learning of regularities in the world, and there are no such regularities available for philosophical conclusions. Philosophical intuitions, even one's own, are not to be trusted, and occasional agreement among philosophers in their intuitive judgments can simply be a sign of social influence rather than veracity. Hence the pure-reason approach to philosophy is as unpromising as the religious approach.

Analytic

For the past century, the dominant style of philosophy in the English-speaking world has been analysis based on the careful study of language and logic. On this view, philosophy differs from science because of its concern with conceptual issues rather than empirical ones.

One of analytic philosophy's major tools has been conceptual analysis, which attempts to give definitions of key ideas such as knowledge. But there are four reasons why this form of definitional analysis needs to be abandoned. First, it always fails because of the inevitable generation of counterexamples that show that any particular attempt to provide necessary and sufficient conditions is inadequate. Second, it often turns out to be circular because the definition of one concept relies on other concepts that may be themselves defined only using the target concept. Third, conceptual analysis can be unduly conservative because it assumes that the concepts used by ordinary people are enduring, whereas improved scientific knowledge often shows that concepts need to be radically revised or even abandoned. Fourth, there are many experimental psychological studies that support the importance of exemplars, typical features, and explanations as being far more important to the structure of concepts than definitions. These findings support the use of three-analysis as a tool of conceptual analysis instead of trying to provide necessary and sufficient conditions.

According to Michael Dummett, the fundamental axiom of analytic philosophy is that the only route to the analysis of thought goes through the analysis of language. A rudimentary understanding of psychology and neuroscience makes this axiom dramatically implausible, because language is only one aspect of thought, which also includes perception, imagery, emotion, action, and nonverbal communication. Language is certainly important, but it is only one facet that needs to be taken into account in a full theory of mind and knowledge.

Some people view naturalistic philosophy as part of analytic philosophy, but I think they are fundamentally different in their assumptions and methodology. To clarify the difference between analytic and natural philosophy, the following is a list of 11 dogmas that are often assumed by analytic philosophers but rarely explicitly defended. For each, I state the natural alternative.

1. The best approach to philosophy is conceptual analysis using formal logic or ordinary language. Natural alternative: Investigate concepts and theories developed in relevant sciences. Philosophy is theory construction, not just conceptual analysis.
2. Philosophy is conservative, analyzing existing concepts. Natural alternative: Instead of assuming that people's concepts are correct, develop new and improved concepts embedded in explanatory theories. The point is not to interpret concepts but to change them.
3. People's intuitions are evidence for philosophical conclusions. Natural alternative: Evaluate intuitions critically to determine their psychological causes, which are often more tied to prejudices and errors than truth. Do not trust your intuitions.
4. Thought experiments are a good way of generating intuitive evidence. Natural alternative: Use thought experiments only as a way of generating hypotheses and evaluate hypotheses objectively by considering evidence derived from systematic observations and controlled experiments.
5. People are rational. Natural alternative: Recognize that people are commonly ignorant of physics, biology, and psychology and that their beliefs and concepts are often incoherent and formed by irrational processes such as wishful thinking. Philosophy needs to educate people, not excuse them.
6. Inferences are based on arguments. Natural alternative: Whereas arguments are serial and linguistic, inferences operate as parallel neural processes that can use representations that involve visual and other modalities. Critical thinking is different from informal logic.
7. Reason is separate from emotion. Natural alternative: Appreciate that brains function by virtue of interconnections between cognitive and emotional processing that are usually valuable but can sometimes lead to error. The best thinking is both cognitive and emotional.
8. There are necessary truths that apply to all possible worlds. Natural alternative: Recognize that it is hard enough to figure out what is true in

this world, and there is no reliable way of establishing what is true in all possible worlds, so abandon the concept of necessity.
9. Thoughts are propositional attitudes. Natural alternative: Instead of considering thoughts to be abstract relations between abstract selves and abstract sentence-like entities, accept the rapidly increasing evidence that thoughts are brain processes.
10. The structure of logic reveals the nature of reality. Natural alternative: Appreciate that formal logic is only one of many areas of mathematics relevant to determining the fundamental nature of reality. Then we can avoid the error of inferring metaphysical conclusions from the logic of the day.
11. Naturalism cannot address normative issues about what people ought to do in epistemology and ethics. Natural alternative: Adopt a normative procedure that empirically evaluates the extent to which different practices achieve the goals of knowledge and morality.

How such a normative procedure works is explained later in this chapter as part of a broader description of natural philosophy.

Phenomenology

In contrast to the primacy of analytic philosophy in Great Britain and North America, the dominant philosophical approach in continental Europe and South America for the past century has been phenomenology. Originating with the German philosopher Edmund Husserl, phenomenology is the study of conscious experience and of things as experienced. At first glance, this project sounds like psychology, but Husserl sharply separated his method from anything scientific or experimental. He thought that pure reflection and attention could lead to an understanding of the essence of experience that is much more fundamental than science could provide.

From the perspective of modern cognitive science, phenomenology's reliance on introspection and pure reflection is as naïve as analytic philosophy's assumption that the study of language is central to the understanding of thought. Thousands of experimental studies in psychology and neuroscience have shown that conscious, verbal experience is only a tiny part of the mental processing carried out by the brain. At best, phenomenology can provide a description of some superficial aspects of thought that cry out for deeper explanation. At worst, phenomenology leads to conclusions that are demonstrably false, such as the proclamation of existentialists like Jean-Paul Sartre that people are radically free and can choose their essences.

One reason for the appeal of phenomenology as a philosophical method is that it can address profound life questions concerning meaning and death, in ways that seem more insightful than the obsession of analytic philosophy with ways of speaking. I will show that natural philosophy can also pursue profound questions about meaning by embracing connections with empirical information rather than relying on pure reflection.

Like analytic philosophy, phenomenology has its dogmas. One is the unspoken assumption that there is a correlation between obscurity and profundity, so that the execrable writing of Heidegger (whose prose was as bad as his politics) looks like a virtue rather than a vice. This assumption fits with the dogma of confusing history and hagiography, so that the writings of the originators of phenomenology, especially Husserl and Heidegger, can be mined endlessly for nuggets of wisdom. The key dogma is that philosophy can be done a priori by a transcendental reduction of consciousness, superior to science because introspection and self-reflection alone can yield important insights into the nature of the mind. But trying to understand the mind without behavioral and neural experimentation is like trying to do astronomy without telescopes or biology without microscopes. The observations assembled by phenomenologists fall far short of the standards of evidence described in chapter 3.

Postmodernism

Since the 1960s, another philosophical method has been influential in France and in North America in fields such as literature, anthropology, and cultural studies. Postmodernism rejects philosophical assumptions about the nature of reality and objectivity, claiming that everything is socially constructed on the basis of power relations. Postmodernism is clearly at odds with natural philosophy in rejecting the idea that science or any other general approach can tell us much about how knowledge is achieved. Postmodernism is appealing to people skeptical of contemporary society, which does grant science a privileged role in amassing evidence and technology. To critics of science and capitalism, it seems politically progressive and culturally insightful to allow for a free range of vague speculations about social relations.

I share the concern with casting a skeptical eye on contemporary social structures while attempting to explain the full range of cultural activities, but will show that these enterprises can be pursued much more productively from a naturalistic perspective. For example, *Mind–Society* contains a detailed account of power based on human emotional interactions, and chapter 9 explains the arts by drawing on theories of brain, mind, and society. The best way to counter the nihilistic tendencies of postmodernism is to show that natural philosophy, allied with the cognitive and social sciences, can pursue the cultural aims of postmodernist thinkers with greater critical insight and practical effectiveness.

Naturalism

The approach to philosophy as closely tied to scientific investigations is sometimes called "naturalistic philosophy" or "philosophy naturalized," but I like the more concise term *natural philosophy*. Before the words "science" and "scientist" became common in the nineteenth century, researchers such as Newton described what they did as natural philosophy. I propose to revive this term to cover a method that ties epistemology and ethics closely to the cognitive sciences and ties metaphysics closely to physics and other sciences.

This natural philosophy is not new, for it has been practiced in various ways by such distinguished philosophers as Thales, Aristotle, Epicurus, Lucretius, Francis Bacon, John Locke, David Hume, John Stuart Mill, Charles Peirce, Bertrand Russell, John Dewey, W. V. O. Quine, and Thomas Kuhn. There are also many contemporary philosophers making progress on problems concerning the nature of knowledge, reality, and ethics without succumbing to the dogmas of analytic philosophy and phenomenology. Philosophy needs to be extraverted, directing its attention to real-world problems and relevant scientific findings, not introverted and concerned only with its own history and techniques.

The following are some central principles of naturalistic philosophy. They are not to be taken as dogmas or a priori truths but rather as heuristics for guiding development of philosophical theories that would justify naturalism by virtue of its greater explanatory power and practical import compared to alternative approaches.

1. Philosophy can do without supernatural entities such as gods, souls, heavenly ideas, and superhuman meanings.
2. Philosophy is tightly interconnected with science.
3. Philosophy differs from science in being more general, ranging across all of the sciences, and in being more normative, concerned with how the world can be made better.
4. Philosophical theories are to be evaluated on the basis of coherence with evidence rather than intuition, thought experiments, and internal consistency.

5. Values are interconnected with facts.
6. The challenge is to show how philosophy can connect with science to answer fundamental questions about knowledge, reality, morality, meaning, and the arts.

ELEMENTS OF NATURAL PHILOSOPHY

My own approach to natural philosophy builds on new theories of mind and society. I next review the multidisciplinary method used in *Brain–Mind* and *Mind–Society* and address the challenging question of how naturalism can deal with normative questions about how people and the world ought to be. A key to forging the connection between values and facts is a precise conception of coherence that challenges philosophical assumptions about linear inference akin to deductive proof. I outline how natural philosophy aims for coherence with scientific accomplishments and then examine the relations between this approach and the recently popular movement of experimental philosophy.

The Social Cognitive-Emotional Approach

My approach to natural philosophy draws repeatedly on a new theory of neural cognition described in chapter 2. Chris Eliasmith's Semantic Pointer Architecture is capable of accounting both for perceptual, embodied aspects of thinking and for abstract aspects that go beyond the senses. In line with current understanding of how the brain works, this theory of cognition fits intimately with emotions, because perception, inference, and emotion are all explained by semantic pointers, a special kind of neural process that integrates verbal and sensory-motor representations. For example, semantic pointers for emotions are patterns of neural firing that combine internal perceptions of bodily changes with judgments about goal accomplishment.

The semantic pointer theories offered are mechanistic in that they describe the operations of the brain as resulting from neural parts whose connections and interactions produce regular changes. But the mechanisms are sufficiently complex to accommodate the emergence of meaning, consciousness, and social interactions. Chapters 3, 4, and 5 provide a thorough discussion of mechanism, explanation, and emergence.

The scientific theories needed for philosophical purposes go beyond those of individual psychology and neuroscience to include a mechanistic theory of social interactions. Philosophy since Hobbes and Descartes has focused on individuals, but we will see the benefits of pursuing philosophical questions in ways that recognize the intimate connections between individual minds and their social context. Unlike postmodernism, natural philosophy is not concerned with vague ideas about social construction but instead draws on accounts of how mental mechanisms and social mechanisms depend on each other. Societies do not reduce to minds, and minds do not reduce to societies, because thinking and social communication are interdependent.

Two mechanistic systems are interdependent if changes in each system causes changes in the other. We will see that interdependence rather than reduction is the best way to think of the relation between the cognitive and the social, between cognition and emotion, between philosophy and science, and between descriptive facts and normative values. For example, social norms are tied to cognitive rules and emotional reactions based on needs and appraisals.

The major sticking points for natural philosophy are mind, morality, and meaning, which some philosophers think are forever immune from scientific explanation. But it used to be thought that science was equally inept at explaining the universe, life, and religion, all of which have progressively succumbed to scientific explanation. Natural philosophy can thrive using a full set of social, cognitive-emotional mechanisms extended to handle mind, morality, and meaning. Sorting out mind is the first step, but connecting it to social mechanisms is crucial for generating explanations of morality, meaning, justice, and beauty.

In *Mind–Society*, I use the term *social cognitivism* to describe my approach to the social sciences and professions, and it also sums up my variety of naturalistic philosophy. This approach is cognitive because it relies heavily on cognitive neuroscience (including emotions) to explain individual behavior, but it is also social because communicative interaction affects many thoughts and behaviors. We will see that philosophical questions about knowledge, reality, morality, justice, meaning, and beauty are inherently social, dependent as much on human interactions as on individual thinking. Hence social cognitivism extends to natural philosophy and the arts.

Normative Questions

The standard philosophical objection to naturalism is that it is incapable of dealing with normative issues about what ought to be. But natural philosophy can overcome this objection by applying the following *normative procedure*.

1. Identify a domain of practices, such as knowledge or ethics.
2. Identify candidate norms for these practices, such as logic or utilitarianism.

3. Identify the appropriate goals of the practices in the given domain, such as truth or human welfare.
4. Evaluate the extent to which different practices accomplish the relevant goals.
5. Adopt as domain norms those practices that best accomplish the relevant goals.

The trickiest part of this procedure is step 3, because it requires value judgments about appropriate goals, but later chapters provide reasonable ways of accomplishing goal evaluation with respect to knowledge, ethics, justice, meaning, and art.

People are rational when their inferences and actions conform to the relevant norms justified by the normative procedure. *Mind–Society* provides numerous cases of irrationality in personal relationships, politics, economics, and international affairs.

Facts (about how things are) and values (about how things ought to be) are different, but they are interconnected, as shown in Figure 1.1. Descriptions of the world in the form of empirical evidence and explanatory theories can provide information about available practices and goals, especially about the extent to which different practices accomplish different goals. This figure does not display connections that run in the other direction, from the normative to the descriptive, but these are discussed in chapter 3 in connection with values in science.

Chapter 6 argues that ethical values are emotional attitudes, where emotions are in part cognitive judgments that depend on facts concerning how people can accomplish their goals. So values are in part factual, at the same time that determining what is factual requires cognitive processes that invoke values such as

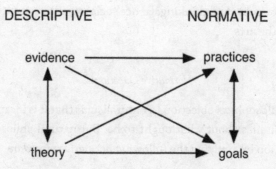

FIGURE 1.1 How descriptive information can be relevant to normative conclusions. Arrows indicate inferential relevance. Reprinted from Thagard 2010b by permission of Princeton University Press.

truth and explanation. Hence facts and values are linked by a kind of nonvicious circularity that can be captured by the idea of coherence presented in the next section.

The normative procedure provides an alternative to dogmatism and subjectivism. Dogmatism relies on religious doctrines or alleged a priori truths to provide an incontrovertible basis for judging what ought to be, but these dogmas are always challengeable. Seeking absolute foundations leads to subjectivism, the skeptical view that there are no objective normative standards at all, so that judgments about what ought to be are merely expressions of individual or cultural opinions and attitudes.

In contrast, my normative procedure allows for the assessment of what ought to be by connecting judgments of value with judgments of facts, without attempting to deduce values from facts. In this way, natural philosophy can contribute to plausible normative accounts of what ought to be done in the development of knowledge, morality, meaning, and art. Values are neither derived nor eliminated but rather assessed by coherence with facts.

Coherence

The key to avoiding both dogmatism and relativism is a theory of coherence that allows reasonably interconnected assessment of both facts and values. Coherence has to be much richer than mere consistency, because there can be multiple systems of belief that are equally consistent but incompatible. Moreover, coherence has to be broad enough that it can cover not only beliefs but also value judgments. The required theory of coherence results from understanding it as a process of parallel constraint satisfaction.

For example, suppose you have an important decision to make, such as what career to pursue. Such decisions face negative constraints such as the difficulties of being both a lawyer and an artist. They also face positive constraints such as that being a lawyer will make you more money and being an artist might bring you more emotional satisfaction. Parallel constraint satisfaction means that you have to figure out how to weigh these actions and goals all at once, including the possibility of changing your mind about the importance of different goals. In logic and language, inferences are serial, one step at a time, but decision making requires you to consider all the constraints at one, which is not as difficult as it sounds because the brain is a parallel processor with billions of neurons interacting with each other.

The following is an informal characterization of coherence problems, which can also be specified more exactly using mathematical and computational analysis.

Coherence requires assessment of elements, which are mental representations such as beliefs, concepts, images, actions, and goals. The elements can cohere (fit together) or incohere (resist fitting together) in various ways. If two elements cohere (e.g., because one explains or facilitates the other), then there is a positive constraint between them. If two elements incohere (e.g., because they are inconsistent or otherwise incompatible), then there is a negative constraint between them. The elements need to be divided into ones that are accepted and ones that are rejected. A positive constraint between two elements can be satisfied either by accepting both elements or by rejecting both elements. A negative constraint between two elements can be satisfied only by accepting one element and rejecting the other. A coherence problem consists of dividing elements into accepted and rejected ones in a way that satisfies the most constraints.

Decision making is a coherence problem that requires selecting a set of actions and goals that fit together, with positive constraints concerning which actions accomplish which goals and negative constraints concerning which actions and which goals are incompatible with each other. For example, a decision about where to go on vacation assesses different actions, such as going to a beach or the mountains or a big city, against various goals such as relaxation, fun, and cultural enrichment. The negative constraints are the difficulties of going simultaneously to the beach, the mountains, and a big city, while the positive constraints are the extent to which each of these options accomplishes the goals. Figure 1.2 displays a vacation decision as a coherence problem. The dotted line indicates a negative constraint based on the difficulty of going to both a beach and a big city (with a few exceptions like Barcelona). The solid lines indicate positive constraints such as that the beach is relaxing and the city is culturally exciting. Later chapters show how coherence as constraint satisfaction works to assess competing beliefs, ethical judgments, and aesthetic evaluations.

The repeated application of coherence methods unifies natural philosophy, applying to all its major branches: epistemology, metaphysics, ethics, and aesthetics. But the coherence of natural philosophy is not merely methodological, because

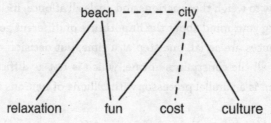

FIGURE 1.2 A vacation decision as a coherence problem, with positive constraints shown by solid lines and negative constraints shown by dotted lines.

the philosophical theories constructed for each of these branches need to be coherent each other, not just with the method of coherence and the relevant scientific knowledge. This book displays the fine fit among the following philosophical positions: multilevel materialism in philosophy of mind, reliable coherentism in epistemology, scientific realism in metaphysics, needs-based consequentialism in ethics, needs sufficiency in justice, and neural expressivism in aesthetics.

The coherence approach pursued in this book has numerous philosophical advantages. First, it fleshes out Sellars' *hangs together* metaphor for philosophy. Hanging together results from relations between different elements that generate an overall coherence judgment based on parallel constraint satisfaction. Second, coherence shows how to live with the unavoidable uncertainty that has driven generations of philosophers to various forms of aggressive dogmatism and despairing relativism. Coherence judgments will always be tentative and fallible, subject to later revision. Nevertheless, they can be satisfying and reassuring if there has been a systematic attempt to consider all relevant elements and constraints.

We will repeatedly see the bearing of philosophy on important questions in the cognitive and social sciences, but relevance also runs in the opposite direction. In contrast to the individualist bias of most philosophy, social interactions are important for answering basic questions in all of philosophy, from epistemology to aesthetics. Humans are inherently social in ways that strongly affect the development of knowledge, morality, justice, and art.

Other implications of science for philosophy include appreciation that standard conceptual analysis depends on an empirically false theory of concepts and needs to be supplanted by three-analysis. Cognitive and social sciences provide explanations of why metaphysical doctrines like theism are so appealing and of why so many philosophers are afraid of naturalism.

Coherence in the precise sense of parallel constraint satisfaction plays an important role in the entire book. Chapter 2 describes the mind as a coherence engine that smoothly integrates semantics, syntax, and pragmatics, in contrast to the syntactic view of mind as using logical inference. Coherence is not just a matter of relations among beliefs but can also cover other kinds of mental representations including images and emotions.

Similarly, the approach to knowledge in chapter 3 relies heavily on coherence rather than on foundations. Nevertheless, the discussion of reality in chapter 4 rejects the coherence theory of truth in favor of the view that truth is correspondence to an independent reality. Chapter 5 shows how explanations connect knowledge with reality. Chapters 6 and 7 deal with issues concerning morality and justice, extending coherence to cover emotions and emphasizing coherence with human needs.

Chapter 8 treats both the meaning of language and the meaning of life as coherence problems. Beauty is analyzed in chapter 9 as primarily resulting from emotional coherence, and other emotional responses to the arts also depend on coherence between cognitive appraisals and physiological responses. The final chapter on future philosophy reviews how the overall coherence of all of these theories furnishes a full system of philosophy and sketches potential applications to problems about free will, mathematics, and nonhuman minds.

Experimental Philosophy

Data-driven experimental philosophy is an important twenty-first-century movement that is compatible with my more theoretical approach to natural philosophy. Experimental philosophy makes valuable contributions in two main ways. First, it provides an effective antidote to assumptions of analytic philosophy and phenomenology that thought experiments and introspection provide insights into how things are and how they ought to be. Instead of relying on the solitary intuitions of a philosopher about stories that the philosopher made up, experimental philosophy consults the reactions of numerous people in populations more culturally diverse than are usually found in philosophy departments. Diversity in philosophical intuitions has been found for ethnicity, gender, personality, philosophical background, and age. Experimental philosophy thus serves to undermine the dogmas of analytic philosophy criticized earlier.

Second, experimental philosophers valuably extend the range of data relevant to assessing philosophical theories. Psychologists do abundant experiments, but their concerns are not always philosophical, and philosophy should not wait for psychologists to amass evidence relevant to epistemology, metaphysics, and ethics. So experimental philosophy is a valuable source of additional information for the development of theories in natural philosophy.

However, as currently practiced, experimental philosophy has several limitations that undercut its relevance to natural philosophy. First, almost all published results in experimental philosophy are surveys in which people, ranging from undergraduates to paid contributors on Amazon's Mechanical Turk, are asked to give their reactions to various scenarios. Psychologists have long been trained to look critically at the results of such surveys because of biases such as the tendency of people to tell the experimenter what the experimenter wants to hear. Psychologists use a variety of techniques such as deception and reaction time measures that detect aspects of thought not reachable by survey questions alone.

Second, with rare exceptions, experimental philosophy has avoided the experimental techniques of cognitive neuroscience, which are increasingly proving relevant to developing deeper theories in cognitive, social, development, and clinical psychology. Experimental psychology has learned that it cannot ignore the brain, and experimental philosophy needs to gain similar insights by means of brain scans and other neuroscientific methods.

Third, the current findings about what people think obtained by experimental philosophy are of limited use in developing and evaluating philosophical theories that are general and normative. Finding out what people think about minds is interesting for fields such as developmental psychology but has small relevance for understanding how minds actually work, just as finding out what ordinary people think about forces and life has little relevance for physics and biology. Decades of research in psychology and neuroscience show that people are astonishingly ignorant of the mental mechanisms that produce thought, so their naïve judgments are of little use in figuring out the nature of mind, knowledge, reality, and morals.

All three of these objections can be overcome by more sophisticated work in experimental philosophy. Philosophers can conduct more illuminating experiments using techniques well understood in psychology and neuroscience. Their experimental results will not have direct implications for finding answers to philosophical problems, but they can help point to better theories of mind that will have such implications. When enhanced experimentally and theoretically, experimental philosophy can play a valuable constructive role in natural philosophy that goes beyond its already useful debunking role in undermining the introspective methods of analytic philosophy and phenomenology.

OVERVIEW OF THIS BOOK

Ideally, a system of philosophy would be presented all at once and evaluated by parallel constraint satisfaction that determines how well the contributing theories fit with each other and with scientific knowledge. But human communication is serial, so a philosophical system needs to unfold one part at a time. All branches of philosophy make assumptions about the nature of mind, so chapter 2 starts by proposing multilevel materialism as the theory of mind that fits best with current evidence from psychology and neuroscience. The materialist conclusion that all mental processes are brain processes is supported by the explanatory successes of the Semantic Pointer Architecture. This form of materialism is multilevel because of the molecular, psychological, and social mechanisms that interact with neural ones.

Chapter 3 addresses central questions about the nature of knowledge, arguing that the best epistemology is reliable coherentism. Coherence rather than indubitable foundations is the route to knowledge, but coherence must be constrained by reliable evidence. This theory of knowledge goes beyond individual minds because knowledge usually develops in groups of people. The integrated cognitive-social theories developed in *Brain–Mind* and *Mind–Society* provide the basis for an epistemology of both individuals and groups. Although this theory of knowledge is tied to facts about brains, minds, and societies, it also uses value-related coherence assessments to justify it as objectively normative, not just descriptive.

The theories about mind and knowledge developed in chapters 2 and 3 point to a general theory of reality in chapter 4. The materialist theory of mind defended in chapter 2 undercuts philosophical theories such as idealism and dualism that try to make mind primary. As the best approach to metaphysics, I defend scientific realism, the view that science aims and sometimes succeeds at discovering how the world is. Then ontology—the study of what fundamentally exists—should draw heavily on the best available scientific theories, although they cannot be expected to address philosophical questions at full-level generality. This approach to reality fits well with the coherentist approach to knowledge but rejects the view that truth is just coherence rather than correspondence to reality.

Chapter 5 explores the interconnections of epistemology and metaphysics by tackling questions about explanation, causality, reduction, and emergence. It argues that explanation in the cognitive and social sciences are usually descriptions of causal mechanisms. The relation between explanatory levels is typically emergence rather than reduction, because wholes are not always just the sum of their parts.

Chapter 3 touches on normative issues connected with the justification of knowledge, but chapters 6 through 9 are more intensely normative. Chapter 6 develops a theory of morality that answers fundamental ethical questions concerning an objective basis for right and wrong. It uses semantic pointer theories of concepts and emotions to generate new accounts of values and ethical judgments, aiming for both psychological plausibility and normative justification. Like the theory of knowledge developed in chapter 3, the theory of morality is fundamentally social, considering people as interacting to form groups, not just as isolated agents.

Chapter 7 on justice is even more social as it concerns public life, particularly the operation of the state. It addresses descriptive and normative questions about justice in modern, complex societies, particularly what roles governments should play in pursuing social justice. Using both facts and values, social democracy is defended as the most legitimate approach to just government.

Chapter 8 discusses meaning, addressing empirical issues about meaning in language and prescriptive issues about the meaning of life. Surprisingly, questions about the meaning of language and the meaning of life turn out to have similar answers based on construing meanings as complex processes rather than things. The semantic pointer theory of mind shows how to integrate the multiple dimensions that go into achievement of meaning in language and thought, showing how representations can be meaningful through interactions with other representations, the world, and other people. Analogously, the semantic theory of emotions shows how to integrate the different aspects of existence that make people's lives meaningful, including love, work, and play.

Chapter 9 turns to aesthetics, the branch of philosophy concerned with questions about the nature and value of the arts. It shows how painting and music depend on emotional coherence among semantic pointers. Judgments of beauty and other kinds of aesthetic evaluation all makes sense, descriptively and normatively, through a general theory of mind, knowledge, and reality.

Finally, chapter 10 reviews how the individual theories developed for different branches of philosophy all combine into a coherent whole. It also sketches how natural philosophy might deal with important but unresolved issues concerning free will, mathematics, and nonhuman intelligence. One of the major stumbling blocks for materialist theories of mind, knowledge, and reality is the puzzling nature of mathematics. I outline how mathematical knowledge might be understood by combining the semantic pointer theory of mind from chapter 2 with the coherence theory of knowledge from chapter 3. Solving the problem of free will depends on future discoveries about how the brain connects conscious thought with actions. Similarly, difficult moral questions about the mental capacities of machines and nonhuman animals will require future progress in both science and philosophy.

SUMMARY AND DISCUSSION

In Chekhov's play *The Seagull*, a character remarks: "How easy it is to be a philosopher on paper and how hard it is in life." Actually, it is not that easy to be a philosopher on paper either. Philosophy would even be impossible if its goal were to establish eternal, indubitable truths that everyone accepts. But also challenging is the more modest goal of coming up with general and normative hypotheses that are coherent with available scientific knowledge. This book rises to the challenge by developing interconnected theories of knowledge, reality, morality, justice, meaning, and the arts.

C. P. Snow lamented the two separate cultures of the sciences and the humanities, but this book tries to bring them together by linking the study of philosophy and the arts to the cognitive and social sciences. My aim is not to reduce the humanities to the sciences but rather to display fertile interconnections that show that philosophical questions and artistic practices can be much better understood by bringing to bear scientific knowledge about how minds operate and interact in social contexts. The sciences and the humanities are interdependent, because the natural and social sciences cannot avoid questions about methods and values that are primarily the province of philosophy.

My method for pursuing philosophy is very different from more familiar philosophical approaches such as analytic philosophy and phenomenology. Little can be learned about mind and reality by introspection, thought experiments, storytelling, and narrow attention to language and logic. Instead, we can proceed by identifying key philosophical questions, canvassing alternative answers, and evaluating answers on the basis of empirical and theoretical knowledge derived from the sciences. The value of this method cannot be established in advance, a priori, but requires systematic investigation to establish theories that cross all the main concerns of philosophy and the arts.

Natural philosophy has a long and venerable history, going back to the ancients such as Thales, Aristotle, and Epicurus. My version is unusual because I connect philosophical issues with current theories in psychology and neuroscience. The theories employed are not narrowly cognitive but also integrate emotions as crucial parts of thinking. Rather than focus just on individuals as usually happens in Western philosophy, I treat all philosophical questions as inherently social, from philosophy of mind through aesthetics.

The cognitive-emotional-social outcomes subdue the traditional fear that mixing philosophy with empirical science gets in the way of establishing objective truths about knowledge, reality, and morality. These results are only provisional and tentative, because philosophy and the arts are subject to the same attitude that Donald Hebb recommended for science: a good theory is one that holds together long enough to get you to a better theory. Because scientific knowledge advances, and because understanding in the humanities needs to move with it, knowledge about philosophy and the arts should also aim for ongoing progress instead of for eternally established answers.

Philosophy thereby directs its attention to real-world problems and relevant scientific findings rather than just its own history and techniques. But philosophy is not replaceable by science, because cutting-edge science invariably encounters general questions about knowledge such as what justifies a theory, as well as general questions about reality such as what kinds of entities exist. Moreover, human uses of science always encounter normative questions about how science can be used to improve society or make it worse and how people can think better about science, technology, and society. Hence philosophy and science are not merely continuous with each other but interdependent.

Rather than diminish philosophy, my goal in this book is to show its importance for diverse human enterprises, including science, politics, the arts, and everyday life. Natural philosophy can draw on the sciences to dramatically increase understanding of fundamental issues concerning mind, meaning, and morality. Many philosophers have fled from naturalism because of its perceived threat to objectivity, truth, and faith, but only faith is at risk. Philosophy has tended to be mind-blind and even psychophobic, but chapter 2 presents a theory of mind with revolutionary implications for epistemology, metaphysics, ethics, and aesthetics.

NOTES

On naturalism and other philosophical methods, see Bashour and Miller 2014; Cappelen, Gendler, and Hawthorne 2016; Daly 2015; and Haug 2014. The variant of naturalism that emphasizes neuroscience originated with Patricia S. Churchland 1986, 2002, 2011; see also Paul M. Churchland 2007. Thagard 2009 discusses the relation between philosophy and cognitive science.

Hawking's dismissal of philosophy is in Hawking and Mlodinow 2010. My response is in a blog post: https://www.psychologytoday.com/blog/hot-thought/201011/is-philosophy-dead.

For references to Chris Eliasmith's Semantic Pointer Architecture, see the notes to chapter 2.

The method of three-analysis originates in this *Treatise*. It is based on the theory of concepts in Blouw, Solodkin, Thagard, and Eliasmith 2016; see also *Brain–Mind*, chapter 4. I indicate concept names by italics: the word "dog" stands for the thing dog represented in the mind by the concept *dog*.

For the history of the "nonsense is nonsense" quote, see http://leiterreports.typepad.com/blog/2005/05/drebenized.html.

Sellars 1962 (p. 1) and Dennett 2013 (p. 70) describe philosophy as aiming to understand how things hang together.

The prototypical religious philosopher is Thomas Aquinas. For a critique of faith as a source of knowledge, see Thagard 2010b.

Dummett 1993 (p. 128) states his axiom of analytic philosophy. Williamson 2007 defends philosophical analysis. Thagard 2014e describes the limitations of thought experiments. On intuition, see Osbeck and Held 2014. Schwitzgebel and Cushman 2015 show that trained philosophers have some of the same cognitive biases as ordinary people.

For introductions to phenomenology, see Smith 2013 and Gallagher and Zahavi 2012. On the importance of unconscious cognitive processes, see Kihlstrom 1987 and recent textbooks in cognitive neuroscience such as Banich and Compton 2018. Wolin 2016 examines Heidegger's Nazism.

Important contemporary naturalistic philosophers include William Bechtel, Mario Bunge, Patricia Churchland, Paul Churchland, Carl Craver, Lindley Darden, Daniel Dennett, Peter Godfrey-Smith, Alvin Goldman, Daniel Hausman, Joshua Knobe, Robert McCauley, Philip Kitcher, David Papineau, Peter Railton, Alexander Rosenberg, Eliot Sober, Miriam Solomon, Kim Sterelny, Stephen Stich, John Turri, and William Wimsatt. Read their books.

The normative procedure and Figure 1.1 are from Thagard 2010b (p. 211).

Coherence as constraint satisfaction is from Thagard 1989, 1992, 2000, 2012c. For the math, see Thagard and Verbeurgt 1998.

For experimental philosophy, see Sytsma and Buckwalter 2016 and Sytsma and Livengood 2016. Diversity in philosophical intuitions is summarized by Colaço, Buckwalter, Stich, and Machery 2014. Neuroscientific methods are applied to experimental philosophy in Jenkins, Dodell-Feder, Saxe, and Knobe 2014.

The Hebb quote about a good theory is reported by Mintzberg 2009 (p. 43).

Wilson 1998 wants to unify knowledge in the humanities and sciences using evolutionary biology, but theoretical neuroscience suggests a much broader and deeper account of cultural learning and the interactions between mental and social mechanisms, as shown in *Mind–Society*.

PROJECT

Write a history of natural philosophy from Thales to today that describes the coevolution of science and philosophy. Do three-analyses for the major branches of philosophy, including epistemology, metaphysics, ethics, and aesthetics.

2

Mind

MENTAL PROCESSES

Why start with the mind when building a system of natural philosophy? It might seem more appropriate to start with a theory of knowledge and then use it to figure out what we know about thinking. Perhaps we could start with metaphysics as the theory of reality, since minds are part of reality. From a coherence point of view, it does not matter where we start, because no branch of philosophy is the axiomatic foundation for the rest. What matters is how well all of the philosophical theories fit with each other and with scientific knowledge.

Strategically, however, it is best to start with mind because a strong theory of mental processing has major implications for all other branches of philosophy. The theory of knowledge requires an appreciation of what does the knowing, namely minds. General theories of reality all hinge on theories of mind, for example concerning whether mind is an original aspect of the universe or just a relatively recent addition. Questions about morality and justice depend partly on fundamental human needs, including psychological ones that connect with theories of mind. Aesthetics theories also presuppose theories of mind needed to make sense of beauty and other aesthetic reactions.

Accordingly, I offer a philosophical theory of mind, in line with the much more detailed psychological and social theories in *Brain–Mind* and *Mind–Society*. The philosophical procedure in chapter 1 requires (a) identifying important questions that a theory of mind should be able to answer and (b) reviewing the major theories that philosophers have proposed to deal with them. I then argue for a specific form of materialism that connects mind to brain using Chris Eliasmith's Semantic Pointer Architecture (SPA) for cognitive neuroscience. This theory shows that the best available explanation of the full range of mental processes comes from neural mechanisms that operate in concert with molecular, mental, and social mechanisms, justifying the philosophy of mind that I call "multilevel materialism."

Multilevel materialism defends the following principles:

1. Mind, like everything else, consists of matter and energy.
2. Explanations of mind are mechanistic, with mental processes operating through interconnected and interacting parts that produce regular changes.
3. Molecular, neural, mental, and social mechanisms all contribute to mind.
4. Mechanisms at different levels interact through neural mechanisms, which are therefore central to explanations of mind.

Table 2.1 summarizes the four levels of mechanisms.

The centrality of neural mechanisms is not enough to justify the simple identity that the mind is the brain, but it does sustain the claim that all mental processes are brain processes. Molecular mechanisms relevant to brain operations include neurotransmitters, protein folding, and hormones, all of which affect neural firing. Psychological explanations invoke mental representations such as concepts and images, which *Brain–Mind* argues are best explained by neural mechanisms involving semantic pointers. *Mind–Society* provides examples of molecular-social interactions, such as when alcohol helps people party and public stress raises cortisol levels. These interactions are all mediated by neural firing. Alcohol affects neurotransmitters including dopamine and GABA, thereby changing neural firing

TABLE 2.1

Four Levels of Mechanisms Relevant to How the Mind Works

Level	Parts	Interactions	Changes	Results
Social	people	communication	group behavior	social practices
Mental	mental representations	processes such as inference	shifts in representations	thought, action
Neural	neurons	electric/chemical transmission	neural firing patterns	representations, inferences
Molecular	genes, proteins, neurotransmitters	chemical reactions	neural properties	neural firing

that changes behavior. Public stress such as oral presentations increases cortisol levels via perceptions of the audience accomplished by neural firing.

The multilevel mechanisms approach to philosophy of mind is open to many objections. I respond to the common reply that we can easily imagine that the mind is different from the brain, specifically that the mind operates independently of semantic pointers. Second, I consider the view that minds should not be explained by neural mechanisms because minds are embodied, extended, embedded, and enactive. Third, I show that the social characteristics of mind are compatible with the claim that mental processes are neural. Finally, I deal with the biggest obstacle to materialist theories of mind by explaining consciousness as resulting from brain mechanisms for semantic pointer construction and competition.

ISSUES AND ALTERNATIVES

Like theories in science, theories in philosophy should be able to explain important phenomena, dealing with issues that can be expressed as questions, such as the following for the philosophy of mind:

1. What is mind?
2. What is the relation between mind and body and how do they affect each other?
3. How can conscious experience be explained?
4. Why do people act as they do?

These four issues are crucial to many other philosophical questions concerning the nature of reality, morality, and art.

Over the 3,000 years of the history of philosophy, numerous answers to these questions have been proposed. Historically and today, the most popular philosophy of mind has been dualism, the view that people consist of two fundamentally different things kinds of thing: a spiritual mind and a physical body. This view is entrenched in the worldviews of major religions including Christianity and Islam and has been defended by major philosophers such as Descartes and Kant.

Dualism keeps mind separate from matter, but the philosophical doctrine of idealism goes even farther by claiming that everything that exists is fundamentally mind. On this view, there is no division between mind and matter, because body and all other aspects of matter are fundamentally mind-dependent. Advocates of the primacy of mind included notable philosophers such as Berkeley and Hegel but only a few contemporary ones.

A currently popular version of dualism with a hint of idealism is panpsychism, according to which everything in the universe has some degree of mental character to it, even the atoms in a rock sitting at the bottom of the ocean. This view gains some plausibility from the problem of explaining consciousness, through the hope that we can explain how human minds end up with a lot of consciousness because there is a little consciousness in everything. Panpsychism loses its appeal if consciousness arises just from neural mechanisms.

A rarer view tries to moderate between idealism and materialism by suggesting that there is only one kind of thing in the world that it is neutral between mind and matter, both of which are manifestations of something else. The main problem with this neutral monism is that it ignores the abundant evidence for the existence of matter and energy from the natural sciences. Another problem is it has no answer to how aspects of mind and aspects of matter result from some completely unknown neutral underpinning.

Most naturalistic philosophers adopt some version of materialism based on the hypothesis of physics and chemistry that the universe consists of forms of matter and energy, convertible into each other because energy equals mass multiplied by the square of the speed of light, $e = mc^2$. One extreme version, eliminative materialism, says that mind simply does not exist and should be eliminated from scientific theories. Behaviorists, for example, said scientific psychology should only consider how environmental stimuli lead to behavioral responses, with no need to consider the mind as a ghostly intermediate. More recently, advances in understanding of neural processing have been taken by some philosophers to suggest that old concepts of mind such as *belief* and *desire* can simply be abandoned as irrelevant to a rigorous account of mind.

However, contemporary neuroscience works hard to explain many aspects of mind such as perception, memory, emotion, and inference. These explanations provide provisional support for the mind–brain identity theory, which says that all mental processes are brain processes. To make this plausible, neuroscience has to identify more than correlations between mental activity and the brain activities found in brain scanning experiments, providing neural mechanisms that can explain all mental operations. The new Semantic Pointer Architecture provides a strong candidate for specifying these mechanisms, including the formation of beliefs and desires (*Brain–Mind*, chapters 4 and 7). Then the concepts *belief* and *desire* become modified and absorbed into cognitive science, not eliminated.

Materialism also comes in nonneural forms. Functionalism claims that there is no point in looking for identities between mental processes and brain processes because mental operations can be physically realized in various ways. For example, if a driverless car decides to make a right turn using its computer, then decisions are not brain processes because computer mechanisms are just as adequate for making decisions. However, the underlying mechanisms for computer decisions and human decisions are so different that it is better not to suppose that there is some general mental process of decision making to be characterized computationally rather than neurologically. So it remains appropriate to identify human decisions with brain mechanisms if these can be plausibly specified.

Mind–brain identity is often considered a form of reductionism, which in its most extreme form maintains that the mechanisms responsible for mental processes are fundamentally those described by physics. However, *Brain–Mind* makes the case that there are four different levels of mechanisms relevant to understanding human minds: molecular, neural, mental, and social. This multilevel materialism rejects the assumption that all explanations of mental processes need to be at the molecular level, let alone at the physical level of atoms and subatomic particles. The claim that all mental processes are brain processes needs to be adjusted to recognize the explanatorily relevant operations of molecules such as dopamine and serotonin, of psychological operations such as inference, and of social operations such as communications among individuals.

The differences among these philosophical positions is clarified by considering their answers to the question of what is consciousness. For dualism, consciousness is a property of nonmaterial souls. According to idealism, consciousness is a part of the universe because the universe is mind. For panpsychism, everything has at least of bit of consciousness, whereas neutral monism states that consciousness results from processes that are neither mind nor matter. Eliminative materialism holds that consciousness does not exist, and functionalism says consciousness is a computational process that could be shared by entities without brains. Mind–brain identity holds that consciousness is a brain process. According to multilevel materialism, consciousness results from interacting neural, molecular, mental, and social mechanisms. How does the interactive firing of neurons help to explain thought?

NEURAL MECHANISMS

My overall argument is an inference to the best explanation: we should accept the hypothesis that mind consists of multilevel mechanisms because it provides a better explanation of the full range of mental phenomena than other philosophies of mind. The key mechanisms used in explaining the broad range of mental phenomena are taken from Chris Eliasmith's Semantic Pointer Architecture.

Mechanisms are systems of interconnected parts whose interactions produce regular changes. A mechanistic explanation is one where observed phenomena are explained as resulting from the changes produced by interacting parts. As chapter 5 recounts, such explanations operate powerfully in all areas of science, including physics, chemistry, biology, psychology, and the social sciences. For example, to explain how the heart pumps blood, look at its interconnected parts that include chambers, valves, veins, arteries, and blood. The interactions of these parts enable the heart to keep blood flowing through the circulatory system and perform the valuable function of conveying energy to all parts of the body. In turn, the parts of the heart function because they consist of cells with biochemical mechanisms.

In brain science, the most important parts are the more than 80 billion neurons connected to each other by axons, dendrites, and synapses. Neurotransmitters, hormones, and other brain cells called glia enable them to interact with each other and other parts of the body. Neurons depend on molecular mechanisms such as protein folding occurring within neurons and transmission of chemicals such as glutamate between neurons. The challenging question is how these neural and molecular mechanisms can give rise to thinking in humans and other animals.

The claim that all mental processes are brain processes is an explanatory identity, a kind of hypothesis common in the history of science: for example, air is a mixture of gaseous elements (primarily oxygen and nitrogen), and fire is rapid combustion involving combination with oxygen. These hypotheses identify ordinary things with scientific entities and processes in a way justified because the identities explain empirical phenomena.

The only way to justify multilevel materialism is to enumerate mental processes and show that each of them is explained by brain mechanisms. We can quickly generate a list of familiar mental processes that are candidates for explanation, including perception, imagery, memory, learning, problem solving, decision making, and language use. A step toward identification would be correlational, for example using brain scanning to show that whenever a particular mental process occurs there is activity in brain areas. But a much deeper identification requires finding brain mechanisms that are causally responsible for different aspects of the mental process to be explained.

The major intellectual problem is to figure out collections of billions of neurons can generate complex phenomena such as perception and language. The history of neuroscience has had a series of breakthroughs concerning mechanisms that show how brains can support thought, including the following:

1889: The realization by Ramón y Cajal that brains consist of cells, later called neurons, that are the most important parts of brain operations.

1897: The proposal by Charles Sherrington that neurons are connected by synapses that provide the main interactions between them.

1907: The hypothesis of Louis Lapicque that neurons fire (spike) by integrating input current from other neurons.

1949: The cell assembly hypothesis of Donald Hebb that neurons can work together to represent the world.

1949: The proposal by Hebb that systems of neurons learn by increasing the synaptic strengths between them when the neurons fire simultaneously.

1986: The demonstrations by Rumelhart, McClelland, and their collaborators that numerous psychological phenomena can be accounted for by sophisticated neural networks.

These advances still leave a big gap between the operations of neurons and high-level mental processes such as problem solving and creativity. Fortunately, a recent breakthrough in theoretical neuroscience has narrowed this gap substantially.

SEMANTIC POINTERS

Chris Eliasmith is a Canadian theoretical neuroscientist and philosopher who developed the Semantic Pointer Architecture as a general theory of how the brain accomplishes mental tasks such as problem solving and learning. In cognitive science, an architecture is a general proposal about the structures and processes that produce thinking. Semantic pointers are neural processes that represent the world and interact with other representations by compressing patterns of neural firings, including ones from sensory-motor inputs.

What Are Semantic Pointers?

Begin with the familiar idea of a representation as something that stands for other things. The most familiar representations are words and pictures, for example the word "wine" that stands for a drink made from fermented grapes, and a picture of a glass filled with red liquid that also stands for wine. More theoretically, cognitive science explains how thinking works by supposing that minds contain mental representations such as images, concepts, beliefs, and desires. For example, you have the concept *wine* in your head that is associated with sensory images of what it looks like and how it tastes. Cognitive neuroscience proposes that these mental representations are processes in the brain requiring the interactions of thousands, millions, or even billions of neurons.

The simplest kind of neural representation would be a single neuron that fired fast or slow depending on the presence of some stimulus, such as a glass of red wine. The neuron would function as a red wine detector if it fired fast when red wine is present but very slow when red wine is absent. We could then describe the neuron as representing wine.

This description is too simple for two main reasons. First, neurons do not simply fire fast or slow but with different patterns, just as a drummer can produce different kinds of rhythms. The song fragments "happy birthday to you" and "God save our gracious queen" are each six beats, which can be played fast or slow, but they also are played with different rhythms. Similarly, neurons can fire in different patterns, not just fast or slow.

Second, the brain rarely uses single neurons to stand for anything alone but rather uses groups of neurons, just as a drummer is normally a participant in a whole band that involves other musicians. The neural representation of red wine is captured in the brain by thousands or millions of neurons, each of which is capable of firing faster or slower, using many patterns. Collectively, the number of patterns that can be produced by a group of neurons is enormous: a group of 100 neurons, each of which is capable of firing from 1 to 100 times per second, can collectively produce more than a 100^{100} patterns, more than the number of stars in the universe. In general, a neural representation is a pattern of firing in a group of neurons, just as a musical piece is a temporal pattern of notes produced by musicians.

Semantic pointers are a special kind of neural representation built out of simpler ones, a complicated pattern built from patterns. Expanding the musical analogy, consider a musical performance where a symphony orchestra, with players of instruments such as violins and trumpets, is accompanied by a choir consisting of human singers. Each of the groups on its own can perform impressively, but the combination of orchestra and choir, for example in Beethoven's *Ninth Symphony*, can execute pieces that neither can do on its own. Similarly, semantic pointers are neural representations that combine other representations. Just as the orchestra and choir required a director to bring them together and coordinate them, so semantic pointers require neural processes that can bring other neural representations together into more complex ones.

Consider, for example the concept of red wine, which combines the perceptual concept *red* with the more diverse concept *wine*. Red could just be narrowly represented by patterns of firing in a group of neurons in visual brain areas such as

V4, but *wine* is a concept that connects verbal information, such as that wine is a drink made from fermented grapes, with perceptual information about the taste, feel, and smell of wine.

My musical analogy has a director who combines and coordinates the orchestra and the choir, but the brain does not have a director to combine concepts. Instead, such functions need to be carried out collectively by groups of neurons. Semantic pointers are formed by neurons capable of accomplishing a mathematical function called convolution that binds patterns of neural firing into new patterns. It is as if the orchestra and the choir were brought together by a whole committee of directors and vocal coaches who interacted with each other and the members of the orchestra and choir. The committee would serve to bring those two groups together and to coordinate them to produce the new powerful entity of the symphonic choir. The mathematics of convolution are too technical to describe here, but you can think of it as on operation that braids representations together like strands of hair or rope. For a rigorous, mathematical explanation of binding and semantic pointers, see Eliasmith's *How to Build a Brain* and *Brain–Mind* (chapter 2, appendix).

The semantic pointer for wine combines simple representations of verbal information with perceptual information about the look, feel, or taste of wine. Semantic pointers can also bind motor information such as what it feels like to lift a glass of wine in a toast and emotional information about the enjoyment or discontent of drinking wine. The construction of these new patterns can proceed repeatedly, building up more and more complicated structures. We can go from *wine* to *red wine* to *fruity red wine* to *exquisitely robust red wine made from Cabernet Sauvignon grapes*.

The power of semantic pointers is that they show how neural representations can both be formed by sensory inputs and also combined into new structures that go beyond the limitations of the senses. Compare the making of mayonnaise, which ends up as thick and creamy even though its ingredients are not (oil, egg yolk, and vinegar or lemon juice). Whipping the oil and vinegar with egg yolk emulsifies them to produce the emergent properties of being thick and creamy that are not found in the ingredients. Analogously, binding patterns of neural firing into semantic pointers produces new patterns with novel representational capacities. Once formed, semantic pointers can be used for many inferential purposes such as explanation and problem solving, just as mayonnaise can further be combined with other ingredients such as garlic and mustard and used to dress salads or meats. See chapter 5 for a full account of emergence.

Semantic pointers as neural processes are much more than just words in the head or words in the heavens, which are the two ways in which philosophers tend to think of mental representations such as concepts. There is nothing supernatural about them, as they are biological processes that are formed by mechanisms of neural firing. Moreover, they are much more than just words, because they can incorporate multimodal sensory-motor information that provides contact with the world through physical interactions. (I use the term "multimodal" to cover the full range of mental representations: verbal, external and internal senses, motor actions, and emotions.)

A concise definition, flawed as all definitions are, could be: semantic pointers are multimodal neural processes that explain cognition and emotion as results of binding of simpler representations. More useful than a definition is the three-analysis of the concept *semantic pointer* in Table 2.2.

Semantic pointers are semantic in three ways. Philosophers have long noted that meaning is a matter both of the internal relations of representations to each other and the external relations of representations to the world (chapter 8). For example, the concept *wine* has important relations with other concepts such as *drink, alcohol,* and *grape,* but also stands for a liquid observed in glassfuls.

But *wine* also has sensory-motor characteristics, such as its color, smell, mouth feel, drinking motion, and emotional response, connecting it to the world and to the body. Semantic pointers show how to combine both these aspects of the relations of representations to representations and the relations of representations to the world. Representation of the world comes about because the senses interact

TABLE 2.2

Three-Analysis of the Concept *Semantic Pointer*

Exemplars	Images (a glass of red wine), concepts (*wine*), beliefs (drinking wine is healthy), desires (wanting a glass of wine), and emotions (regretting drinking a whole bottle of wine)
Typical features	Pattern of neural firing, produced through binding by convolution
	Connections to other semantic pointers
	Connections to the world by sensory and motor interactions
	Capable of being decompressed into sensory-motor components
	Used in communication
Explanations	Explains: how neural groups can perform cognition, emotion, behavior, communication, social interaction
	Explained by: neural mechanisms that include firing, excitation, and inhibition of neurons, tuning of neurons to environments, and binding by convolution

with the world, for example when light bounces off the wine in a glass and stimulates your retina to produce the perceptions of a red liquid. Moreover, the motor aspects of the representation result from using your body to do things with the wine, such as pouring it into a glass and drinking it. The word "pointer" in "semantic pointer" indicates that the neural pattern for *wine* can be decomposed into (point to) its sensory constituents such as color and taste.

Semantic pointers are also meaningful in a third way, in that they are employed in communication. As philosophers such as Wittgenstein have emphasized, meaning is not just an individual matter of a single speaker but requires interactions among people sharing a language. *Mind–Society* (chapter 3) describes how communication of semantic pointers occurs not just through verbal utterances but also through many kinds of nonverbal behaviors such as facial expressions, gestures, and body language. Hence semantic pointers cover meaning in all three crucial ways: as relation to the world, as relation to other representations, and as a social process in which people interact to share meanings by transfer of semantic pointers from one head to another.

What do semantic pointers do in the brain? When they combine sensory-motor and other information into more powerful representations, they do not throw the contributors away but retain the ability to reconstruct the bodily information that went into them. For example, when the concept *wine* is formed by combining information about words, perceptions, and body actions, the resulting semantic pointer does not just convert it into some word-like structure. Rather, the neural firing process that constitutes a semantic pointer can be used to retrieve some approximation of the sensory-motor information that went into it, just as the combined orchestra and choir in the music analogy can be decomposed into the orchestra and the choir. Semantic pointers are unlike the orchestra and choir, however, in that they cannot be perfectly recovered. The sensory-motor information that goes into a pointer may be only loosely reproduced, just as the compressed music file that plays on a computer is only an approximation to the full sound that was generated by musicians.

Much more rigorous accounts of semantic pointers can be given with biological and mathematical details. My aim here is instead to provide a simplified, accessible account because of the relevance of semantic pointers to philosophical questions including the arts (chapter 9). I outline how semantic pointers are important for solving the mind–body problem by helping to make it plausible that mental processes are brain processes, because semantic pointers are neural mechanisms that potentially explain a broad range of mental functions. If these explanations are more plausible than alternative explanations, then we have support for the generalization that all mental processes result from neural mechanisms. Not all neural processes require semantic pointers, because there are simple operations such as the control of breathing by the cerebellum that do not need rich representations.

The broad range of explanatory successes of semantic pointers are thoroughly described in *Brain–Mind* and the many publications of Chris Eliasmith and his collaborators. Here I only sketch the explanation applied to three kinds of mental process: concepts, emotions, and inference.

Concepts

Concepts are mental representations roughly corresponding to words. Concepts are an important part of psychological theories because of their role in learning, classification, problem solving, and language. Although a few concepts such as *object* and *face* may be innate, children have to acquire thousands of new concepts such as *dog*, *tree*, and *peanut butter sandwich*.

Once acquired, concepts can be used for numerous purposes such as classifying an object in the world as a dog rather than a cat and for figuring out how to get an animal to do something once it is recognized as a dog. Words and concepts usually overlap, but we can have sensory concepts without words and nonsense words without concepts. Concepts are an important part of linguistic processing because all sentences require understanding their constituent concepts. Concepts even allow us to understand sentences never heard before, such as: "The dog ran up a tree after being threatened with a peanut butter sandwich."

The classical view of concepts as definable by necessary and sufficient conditions has been discredited by philosophical and psychological challenges. Philosophers from Plato to Wittgenstein to Hilary Putnam have criticized the classical view, and psychologists have amassed substantial experimental evidence that the classical view is not sustainable. But there have been problems in developing an alternative theory, because some experiments suggest that concepts are prototypes consisting of typical features, while other experiments suggest that concepts are sets of examples or are distinguished by their explanatory roles. The semantic pointer theory of concepts shows how all three of these functions—exemplars, typical features, and explanations—can be accomplished by the same neural mechanisms that accomplish binding of representations into semantic pointers. The adequacy of these mechanisms is shown by computer simulations that implement the mechanisms in a mathematically precise way and reproduce some of the key experimental results.

Because no other current theory of concepts can explain this breadth of results, the method of inference to the best explanation justifies the conclusion that concepts are semantic pointers, from which it follows that concepts are brain processes. Chapter 3 defends the legitimacy of this kind of inference.

The new theory of concepts implies that conceptual clarification should employ the method of three-analysis illustrated in Table 2.2 rather than definitions that provide strict necessary and sufficient conditions. It might seem unfortunate that three-analysis does not always provide a means to settle hard cases about what falls under a concept, for example whether viruses are alive. But this limitation merely reflects the flexibility of concepts that the definitional view tries to conceal. Viruses have some of the typical features of life, such as ability to reproduce, but lack others such as metabolism.

Equally detailed explanations using computer simulations to connect theory with data have been given for mental processes involving perception, arm movement, intention, emotion, creativity, consciousness, and other aspects of mind. Some other explanations are currently much more sketchy and provisional, for example concerning mental imagery and language. We are therefore not yet able to give a strong argument that would jump from an exhaustive listing of mental processes and semantic pointer explanations to the conclusion that all thinking is semantic pointer processing. Rather, my argument is inductive, relying on the fact that many important mental processes have been shown to be explainable in terms of semantic pointers, so that even the most complex kinds of thinking can be explained as neural processes.

You might be worrying that the idea of semantic pointer is vacuous, just another term for mental representation. This worry fails to recognize that the semantic pointer concept is far more precise than the vague and general one of mental representation. First, it relies on specific biological mechanisms including parts such as neurons, interconnections such as synapses, interactions such as transmission of firing capability from one neuron to another by neurotransmitters, and binding of patterns of activity into more complex ones using convolution. Hence semantic pointers explain how mental representations such as concepts work, just as combination with oxygen explains how combustion works.

Second, not all mental representations are semantic pointers, because some low-level perceptions resulting from stimulation of receptors in the eye, nose, and so on are not semantic pointers, which require binding and compression. I also mentioned that there are patterns of firing in the brain that are neither representations nor semantic pointers, because they are simpler processes involved in breathing, balance, and digestion that are biological without being representational.

Emotions

Philosophers and psychologists have long debated the nature of emotions such as happiness. Are they states of supernatural souls, cognitive judgments about goal satisfaction, or perceptions of physiological changes? Semantic pointers explain how brains generate emotions through a combination of cognitive appraisal and bodily perception.

Suppose that something really good happens to you today: you win the lottery, your child gets admitted to Harvard, or someone you like wants to have lunch with you. Naturally, you feel happy, but what does this happiness amount to? On the traditional dualist view of a person, you consist of both a body and a soul, and it is the soul that experiences mental states such as happiness. This view has the appealing implication that you can even feel happiness after your body is gone, if your soul continues to exist in a pleasant location such as heaven. Unfortunately, there is no good evidence for the existence of the soul and immortality, so the dualist view of emotions has little going for it besides wishful thinking.

There are currently two main scientific ways of explaining the nature of emotions. According to the cognitive appraisal theory, emotions are judgments about the extent that the current situation meets your goals. Happiness is the evaluation that your goals are being satisfied, as when winning the lottery solves your financial problems and being asked out holds the promise of satisfying your romantic needs. Similarly, sadness is the evaluation that your goals are not being satisfied, and anger is the judgment aimed at whatever is blocking the accomplishment of your goals.

Alternatively, William James and others have argued that emotions are perceptions of changes in your body such as heart rate, breathing rate, perspiration, and hormone levels. On this view, happiness is a kind of physiological perception, not a judgment, and other emotions such as sadness and anger are mental reactions to different kinds of physiological stages. The problem with this account is that bodily states do not seem to be nearly as finely tuned as the many different kinds of emotional states. Yet there is undoubtedly some connection between emotions and physiological changes.

Semantic pointers show that these theories of emotion—cognitive appraisal and physiological perception—can be combined into a unified account of emotions. The brain is a parallel processor, doing many things at once. Visual and other kinds of perception are the result of both inputs from the senses and top-down interpretations based on past knowledge. Similarly, the brain can perform emotions by interactively combining both high-level judgments about goal satisfactions and low-level perceptions of bodily changes. The judgments are performed in the prefrontal cortex, which interacts with the amygdala and insula that process information about physiological states. Hence happiness can be a brain process that simultaneously makes appraisals and perceives the body.

The integration is naturally performed by semantic pointers that bind representations of three or four constituents: the situation that provoked the emotion, the physiological response, the cognitive appraisal of the goal relevance of the situation, and (in a few species such as humans with large brains) the self. For example, your happiness that a good friend is coming to visit is the semantic pointer that binds neural representations of the visit (which can be visual as well as verbal), the appraisal that your friend helps to satisfy your social needs, the physiological changes such as elevated heart rate, and yourself.

This account of emotions provides precise, mechanistic explanations of a wide range of emotional phenomena, such as emotional shifts and the occurrence of mixed emotions. It also extends to cover emotional consciousness. The semantic pointer theories of concepts and emotions combine to generate a new account of values developed in chapter 6.

Inference and Coherence

To be intelligent, people need to do much more than apply concepts and feel emotions. Problem solving requires inference in which new representations are derived from old ones, as in the logical form of inference modus ponens: if Alice is in the library, then she is studying; Alice is in the library; therefore, she is studying. In logic and computer science, inference is just a matter of syntax, manipulating symbols whose meaning does not matter.

In contrast, the brain is not just a syntactic engine but simultaneously integrates syntax with semantics (meaning) and pragmatics (purpose and context). Semantic pointers elegantly accomplish this integration because they get their meaning from the sensory-motor, inferential, and social processes already described. Computer simulations show that they can contribute to modus ponens and other kinds of logical inference. They handle pragmatics because emotions take goals into account in cognitive appraisals and because neural firings are adjustable based on contexts provided by environmental inputs and ongoing inference.

Moreover, because semantic pointers operate through the parallel processing of billions of neurons connected by excitatory and inhibitory synaptic connections, they naturally can handle the process of parallel constraint satisfaction described in chapter 1. The elements in a coherence problem such as concepts and beliefs are neurally represented by whole semantic pointers, not by individual neurons: there is no single neuron for the concept *dog* or the belief that dogs have ears. Positive and negative constraints are captured by excitatory and inhibitory links among the many neurons that make up the semantic pointers. Parallel constraint satisfaction is accomplished by neural processing that performs semantic pointer competition, a mechanism described later in relation to consciousness. Hence semantic pointers can explain both serial and parallel inferences. Chapter 3 details how coherence can serve to evaluate and justify knowledge claims.

INFERENCE TO THE BEST EXPLANATION TO MULTILEVEL MATERIALISM

The last three sections are just samples of the more detailed explanations available for mental phenomena using the Semantic Pointer Architecture. Taken together, these explanations support the following argument.

1. SPA provides a better explanation of the operations of mind than any other available theory.
2. Therefore, tentatively and approximately, we can accept SPA as the best currently available account of mind.
3. Because semantic pointers are brain processes, we can therefore infer that all mental processes are brain processes.

Step 1 requires considering alternative explanations for mind. Inference to the best explanation is always comparative, and it might be that although SPA is the best current scientific account of mind, it does not provide the best metaphysical account of what is deeply real about the mind. Let us look at the alternatives mentioned earlier.

The semantic pointer version of neural materialism is superior to mind–body dualism in several respects. First, it provides detailed explanations of mental phenomena by describing rich biological mechanisms. All dualism can say is that somehow, mysteriously, the soul manages to carry out mental tasks. Second, neural materialism is simpler than dualism, because it hypothesizes the existence of only one fundamental kind of thing, matter, rather than two different kinds of entities, matter and spirit.

Simplicity is never a virtue in itself, because sometimes more complicated theories are needed in order to explain the full range of phenomena. Modern science postulates 118 elements, many more than the 5 proposed by the ancient Greeks and Chinese. Nevertheless, neural materialism deserves credit for simultaneously explaining more than dualism while assuming less. The main phenomenon that dualism is supposed to be able to explain better than neural mechanism is consciousness.

Third, materialism has no problem figuring out how mind and body can affect each other, because the brain is part of the body, and the physical interactions via the senses and nervous system are well known. In contrast, mind–body dualism has never managed to explain how spiritual souls and material bodies can influence each other.

Mind–brain identity also has numerous advantages over idealism, the view that mind is fundamental in the universe. There is abundant astronomical evidence that matter has been around for at least 13 billion years, whereas there is no evidence that mental phenomena have been around for longer than the less than 1 billion years that animals have existed on earth. Explanations of physical phenomena operate successfully without any mention of mind, and the Semantic Pointer Architecture shows that no special mind stuff is needed to explain complex mental phenomena. Hence idealism is implausible given everything that is known about the operations of the mind and the physical world. Similarly, panpsychism, the claim that everything in the universe has an element of mind to it, is pointless unless there really is something about consciousness that requires everything to have a bit of consciousness in it. The semantic pointer theory of consciousness shows the dispensability of the behaviorally implausible view that atoms and rocks are even slightly conscious.

The semantic pointer version of neural materialism also shows the dispensability of other forms of materialism. Ruthless reductionism that tries to make all materialist explanations operate at one level, such as subatomic physics or molecular biology, is implausible because of the need for explanations operating with mechanisms at different levels. *Brain–Mind* and *Mind–Society* show how neural mechanisms involving semantic pointers mesh naturally with three other levels of mechanisms: molecular, mental, and social. Psychological processes are all neural, but we only understand neural processes by considering their molecular underpinnings, their psychological results, and their social interactions.

Therefore, I am not advocating a mind–brain identity theory. Mind–brain identity is just a slogan for the claim that mental processes are brain processes, which is better formulated as: mental processes are brain processes resulting from neural, molecular, mental, and social mechanisms. The implications of this multilevel version of materialism are discussed in chapter 5 on explanation.

Another version of materialism, functionalism, rejects brain-based materialism because mental states can be realized in many different ways, for example in digital computers and possibly in life forms from other planets that are very different from humans. Although semantic pointers can be effectively simulated in computers, their sensory-motor components will differ enormously between biological and mechanical systems: electronic robot arms and sensors differ markedly from human perception and interaction with the world that operate as a result of hundreds of millions of years of biological evolution. Therefore, it is unlikely that machines will ever have the same physical instantiation as humans, so their minds will be very different. Chapter 10 provides a more detailed discussion of human–machine differences.

Therefore, just as the Semantic Pointer Architecture wins the inference to the best explanation as a scientific theory of mind, multilevel materialism based on it wins the inference to the best explanation as a metaphysical theory. I provided evidence-based answers for two of the fundamental philosophical questions about minds: mental processes are brain processes, and mind and body interact because brains are parts of bodies. Still unanswered are fundamental questions about action and consciousness.

PHILOSOPHICAL OBJECTIONS

Semantic pointers are brain operations that combine verbal, sensory, motor, and emotional information using mechanisms of neural firing and binding by convolution. Despite the explanatory power of the Semantic Pointer Architecture, some philosophers would still insist that it just cannot be a full theory of mind. Let us now look at the kinds of objections that philosophers have made to brain-based theories.

We Can Imagine Minds Without Semantic Pointers

First, consider an argument that has long been a favorite of antimaterialist philosophers, from Descartes to David Chalmers. Minds cannot be brains, even brains construed as semantic pointers, because we can easily imagine having a mind that completely lacks a brain, a body, or anything to do with semantic pointers. Similarly, we can easily imagine our bodies and brains being just as they are but lacking consciousness. Therefore, minds are not brains.

That this argument is a horrendous non sequitur is evident from considering other claims about explanatory identities. We can easily imagine that lightning is not electrical discharge and that combustion is not combination with oxygen, but these acts of fancy are completely irrelevant to the well-founded scientific inferences concerning lightning and combustion. Similarly, imagining the usual physical processes of lightning but without flashes of light, or the usual rapid combination of things with oxygen but without burning, is an exercise in scientific ignorance rather than metaphysical insight. The experimental evidence explained by these identity hypotheses is substantial, and the alternative hypotheses are weak, so the explanatory identifications are thoroughly justified as inferences to the best explanation.

Analogously, because the Semantic Pointer Architecture successfully explains so many mental phenomena as the result of neural mechanisms, the hypothesis that mind is brain-based is not undercut by ignorance-based imagination. Thought experiments that try to use intuitions as a basis for philosophical conclusions are a combination of wishful thinking and circular reasoning, arguing for a desired conclusion merely on the basis of a story that is concocted to support the inference.

Like all scientific hypotheses, the claim that the SPA is the best explanation of brains and minds might turn out to be wrong, in keeping with the fallibility championed in chapter 3. New evidence and better theories may lead to its obsolescence. But based on current evidence, the SPA and the attendant materialist view of mind are reasonable inferences, regardless of whether some philosophers can imagine that they are false.

Minds Are Embodied, Extended, Embedded, and Enactive

Since the 1980s, there have been two main contending approaches to cognitive science. The symbolic approach dating from the 1950s considers intelligence to be the result of computational processes of symbol manipulation of the sort familiar in language and logic. Alternatively, the connectionist approach dating from the 1980s considers intelligence instead as emerging from the interactions of neurons via synaptic connections. Marvelously, the Semantic Pointer Architecture shows how to synthesize these two competing approaches, gaining both the flexibility of connectionist operations in neural networks and the computational power of symbol manipulation.

Some philosophers have maintained that there is now available a third major approach to cognitive science, claiming that minds are embodied, extended, embedded, and enactive (4e cognition). But all of these aspects of mind can be accommodated within the SPA.

The claim that thinking is embodied has both a moderate, plausible version and an extreme version that cannot explain human mental capacities. The extreme form says that, because human thinking is enormously influenced by the kinds of bodies we have, cognitive science can dispense altogether with the notion of mental representation and instead explain human action merely in terms of the ways in which bodies operate in the world.

The failure of behaviorist psychology in the 1950s shows that this kind of radical embodiment is implausible. Even the behavior of rats in mazes cannot be explained without hypotheses about the mental maps that they use to navigate complex environments, and there are countless other phenomena involving perception, planning, problem solving, and language that have needed mental representations for their most plausible explanations. If embodiment required abandonment of the theoretical tool of mental representations, it would mark a desperate retreat in cognitive science rather than an alternative approach.

However, embodiment has a moderate version that is plausible and important. The symbolic approach to cognitive science made thinking too much a matter of language manipulation, with words disconnected from sensory and motor processes. Much experimental work on perception, concepts, emotion, and action suggests that human bodies do have a major influence in the kinds of representations we have. People think with words but also with images that are visual, auditory, tactile, olfactory, gustatory, and kinesthetic.

Moreover, images are important for internal senses such as pain, heat, balance, location in space, and emotions. The use of imagery depends crucially on people's sensory and motor system, so we need to think of mental representations as having concrete forms attuned to vision and other senses. Moderate embodiment is a valuable enhancement to representational theories of mind, not a replacement.

The Semantic Pointer Architecture elegantly integrates the explanatory advantages of both symbolic and connectionist approaches and similarly integrates moderate embodiment. Semantic pointers can be built out of sensory and motor information that they retain in a compressed form. The concept *wine* is not just a verbal symbol but rather a neural process that arises in part out of sensory and motor experiences with tasting wine.

Similarly, the emotion of happiness that you won a lottery arises from internal physiological sensations and visual representation of the winning ticket. Hence concepts, emotions, and other kinds of mental representation that *Brain–Mind* explains as semantic pointers are embodied, making moderate embodiment a support for the SPA rather than a challenge to it.

Moreover, the combinatorial capabilities of semantic pointers enable them to go beyond embodiment to produce kinds of knowledge that would be impossible with only sensory-motor operations. Theoretical ideas in science such as atom, quark, black hole, atomic bond, gene, and virus go far beyond the senses and therefore transcend embodiment. Similarly, mathematical ideas such as infinity and philosophical ideas such as justice do not reduce to sensory representations.

Embodiment is important because much of human knowledge does come through the senses, but we also must recognize what I call "transbodiment," the capacity of human minds to go beyond the senses to construct ideas that get farther and farther away from sensory experience. *Brain–Mind* shows how combining semantic pointers into novel ones by binding that is recursive (repeatedly building on previous structures) is important for explaining creativity, metaphor, and other human capacities. This flexibility and power is also important for explaining scientific knowledge in chapter 3 and mathematics in chapter 10.

The claims that minds are enactive and embedded in the world are also compatible with the SPA. Minds are enactive in the sense that they are concerned with performing actions, not just contemplative thought. The SPA is not just a purely computational exercise but is already being used to run robots that operate in the world by moving arms. Semantic pointers have also been used to provide a general account of intention, emotion and action, as described in *Brain–Mind* (chapter 9). Mental representations such as concepts can bind information about motor operations that are directly concerned with action.

Similarly, the ability of SPA to use sensory and motor information to control robots operating in the world shows that it accommodates the embeddedness of minds. Action and being-in-the-world are not only compatible with semantic pointers but are explained by it, mechanistically. In contrast, philosophical discussions of the third 4e approach to cognitive science have provided few details about the processes that enable minds to be enactive and embedded. Another sense of embedding, concerning the role of minds in the social world, is discussed later.

Finally, consider the extended mind hypothesis that says that objects in the world function as part of thinking. This claim is what Daniel Dennett calls a "deepity," a declaration that might appear profound at first glance but upon closer inspection turns out to be trivially true or clearly false. Obviously, people use tools to extend their minds, such as notebooks, drawings, computers, and instruments. Semantic pointers include representations of such objects and the representations they contain, along with multimodal rules and other means of interacting with these tools. A multimodal rule is an *if–then* mental representation where the *if* and *then* parts can be sensory-motor or emotional, as discussed in chapter 3.

Commonplace observations about minds operating in the world do not challenge the centrality of brain mechanisms in explaining thinking. Brains obviously connect with the world through sensory-motor operations that are incorporated into the SPA. The objects in the world that assist people with thinking are unlike the systems of interacting neurons that enable people to think and use such tools.

To understand how minds extend into the world, we need to know about neural operations that enable people to get sensory information and act on it as a result of motor processes. The stronger claim, that there is metaphysical importance in the ability of minds to influence and be influenced via neural and sensory processes, adds nothing. My iPhone is part of my mind only by a loose metaphor that garbles normal part–whole relations such as how my brain is part of my body. I use my brain and body to manipulate the iPhone, whereas the iPhone can affect my brain only through my body's sensory apparatus.

Philosophers influenced by phenomenology claim that the body and the world are as important to human action as the brain. They are indeed important, but notice that insects and reptiles also have bodies that operate in the world, without the enormous repertoire of intelligent activities accomplished by humans. This range requires transbodiment as well as embodiment, both of which are supported by semantic pointers. Consciousness may well operate in simpler organisms such as fish, but humans have a sense of themselves in social contexts that requires a very large brain for recursive binding (i.e., bindings of bindings).

In sum, embodied, embedded, and enactive approaches to cognitive science make some important observations about how minds perform with bodies in the world but hardly constitute an alternative approach. These observations can be not only accommodated but also explained within the Semantic Pointer Architecture. The brain is a dynamic system of a special kind, using neural representations to deal with a complex and changing world. Stronger claims such as that embodiment eliminates representation and that extended minds include the world are rhetorical exaggerations.

Minds Are Social

Some philosophers have worried about emphasizing minds as brains because of the important social roles that people play. For example, the important questions about morality discussed in chapter 6 require attention to the social contexts in which people act: how we affect each other and how other people make moral judgments about us. It might seem that the Semantic Pointer Architecture treats minds as individual brains utterly disconnected from the social world.

However, *Brain–Mind* (chapter 12) presents a theory of the self for which social mechanisms are as important as molecular, mental, and neural ones. *Mind–Society* extends semantic pointers from a theory about individual thinkers to cover the full range of human interactions based on both verbal and nonverbal communications. These social mechanisms based on both individual thinking and communicative interaction show how semantic pointers help to explain many important phenomena in the social sciences and professions, from romantic relationships to business leadership. Hence the semantic pointer theory of mind and brain accommodates social operations and explains how they work. Appreciating the neural mechanisms responsible for cognition, emotion, and communication are crucial for explaining the diverse kinds of social interaction that occupy people every day.

CONSCIOUSNESS

The greatest challenge for a materialist theory of mind is explaining conscious experiences that are part of perceptions, emotions, and other kinds of thinking. Behaviorists have even denied that consciousness exists, leaving nothing to be explained. But every human is familiar with experiences such as seeing blue, hearing a guitar, tasting beer, feeling happy or sad, being in pain, getting hot or cold, losing their balance, and thinking about what to do next weekend. Natural philosophy needs to explain consciousness, not ignore it.

Fortunately, neural explanations of consciousness are becoming available, and I review three, based on information integration, neuronal workspace broadcasting, and semantic pointer competition. None of these is yet a full theory of consciousness, but the third is a promising start to a comprehensive, materialist theory of consciousness.

The questions that a theory of consciousness should be able to answer include the following: Why do people have different kinds of conscious experiences? How did consciousness evolve? Why do conscious experiences start and stop? Why does consciousness shift from one matter to another? Are different animals capable of different kinds of consciousness? How is consciousness diverse but unified? What brain mechanisms contribute to consciousness? I do not include on this list hopelessly vague questions such as why is there "something it is like" to be conscious.

Information Integration

In numerous articles and a book, the neuroscientist Giulio Tononi has advocated the information integration theory (IIT) of consciousness. He claims that consciousness in humans and other entities is the result of their ability to integrate information of different kinds. This claim seems plausible at first glance, because our conscious experiences do integrate various kinds of information, as when my being aware of shooting a basketball combines sights, sounds, touches, and muscle movements into a unified experience. Consciousness occurs in anything that has more information as a whole than in its parts.

On closer examination, however, Tononi's IIT has numerous flaws.

1. IIT attributes consciousness to entities such as photodiodes and cell phones that show no behavioral evidence of being conscious. Tononi bites the bullet but clarifies that he does not make the panpsychist claim that everything has a degree of consciousness. Smartphones clearly do integrate information, in that the whole phone ties together signals gathered by various parts, including WiFi, cellular data, camera, microphone, and keyboard. Similarly, countries like Canada have governmental agencies that have more information than their parts such as cities. IIT attributes consciousness to entities that display no behavioral evidence of consciousness. IIT even implies that individual neurons are conscious, because they integrate information beyond their proteins.

2. IIT is vague. Information for Tononi is not the precise mathematical notion of Claude Shannon but a richer meaning-related phenomenon in systems in which there are "differences that make a difference." He means causal differences, but he never says what these are. Integration occurs when mechanisms are not reducible to independent components, but this is useless without some precise characterization of reducibility and independence. An experience is identified as a "conceptual structure," which turns out to be an unspecified probability distribution.

3. IIT's mathematical characterization is flawed. Mathematical formulations are meant to clarify ideas, not obscure them, but it is difficult to understand how to calculate the quantity PHI that Tononi says is a measure of information. Critics conjectured that calculating PHI would be computationally intractable, and Tononi confirms that "the present analysis is unfeasible for systems of more than a dozen elements." Brains have billions of neurons, and modern computers have billions of transistors, so we cannot even begin to calculate their quantity of information integration.

4. The axioms of IIT are not self-evident, contrary to Tononi's claims. The five central axioms of IIT may be true, but they are not self-evident. The first axiom is that consciousness exists, whose lack of self-evidence is clear from the fact that smart people have denied it: behaviorist psychologists have tried to expunge consciousness from science, and some philosophers have also doubted its existence. I think that consciousness does indeed exist, but it takes evidence and theoretical argument to defend this inference. I argue in chapter 3 that there are no self-evident truths. Therefore, the information integration theory of consciousness is too flawed to be accepted.

Neuronal Workspace Broadcasting

Stanislaus Dehaene is a distinguished French cognitive scientist who has done important experimental work on conscious experience using instruments such as functional magnetic resonance imaging to identify the cerebral basis of consciousness. To explain the results of these experiments, he developed a theory that claims that consciousness is information broadcasting within the cortex, arising from a network of neurons that function to share pertinent information across the brain. This theory is much better than Tononi's because it relies on known neural mechanisms rather than a mathematically obscure notion of information. Moreover, it does not ascribe consciousness to objects such as cell phones that provide no behavioral evidence of consciousness such as verbal reports and physical reactions to pain.

Dehaene describes specific operations that consciousness performs. Conscious information is stable, enabling us to hang on to it for a longer time than fleeting subliminal information. Consciousness compresses incoming information from a huge stream of information to carefully selected symbols. The selected information can be routed to other processing stages to perform chains of operations. Hence consciousness functions to broadcast information, within the brain but also between people via language.

Dehaene describes four kinds of brain activity that serve as signatures of consciousness, indicating that a stimulus was consciously perceived. First, when a stimulus generates brain activity that crosses a threshold for awareness, activity spreads to circuits in parietal and prefrontal areas. Second, conscious access appears as a slow wave detected by electroencephalograms. Third, electrodes placed in the brain reveal sudden bursts of high-frequency oscillations. Fourth, synchronization of information occurs across distant brain regions.

These signatures are experimentally observed correlates of consciousness, not causes of it that operate in a set of mechanisms. Dehaene summarizes:

> Conscious perception results from a wave of neural activity that tips the cortex over its ignition threshold. A conscious stimulus triggers a self-amplifying avalanche of neural activity that ultimately ignites many regions into a tangled state. During a conscious state, which starts approximately 300 milliseconds after stimulus onset, the frontal regions of the brain are being informed of sensory inputs in a bottom-up manner, but these regions also send massive projections in the converse direction, top-down, and to many distributed areas. The end result is a brain web of synchronized areas whose various facets provide us with many signatures of consciousness: distributed

activation, particularly in the frontal and parietal lobes, a P3 wave, gamma-band amplification, and massive long-distance synchrony. (p. 140)

Dehaene explains these signatures by proposing that consciousness is brain-wide information sharing using long-distance networks in the prefrontal cortex to select and disseminate information throughout the brain.

These observations are important but fall short of a satisfactory explanation of consciousness for the following reasons. First, they are weak on description of mechanisms of how neural broadcasting can occur across multiple brain areas. Second, workspace broadcasting says nothing about why various conscious experiences are so different from each other. Dehaene's hypothesis that consciousness is just the flexible circulation of information within a network of cortical neurons does not seem to be able to explain why there are different kinds of experiences in perception, emotion, and thought.

Third, the brain can accomplish broadcasting across brain areas without consciousness, raising the question of what else must be added to produce consciousness. Fourth, global broadcasting does not explain the unity of consciousness, for example why experience of the glass of wine ties together its shape, color, and taste.

Semantic Pointer Competition

Dualists, phenomenologists, and some analytic philosophers find it impossible to imagine how brain processes could explain the varieties and character of conscious experience. Their attitude is similar to that of some nineteenth-century biologists who thought that life could never be explained without introducing some special life force that exceeded biological mechanisms. But no one invokes life force anymore, because life is widely viewed as the result of various interacting mechanisms, including cell division, metabolism, digestion, and sexual reproduction. The study of consciousness today is like the study of life in the nineteenth century, when discovery of the relevant mechanisms was just beginning.

Nevertheless, neural mechanisms that help support consciousness are becoming known, including ones that operate more basically than information integration and neuronal workspace broadcasting. In line with the SPA, three mechanisms important for consciousness are neural representation, binding by convolution that produces representations available for subsequent recursive binding, and competition among semantic pointers. This chapter only sketches the semantic pointer theory of consciousness, as details are provided in *Brain–Mind* (chapter 8) and in my 2014 journal article with Terry Stewart that reports computer simulations of phenomena such as shifts in conscious experiences.

Neural representations are patterns of firing in groups of neurons, able to represent things in the world because sensory-motor interactions produce causal correlations between what happens in the world and what goes on the brain. For example, neurons in your brain can represent the red of a glass of wine because they fire systematically in ways caused by the reflection of light from red things into the retina and subsequent neural processing in several brain areas. For simple stimuli, this mechanism already begins to explain how different conscious experiences can occur, because different sensory systems generate different patterns of firing, for example in producing the color red or the taste of wine.

Neural representation by itself would support only simple conscious experiences of one aspect at a time. But conscious experience is often unified, for example when you perceive a glass simultaneously as tall and round and filled with red wine. Such combined experiences require the mechanism of binding that takes different features and makes them into a unified whole. Even newborn chickens are capable of binding features in this way. Neural binding is often explained by synchronization, for example with the neurons that represents *tall* and *round* coordinated to produce a pattern that signifies *tall-round*. But neural synchrony alone fails to explain how the bound representation can be used for additional purposes, for example to include binding of *tall-round* with *red-wine*.

In contrast, binding by convolution produces a combined representation that is susceptible to further binding, allowing additional combinations subject to the limitation that it takes many neurons to perform convolution. Simple animals like fish or small mammals do not have enough neurons to be able to do complicated combinations, so their conscious experience is probably limited to simple features. With larger brains, however, advanced mammals such as humans, chimpanzees, and dolphins can perform recursive bindings that incorporate numerous features and even a sense of self. Binding by convolution explains how conscious experience can be both unified and diverse and also expands the explanation of why different experiences have different characters. It is not just stimulation by the environment that produces different patterns of neural firing but also the binding of different neural representations together.

Brain–Mind (chapter 2) describes how semantic pointers have the capability of *modal retention* not found in other cognitive architectures. Unlike symbols, semantic pointers retain in approximate, compressed form the information resulting from the particular sensory or motor modality that produced them. This information can then be partially regenerated through decompression when semantic pointers are unpacked to produce some approximation to the sensory inputs that went into them. Modal retention is important for explaining why sensation and imagery give rise to a great variety of conscious experiences.

Neural representation and binding are not enough to explain another important aspect of consciousness: it is heavily affected by attentional shifts. At one moment, you may be conscious primarily of a conversation with a friend but then shift dramatically when your phone rings. To explain attention, we can use competition, the same mechanism that the semantic pointer theory of concepts uses to explain classification. We classify animals, for example, by neural competition among concepts such as *dog* and *cat*, and such concepts and other semantic pointers compete by mutual inhibition to enter the narrow space of consciousness.

The formation and functioning of semantic pointers is unconscious, and only limited numbers of these representations will be sufficiently important to break through into consciousness. A situation such as your phone ringing and showing a picture of a close friend enables the neural representation of your friend to outcompete lesser matters you were thinking about.

Emotions are a major part of conscious experiences, in ways explained by competition among semantic pointers that bind representations of situations, appraisals, and physiology. You may be in a bad mood, but when you hear that your favorite sports team just won a big game, your resulting happiness outcompetes other emotions to enter your conscious experience.

Binding by convolution in neural networks accomplishes information integration much more specifically than in Tononi's theory, using neural representations, binding, and attentional competition. It takes all three of these mechanisms to produce consciousness, not just integration, which is an effect of binding rather than the basis for consciousness. It might be objected that the semantic pointer theory of consciousness is parochially restricted to biological systems with brains. But to date, such systems are the only ones that have been found to exhibit behaviors that suggest consciousness such as facial reactions to noxious stimuli, so the restriction is empirically appropriate.

If none of your semantic pointers have neurons with much firing, then you need not be conscious of anything, as happens during some conditions such as dreamless sleep and vegetative comas. The neurons constituting semantic pointers need to be firing with a rate or pattern that exceeds a threshold to be conscious. Events such as falling asleep because of adenosine buildup, receiving anesthetic chemicals, or getting blows to the head can push neurons below this threshold. Consciousness can be regained through fairly local processing in the brain, for example with the stimulus of a phone ringing, rather than through Dehaene's process of broadcasting among brain areas. Broadcasting naturally occurs because of axonal connections between neurons spread throughout the brain, but the broadcasting is an effect of the mechanisms for consciousness, not a cause, just like information integration.

There is currently not enough knowledge to judge whether consciousness evolved by natural selection to improve fitness for survival and reproduction or whether it is just a side effect of neural mechanisms that do increase fitness. Perhaps consciousness was a side effect for simple organisms such as mice but turned out to be advantageous for humans with larger brains that benefit from conscious awareness that supports better learning, teaching, social interactions, and delayed gratification.

Either way, we get a biological answer to the question of why humans ended up with conscious experiences, through the evolution of neural mechanisms that include neural firing, binding, and competition. Explanations of how different experiences arise in people come from combining neural specifics of experiences based on perception, thought, and emotion with these mechanisms. For example, emotional experiences such as feeling happy or sad are explained by bindings of neural representations of bodily states, appraisals, situations, and selves into semantic pointers that outcompete other representations. Perceptual experiences such as color, sound, and taste require different firings and bindings and hence feel differently from emotions. Thus progress is being made on the so-called hard problem of explaining qualitative experiences.

Admittedly, there is still an explanatory gap between brain theory and consciousness, but this gap is shrinking thanks to an increasing understanding of mechanisms such as semantic pointers. The history of science shows that filling explanatory gaps is a long process: it took centuries for water to go from an unanalyzed element to an unexplained compound of hydrogen and oxygen to a substance resulting from bonds between atoms caused by electron mechanisms. Better understandings of neural representation, binding, competition among semantic pointers, and mechanisms to be discovered should continue to fill the explanatory gap for consciousness.

Supernatural Explanations of Consciousness

Empirical and theoretical investigations of the neural basis for consciousness are still in their early days, and it may well turn out that neural representation, binding by convolution, and semantic pointer competition need to be supplemented with other mechanisms in order to explain a broader range of phenomena. Nevertheless, the semantic pointer theory provides a good start in seeing how consciousness is a brain process.

If any of information integration, neural workspace integration, or semantic pointer competition turns out to be the winning scientific theory of consciousness, then the explanatory appeal of dualism and idealism evaporate. Already the attribution of consciousness to a special mind-stuff is a feeble story compared to growing attempts to work out how brains produce consciousness through interactions with the body and the world. Conscious experiences occur because brains have evolved with capacities for representation, binding, and competition, not because consciousness is an original constituent of the universe.

SUMMARY AND DISCUSSION

Dualism, idealism, and various forms of materialism are philosophical theories intended to answer general questions about the nature of mind and its relation to actions and consciousness. I have defended multilevel materialism, which contends that all mental processes are brain processes while allowing the importance of molecular, mental, and social mechanisms that complement neural ones. Eliasmith's Semantic Pointer Architecture provides a good candidate for explaining how the brain has thoughts and conscious feeling. Representation by patterns of firing in groups of neurons, binding of representations into more complex ones by convolution, and competition among semantic pointers serve to produce perception, inference, and consciousness.

Twenty-five hundred years ago, Hippocrates asserted in his work on epilepsy that "from nothing else but the brain come joys, delights, laughter and sports, and sorrows, griefs, despondency, and lamentations." Only in the last century has understanding of the brain grown to make this assertion highly plausible. Only in the last decade has a sufficiently detailed account of the brain developed to provide mechanistic explanations of a broad range of mental and social phenomena.

Philosophical objections to materialism are easily dealt with. The conceivability of minds without brains and of mental processes without semantic pointers is of no relevance to how minds actually operate in this world, just as the conceivability of water that is not H_2O and of H_2O that is not water does not undermine the fact that water actually is H_2O.

Because of their sensory-motor operations, semantic pointers naturally incorporate important aspects of embodiment and action embedded in the world while also enabling minds to transcend the body in order to engage in abstract thought. The operations of minds in social situations requires a theory of communication as the transfer of semantic pointers by verbal and nonverbal means. Minds are extended in the sense that they employ external resources, but brains are similarly extended through their sensory and motor connections with the world, without any need to suppose that external tools are parts of brains or minds.

The major challenge to scientific, materialist theories of mind is the problem of consciousness, of how to make sense of people's qualitative experiences. Introspection and systematic self-observation are a useful start in figuring out what needs to be explained, but deeper explanations require identification of underlying mechanisms. Dualism and idealism rest content with mysteries about how souls and spirits just have consciousness built-in. Panpsychism overextends consciousness to entities that show no evidence of being conscious in the vain hope that bits of consciousness add up to full-blown human consciousness.

In contrast, neural theories are starting to identify neural mechanisms that can explain numerous aspects of consciousness including its diverse characteristics, unity, initiation, attentional shifts, and degrees of complexity across organisms. Consciousness emerges from the interactions of mechanisms, in line with the account of multilevel emergence presented in chapter 5.

The semantic pointer theory of consciousness explains consciousness and numerous phenomena using mechanisms of neural representation, binding by convolution, and semantic pointer competition. The importance of neural mechanisms undercuts assumptions that materialist explanation needs to reduce everything to fundamental operations of physics. Rather, explanation of human behavior can valuably operate at multiple levels, including molecular, mental, and social ones. The identity claim that mind is brain is a useful heuristic because of the great contribution of neural mechanisms to human thought and action, but is simplistic as a metaphysical theory.

The theory of mind developed in this chapter has implications for all of the philosophical questions considered in the rest of this book. Chapter 3 describes how epistemology is altered by an enriched view of the nature of knowing minds. Multilevel materialism with respect to minds is only one aspect of a general metaphysics, but chapter 4 shows how other aspects cohere with it. Ethical questions about morality, justice, and the meaning of life in chapters 6 to 8 depend heavily on understanding of emotions as semantic pointers. Chapter 9 shows that aesthetic questions about beauty and other reactions to painting and music become open to new answers when the mind is viewed as operating with the neural mechanisms of the Semantic Pointer Architecture. Hence a multilevel materialist theory of mind based on how brains operate is the key to connecting traditional philosophical questions.

NOTES

For dualism, see Robinson 2016, Chalmers 2010, and Nagel 2012. For further critique, see Thagard 2010b.

Guyer and Horstmann 2015 review idealism.

Seager and Allen-Hermanson 2010 survey panpsychism.

For an introduction to materialist theories of mind, see Churchland 2013. Ramsey 2013 reviews eliminative materialism.

More evidence for multilevel materialism is in *Brain–Mind* and *Mind–Society*. The importance of explaining the mind at more than one level is also emphasized by McCauley and Bechtel 2001, Bechtel 2017, and Piccinini and Craver 2011. On levels of explanation, see also Findlay and Thagard 2012.

The SPA is in Eliasmith et al. 2012, Eliasmith 2013, and in numerous papers (http://compneuro.uwaterloo.ca/publications.html). For binding by convolution, see Plate 2003 and Eliasmith and Thagard 2001. The chapter 2 appendix in *Brain–Mind* provides a more technical account of semantic pointers and convolution, including comparisons with other kinds of mental representation.

Thagard 2014c discusses explanatory identities. On personal identity, see https://www.psychologytoday.com/ca/blog/hot-thought/201804/are-you-the-same-person-you-used-be.

Finger 1994 provides a history of neuroscience. Important theoretical advances include Hebb 1949 and Rumelhart and McClelland 1986.

Thagard 2012a reviews cognitive architectures. See also the comparisons in the appendix to chapter 2 of *Brain--Mind*.

The semantic pointer theory of concepts is presented in Blouw, Solodkin, Thagard, and Eliasmith 2016. Murphy 2002 reviews much of the evidence for it.

On emotions as semantic pointers, see Thagard and Schröder 2014 and Kajić, Schröder, Stewart, and Thagard forthcoming.

Thagard and Aubie 2008 describe how neural populations can compute coherence.

The limits of philosophical imagination and intuition are discussed in Thagard 2014e.

On embodiment, see Dove 2011, Pezzulo et al. 2011, and Shapiro 2014. The distinction between extreme and moderate embodiment is from Thagard 2012c.

Thompson 2007 and Noë 2009 emphasize enactivism. Schröder, Stewart, and Thagard 2014 provide a semantic pointer theory of emotion and action.

Clark 2008 defends the extended mind. The deepity idea is from Dennett 2013. For a parody called the "extended breath," see https://www.psychologytoday.com/blog/hot-thought/201310/the-extended-breath.

For consciousness as information integration, see Tononi 2004, 2012; Oizumi, Albantakis, and Tononi 2014 (quote from p. 24); and Tononi, Boly, Massimini, and Koch 2016. For critique, see Cerullo 2015, Thagard and Stewart 2014, and http://www.scottaaronson.com/blog/?p=1799.

Dehaene 2014 presents the neural workspace theory of consciousness. The quote is from p. 140.

Thagard and Stewart 2014 present the theory and computational model of consciousness as semantic pointer competition. See also *Brain–Mind*, chapter 8. Dennett 2017 describes consciousness as an "evolved user-illusion," but conscious experience needs to be explained rather than explained away. See chapter 5 on premature elimination.

The Hippocrates quote is from: http://classics.mit.edu/Hippocrates/sacred.html.

PROJECT

Perform three-analyses of the concepts of *dualism* and *materialism* and use the technique of value maps (chapter 6) to indicate their emotional differences.

3

Knowledge

MINDS AND KNOWLEDGE

Knowledge matters. Should parents vaccinate their children against measles, or should they hold off in order to avoid autism? The answer depends on what is known about the effectiveness of vaccines and their alleged links to autism. Should governments take strong measures to reduce greenhouse gases in order to prevent global warming from producing sea-level rises and other calamities? The answer depends on what is known about the causal chain from industrial activity to climate change to disasters. Both science and everyday life demand a theory of knowledge.

A good theory of mind helps to constrain and shape epistemology, the theory of knowledge, but this influence is not because philosophy of mind provides a foundation for the rest of knowledge. In philosophy, as in the rest of thought, there is no pivotal point that provides a foundation for everything else. Rather, the theory of mind is an important contributor to the general fit among theories of knowledge and reality.

This chapter shows how viewing mind from the viewpoint of the Semantic Pointer Architecture provides new perspectives on the most important epistemological questions. What is knowledge? How does it grow? How is it justified? Is there any knowledge at all? To what extent is knowledge social? My answers to these questions differ markedly from the standard ones in analytic philosophy and phenomenology, in keeping with the explanatorily powerful theory of mind in chapter 2.

Some of the conclusions defended in this chapter are as follows. Knowledge is much more than beliefs, because it extends beyond words and sentences to include nonverbal representations that are visual, auditory, tactile, olfactory, gustatory, and visceral. Nonverbal aspects of knowledge are easily accommodated within the Semantic Pointer Architecture, which covers knowledge about how to do things and knowledge of sensory experience. Moreover, emotions, which are often seen as antagonistic to knowledge, can actually be an important part of it, especially with respect to values. Emotions are also important for explaining how knowledge builds up over time, taking into account the passionate pursuit of discovery as well as justification. On the other hand, emotions can also undermine knowledge when personal motivations swamp evaluation of evidence, as has happened with vaccines and climate change. *Brain–Mind* (chapter 6) presents a semantic pointer theory of emotion.

Consonant with the theory of mind, justification is based on coherence, on how pieces of knowledge fit with each other rather than being independently justified. Coherence also shows the implausibility of the skeptical view that people have no knowledge of all. A recent version of skepticism is that there is no objective knowledge because science and other activities are socially constructed, but a sophisticated social account of knowledge development, tied to the operations of individual minds, undermines this kind of skepticism as well.

ISSUES AND ALTERNATIVES

Philosophy is concerned not only with local questions about what is known in particular areas but also with general questions concerning the nature and justification of knowledge. Whereas astronomy is about the stars and medicine is about the human body, epistemology considers knowledge across the full range of science and human experience. It addresses not only what people think about knowledge but also normative questions concerning whether they are justified in how they think and how they can be helped to think better. A good theory of knowledge should provide answers to all of the following questions:

1. What is knowledge? A good answer to this question requires more than a definition, providing insight into the structure of what is known.
2. How does knowledge grow? Knowledge should be viewed dynamically, as the historical process of increasing knowledge over time, rather than the sudden acquisition of a whole system of knowledge. Early humans

100,000 years ago had valuable knowledge about their environments and their social lives, but they did not acquire this knowledge all at once: they attained it through lifetimes of individual and social experiences. Around 6,000 years ago, humans developed powerful new ways of increasing knowledge using writing, mathematics, and science. Epistemology needs to be able to explain not only the growth of knowledge in individuals, before and after these cultural developments, but also the enormous accomplishments of science as a social activity.

3. How is knowledge justified? Gaining knowledge is not merely the acquisition of mental representations such as concepts and beliefs, many of which may be erroneous. Epistemology should provide a theory of what makes some mental representations better than others at capturing the nature of the world. Mention of the world shows that questions about knowledge are interconnected with questions about reality, the topic of the next chapter.

4. Is there knowledge at all? Skepticism cannot be refuted by some conclusive deductive argument, for one could obstinately deny that deductive argument is a source of knowledge. Rather, coherence-based arguments show how both traditional and postmodernist versions of skepticism are implausible.

5. Is knowledge social? Epistemology has usually been concerned with individuals, but postmodernists claim that all knowledge is socially constructed. More moderately, there are identifiable ways in which groups and their values influence the development of knowledge.

A successful theory of knowledge should provide plausible and interconnected answers to all of these questions. The history of philosophy has had various candidates for describing and justifying knowledge. The most intuitive view is that knowledge is like a building that requires a strong foundation on top of which the rest of knowledge can be firmly based. There are two main types of foundation-based epistemologies: rationalists look for a foundation in the products of reason alone, whereas empiricists look for a foundation in sense experience. Unfortunately, attempts to find such foundations have repeatedly failed, threatening to leave abject skepticism as the ugly alternative.

Happily, coherence provides a powerful alternative to foundations, looking for pieces of knowledge that fit together like a puzzle rather than deriving from some solid foundation. Coherence-based philosophers have used alternatives to the foundation metaphors. Knowledge is a cable of interconnecting strands rather than a chain of crucial links. It is a pyramid, a web of belief, or a raft to be repaired piecemeal as it floats along, not a building that has a foundation. These metaphors are suggestive but lacking in psychological detail and explanatory rigor. A much stronger account of coherence, sketched in chapter 1, analyzes coherence as constraint satisfaction.

One of the greatest impediments to an adequate theory of knowledge is the specious pursuit of certainty and necessity. In contrast, coherence allows for the virtuous fallibility of what is believed in science and everyday life, while showing how justification can be based on adequate reliability rather than on absolute certainty. The view I call "reliable coherentism" provides answers to epistemological questions that are easily seen to be superior to those provided by foundationalism and skepticism.

WHAT IS KNOWLEDGE?

The question "what is knowledge?" seems to be a request for a definition, but more than 2,000 years of philosophy has failed to produce a generally accepted definition. The closest is the claim that knowledge is true justified belief, which says that you know something if and only if you believe it, your belief is justified by rational procedures, and the belief is true: the world is as your belief describes it. This definition is not a bad approximation to knowledge but has numerous problems.

Counterexamples

First, clever philosophers since Edmund Gettier have generated numerous counterexamples like the following: Suppose that you believe that there is a green car on the street in front of where you live because you just saw that car. Without you noticing, someone just drove off in that green car but another green car replaced it. Then your belief that there is a green car is justified because you saw one, and it is true because there is a green car, but many philosophers think that this is not really knowledge because the cars were switched. Attempts have been made to add additional conditions to truth, belief, and justification to overcome counterexamples, but the process has not resulted in any standard definition that most philosophers accept.

From the perspective of the semantic pointer theory of concepts described in chapter 2, the failure to find a definition of knowledge as a set of necessary and sufficient conditions is not at all surprising, because very few concepts outside of mathematics have such definitions. Hence it is much more reasonable to look for typical features of knowledge in accord with the three-analysis of knowledge presented later. Then truth, justification, and belief (along with some other conditions such as reliability that philosophers have tried to use to repair the definition) all add up to typical features rather than to necessary and sufficient conditions.

Nonverbal Knowledge

Second, the restriction of knowledge to belief is problematic because it makes knowledge excessively verbal. In analytic philosophy, knowledge is taken as a propositional attitude, a relation between an abstract self and an abstract meaning of a sentence. In contrast, natural philosophy ascribes knowledge to real selves with real mental representations. *Brain–Mind* (chapter 12) presents a multilevel mechanism theory of the self and shows (chapter 2) how beliefs can be built up of semantic pointers, including ones that are primarily sensory-motor rather than verbal. Knowledge can consist of mental representations that have no verbal aspect at all, such as pictures. My mental map of Canada constitutes knowledge of the spatial relationships of the Canadian provinces to the American states without me having any explicit verbal beliefs about which provinces abut which states. I can also have knowledge about tastes, smells, touches, sounds, pains, and so on without converting any of these into verbal, propositional beliefs. We need a conception of knowledge to cover this kind of knowledge-of in addition to the more traditional conception of knowledge as knowledge-that.

Moreover, the assumption that knowledge consists only of beliefs ignores important kinds of knowledge of how to do things, such as row a boat, play a guitar, cook an omelet, knit a sweater, shoot a basketball, fix a computer, and so on. Knowledge-how is also important in science because of the skills needed to use instruments and conduct good experiments. Attempts to reduce knowledge-how to knowledge-that implausibly suppose that all of our visual, motor, tactile, and other kinds of nonverbal information can easily be translated into words. These mental representations, however, are subject to procedures that are very different from the kinds of logical inference that people apply to verbal beliefs. For example, a mental picture allows operations such as scanning, rotation, flipping, and juxtaposition that lack verbal equivalents. Similarly, imagining a song and varying it along dimensions such as tempo and pitch is a nonverbal mental operation. Knowledge-how concerning art, music, sports, and many other areas of procedural, implicit knowledge do not simply reduce to knowledge-that because they are based on nonverbal representations.

Brain–Mind (chapter 5) describes how much of human knowledge is based on multimodal rules built out of semantic pointers. For example, the rule *if you pull* on the oars of a boat, then the boat will move forward* operates in the mind by means quite different from words. Instead, this knowledge-how consists of a nonverbal rule like <*pull-oars*> → <*boat-moves*>. The brackets indicate that the mental representation of this rule requires semantic pointers based on sensory and motor information. The representation of pulling oars is partly motor as you imagine yourself holding and pulling the oars and partly visual as you see your arms and the oars in motion. Similarly, the mental representation for the boat moving combines your proprioceptive sense of motion and the visual image of the boat moving through the water, possibly combined with the sound of the boat against the water.

The arrow indicates that the connection between the two semantic pointers is not the verbal *if–then* but rather a multimodal causal schema of the sort discussed in chapter 5. Applying a multimodal rule is not the result of logical inference using modus ponens (If P then Q, P, therefore Q) but rather a process of matching the first part of the rule and then executing the second part.

Truth and Justification

Third, with the move away from belief as the fundamental carrier of knowledge to a broader range of representations, the notion of truth in the standard definition has to be broadened. Different mental representations such as visual and auditory images do not correspond to the world as neatly as true beliefs are supposed to do but rather correspond to the world in degrees of approximation. My knowledge of the geography of Canada captured by my mental map is only approximately correct, for example concerning just how Vermont adjoins Québec. Therefore, the assumption that knowledge requires truth needs to be broadened to allow more flexible approximation relationships carried by sensory and motor representations.

Similarly, the traditional account of justification needs to be expanded to go beyond verbal inference based on logic as the means of increasing knowledge. Justification as coherence based on parallel constraint satisfaction accommodates many kinds of mental representations, including sensory and motor images as well as verbal beliefs.

Three-Analysis of Knowledge

The net result of these modifications might seem to be a revised version of the definition of knowledge as true justified belief: Knowledge consists of multimodal mental representations that approximately correspond to the world and are coherent with other representations. More succinctly, knowledge consists of reliably acquired and approximately accurate representations of reality. However, such definitions would still be subject to counterexamples such as the green car, so it is much better to move to a three-analysis along the lines of Table 3.1. By following the semantic pointer theory of mind and abandoning the futile search for definitions, this three-analysis answers the fundamental question of what knowledge is.

The three-analysis shown in Table 3.1 is compatible with the recent emphasis in phenomenology and cognitive science on the embodiment of knowledge. Much of our knowledge about the world is embodied in the sense that our mental representations are shaped by the sensory and motor capabilities of our bodies. As chapter 2 showed, the importance of sensory-motor experience is fully recognized within the Semantic Pointer Architecture's support of moderate embodiment, not the extreme version that eliminates mental representation.

However, the Semantic Pointer Architecture also supports transbodiment, the human ability to create mental representations that go beyond body-bound

TABLE 3.1

Three-Analysis of the Concept *Knowledge*

Exemplars	Perceptions (e.g., color and taste of milk)
	Everyday knowledge (e.g., that cows make milk)
	Scientific knowledge (e.g., that cows evolved by natural selection), mathematical knowledge (e.g., $2 + 2 = 4$)
	Knowledge-of (e.g., how milk tastes)
	Knowledge-how (e.g., how to milk a cow)
Typical features	Mental representations: beliefs, images, and nonverbal rules
	Approximate correspondence to the world
	Justification using reliable perception and coherence processes
	Social influences including testimony
Explanations	Explains: the difference between getting the world right and getting it wrong, and our ability to work effectively in the world
	Explained by: reliable and coherent interactions with the world

representations. Some of these transbodied representations are not veridical because people imagine what does not exist, for example leprechauns, unicorns, and angels. But for explanatory purposes in both science and everyday life, people use the creative capacities of semantic pointer binding to generate highly useful concepts such as *natural selection, gravity, electron*, and *mental representation*. Serious science, mathematics, and philosophy would be impossible without the ability to go beyond people's sensory interactions with the world and motor operations to generate concepts and beliefs that transcend the senses. Even in everyday life, people use similarly abstract concepts like *money, fairness,* and *mind* to guide their lives.

The view of knowledge in my three-analysis serves to capture both the value of first-person experience and its limitations. The value comes from how sensory-motor perceptions gives people knowledge *of* the world that is not reducible to verbal knowledge-*that*. But this kind of first-person experience is not the core of knowledge because it is limited with respect to both scope and reliability. For example, when people try to understand their emotions, they can gain some awareness of their experiences but can easily delude themselves about the mental and social causes of their feelings. It takes transbodied knowledge from psychology and neuroscience along with considerable reflection to explain what is actually going on in embodied personal experiences.

THE GROWTH OF KNOWLEDGE

Plato and a few other philosophers thought that all knowledge is innate: you were born with it although you may need some prodding to remember what you know. Much more plausibly, most knowledge has to be acquired, both individually as babies learn about the world and culturally as whole societies increase their general stock of knowledge. Cognitive science and the Semantic Pointer Architecture describe mechanisms by which knowledge grows. After critiquing the superficiality of biological metaphors such as memes, I describe how new concepts and hypotheses are generated by active minds and how they can be transmitted from one mind to another. The growth of knowledge also requires explanation of how new proposals are evaluated and justified.

Biological Analogies

In 1976, Richard Dawkins proposed that we can understand the spread of cultural entities such as ideas by analogy to biological evolution. He named such entities *memes*, by comparison to the genes that enable the transmission and selection of biological traits. The general attempt to model cultural developments on biological ones is called memetics.

I think that memes and memetics are bad ideas because of the substantial differences between biological and cultural evolution. First, the lumping of all cultural entities together as memes neglects the variety and complexity of mental representations that cause behaviors, including concepts, images, beliefs, emotions, and neural networks. Second, the processes by which mental representations are generated and selected are very different from the ones that operate in biological evolution. Superficially, ideas and genes are similar in that both are generated, selected, and transmitted. But the value of an analogy depends on how well it illuminates the domain to be explained, and the comparison with genes only gets in the way of understanding the generation, selection, and transmission of mental representations and resulting behaviors.

Cognitive science has moved beyond vague notions like *idea* to theories about how concepts, rules, images, and emotions are represented in the mind and brain. These representations have different structures, arise from different learning processes, and function differently in inference and other cognitive processes. Lumping them all together as memes is just a distraction from explaining how mental representations spread among groups of people.

The biggest reason to reject memes as an analog of genes is that the processes of generation, selection, and transmission are so different. Genetics makes an enormous contribution to evolutionary theory by explaining how variability arises by random mutations, selection occurs by survival of the fittest, and transmission occurs by parents passing their genes to their offspring. But variation, selection, and transmission in minds and cultures are unlike evolutionary processes.

First, cultural generation of ideas is far more goal oriented than genetic mutation: when new ideas such as the iPhone are generated by combinations of existing ideas, it is usually because people are intentionally trying to solve some recognized problem. In contrast, genetic mutation is independent of the environmental problems faced by the organism in which mutation occurs.

Second, selection of ideas is very different from selection of genes, because it is performed by intelligent people capable of using various criteria to decide whether the new ideas are better than old ones. Emotion plays an important role because people get excited about ideas that are new and valuable. Selection of genes is much more indirect, dependent on the survival of the organisms that carry them.

Third, transmission of ideas is far more rapid and widespread than transmission of genes. No matter how valuable a biological mutation is to a species, it takes many generations before the new gene manages to spread through a population. In contrast, a valuable new idea such as the iPhone can spread to millions of people in a matter of days through social mechanisms of communication built on the cognitive and emotional mechanisms in individual brains.

Hence the concept of memes provides only a superficial understanding of cultural evolution and blocks the way to deeper investigations of (a) the full range of mental representations and behaviors that contribute to cultural evolution and (b) the cognitive and social mechanisms by which these representations are generated, selected, and transmitted. Rather than relying on a shallow biological analogy, anyone interested in the spread of ideas should learn more about cognitive social science. Then the concept *meme* can be rendered extinct by mechanistic neuropsychological theories of the generation and acceptance of mental representations.

Generating Concepts

Perhaps some concepts are innate in human beings, with likely candidates including *face* and *object*. But the vast majority of the concepts that arise in individuals and cultures are learned. The simplest kind of learning is from perceptual experience, for example when you see a hawk for the first time and quickly learn that a hawk is a large bird that flies around. Other concepts can be learned from other senses, such as the sound of a trumpet, the taste and smell of blue cheese, and the feel of wool. Perceptual learning is accomplished by modifying synaptic connections between neurons that encode sensory inputs.

If all concepts were learned from sensory experience alone, then human culture would be highly impoverished. Instead, people can generate concepts that go far beyond sensory experience, including mental properties such as beliefs, religious ideas such as God and spirit, scientific entities such as atoms and genes, and medical categories such as influenza and virus. How are human minds capable of surpassing sense experience in forming concepts?

The best psychological answer is conceptual combination, where people take existing concepts and put them together into new ones. Such combination happens routinely in the generation and comprehension of language, when people put together pairs such as "blue sweater" and "big dog." More interestingly, when people seek explanations they generate combinations of things that they cannot otherwise sense, such as *atom* which is a combination of *particle* and *indivisible*. No one has ever seen an atom, but the ancient Greek philosophers hypothesized their existence in order to explain how matter operates. Modern science has dramatically changed the understanding of atoms, which are not indivisible but rather formed out of particles like electrons and quarks. But the concept *atom* is still valuable because of its explanatory role in scientific theories.

How does the brain accomplish conceptual combination? Concepts understood as semantic pointers are patterns of neural firing that arise from other patterns through the neural mechanism of binding by convolution. If the concept *particle* is one neural pattern and the concept *indivisible* is another, then we get the new concept *indivisible particle* by binding these patterns together. Binding is not simply verbal but can easily incorporate the sensory and motor aspects of semantic pointers. The result, however, is not limited to the senses, because the resulting concepts can concern objects that are not perceivable. Hence the Semantic Pointer Architecture can explain how conceptual combination by convolution produces observable and theoretical concepts, so that knowledge can be both embodied and transbodied.

Generating Hypotheses

Concepts by themselves do not say anything about the world, whereas beliefs can be used to make both specific claims about things (e.g., this hawk is brown) and general claims about relations between concepts (e.g., all hawks are brown). Representations of the world can be made by verbal beliefs and by nonverbal representations such as a picture of a hawk flying in the sky, formed perceptually.

The formation of general beliefs requires inference, for example when you generalize from seeing various hawks that fly to the conclusion that all hawks fly. For theoretical purposes, however, people need to use *abductive* inferences that generate explanatory hypotheses. Abductive inferences sometimes introduce nonobservable properties, for example when biologists explain the capabilities of some birds to fly long distances by supposing that their brains somehow detect the earth's magnetic field. This ability is not directly observable, but abductive inference generates a hypothesis that birds of some species can have a magnetic sense in order to explain their abilities to navigate.

Analogy is sometimes a great help in generating hypotheses. For example, if you know that one species of bird navigates by sensing electromagnetic fields and another bird navigates over similar distances, then you might generate the hypothesis that the second bird also navigates by electromagnetic fields.

Abductive inference and analogy are risky forms of inference, because they take people beyond what can be directly observed. The risk can be worthwhile, however, because of the enormous explanatory gains that can result. We do not have to just stick with the observation that birds fly long distances but can entertain hypotheses that explain how they did it. Going beyond the information given by perception is an epistemic virtue, not a vice, because it contributes to the cognitive goal of understanding why things happen. The aims of knowledge are not just truth but also explanation and practical benefit to human welfare.

Unlike the random mutations that produce new genetic structures, the generation of concepts, beliefs, and explanatory hypotheses is highly focused by the problem-solving activities of people in both science and everyday life. People do not generate concepts and hypotheses about anything they encounter but only about things that matter to them because of relevance to their goals. The primary means by which people detect relevance is emotion, which meshes with cognition to provide ongoing evaluation of situations and potential actions. As *Brain–Mind* and *Mind–Society* show in detail, cognition and emotion work closely together to ensure that thinking and acting cooperate to accomplish people's goals.

Major contributors to this process are the epistemic emotions, such as interest, wonder, curiosity, excitement, enthusiasm, and boredom. These feelings help to determine the questions and observations that people pursue and thereby influence what concepts, beliefs, and hypotheses they generate. For example, James Watson and Francis Crick were fascinated with the structure of DNA and were therefore strongly motivated to investigate it.

Hence emotions are an important part of epistemic rationality, contrary to the widespread view that emotion is antithetical to it. Chapter 6 on ethics explores connections between emotions and rationality in more detail. For knowledge, the important point is that emotions help focus the generation of mental representations to support the values of explanation and practical benefit as well as truth. Unfortunately, emotions can also encourage defective beliefs based on wishful thinking and self-deception.

Social Transmission

For cultures to develop, there needs to be transmission of mental representations from one person to another, including concepts, beliefs, hypotheses, and values. Major cultural accomplishments such as agriculture, the wheel, mathematics, and the printing press did not have to be invented repeatedly but were transmitted across generations by communication mechanisms such as teaching and writing.

Values are important in science because they influence what kinds of problems people choose to think about, shaping the areas of knowledge that will develop. For example, much more research is currently done on the biological causes of cancer than on the much rarer disease of hemophilia.

More controversially, values can also adjust the thresholds for acceptance of different views, when people require more evidence to accept conclusions that are socially dangerous. Bringing global warming down to manageable levels demands disruptive economic changes that can be justified based on values such as preserving human lives, as long as the evidence for climate change is good enough to warrant the disruption. From the semantic pointer perspective, values are neural patterns that combine concepts or beliefs with emotions. For example, the value of health consists of the neural representation of the concept *health* combined with positive emotions such as happiness and desire.

Unlike biological transmission by genes, social transmission can be fast because mental representations can be transferred among people who communicate with each other, directly by conversation and through many technological means such as writing and electronic media. The social transmission of knowledge would be much faster if people could communicate like computers. If you get a new computer, you only need to establish an electronic connection between it and your old computer to transfer the files with great speed. In contrast, human communication is complicated because people pass not only words but also nonverbal information such as emotions and values, carried by diagrams, sounds, gestures, and body language. *Mind–Society* (chapter 3) presents a general theory of communication as semantic pointer transmission that describes the difficult process of approximately replicating mental representations across people. It explains why education is much faster than genetic transmission but much slower than electronic transmission and also applies to other cultural developments such as religion and business.

In sum, the transmission of knowledge among people is a complicated process that combines individual mechanisms of representation and processing with social mechanisms of communication, both verbal and nonverbal. Semantic pointers provide an account both of the range of mental operations working in individuals and of the means by which many kinds of mental representations can be passed among individuals. It thereby explains how the growth of knowledge is both a psychological and a social process. Because cultural development requires transmission of epistemic emotions and values, it is greatly fostered by the establishment of emotional communities such as scientific laboratories and artists' colonies.

JUSTIFICATION

Social transmission of beliefs and attitudes sometimes propagates bunk rather than knowledge, which needs to be justified by reasonable selection of systems of mental representations. Epistemic selection is much more focused than biological selection, which is a slow process favoring organisms that are more capable of obtaining food, avoiding predators, and finding sexual partners. In science and everyday life, people can intentionally scrutinize candidates for knowledge such as beliefs and other representations in order to select the ones that meet standards.

What standards are appropriate? I have already mentioned three goals of knowledge: truth, explanation, and practical usefulness. There is no a priori way of establishing that these are the goals of knowledge production, but the activities of exemplary knowers such as scientists vindicate this list. Scientists do not just aim for a motley collection of truths but rather seek truths that are important, where importance is partly a matter of accomplishing practical goals such as treating disease. The most valuable scientific beliefs are not the miscellany of observations and generalization but rather the theories that provide explanations of many generalizations, such as Einstein's theory of relativity and Darwin's theory of evolution by natural selection. Hence justification of knowledge requires showing that there are methods that accomplish these goals better than alternatives, in line with the normative procedure sketched in chapter 1.

Foundational Justifications Fail

Philosophical views about justification have often been biased by the unusual case of axiomatic mathematics. Euclidean geometry, for example, starts with some basic postulates, such as that all right angles are equal to one another, from which many theorems can be rigorously deduced. Analogously, knowledge in general could be justified by recognizing a few unchallengeable true beliefs and then using valid arguments to generate all other beliefs. The axioms provide a foundation for everything else that is known, as Descartes tried to do with his famous "I think, therefore I am."

Unfortunately, this appealing method does not work even in mathematics, as chapter 10 details. For knowledge in general, the method is disastrous, because the two main ways of finding basic truths fail abysmally. The rationalist version of foundationalism looks for some a priori truths from which all other truths can be derived deductively, generating a system of knowledge that is independent of sense experience. But a priori truths are like fairies and angels, attractive entities that no one has ever been able to find. Even apparently obvious claims such as that a proposition cannot be both true and false have been challenged by some philosophers.

Many philosophers try to use thought experiments to generate a priori truths, telling stories that are supposed to be evidence that some conclusions are intuitively obvious. For example, Descartes tried to establish that the mind is different from the body by imagining minds that do not have bodies. As chapter 2 described, such imaginings are undercut by discoveries about the neural mechanisms that make minds work. In general, philosophical thought experiments are about as reliable a method as reading religious texts such as the Bible or Quran.

Thought experiments can be useful for generating hypotheses, as they often have done for scientists whose influence depends on the accumulation of experimental results arising from observations of the world. On their own, thought experiments are not a reliable source of justification for several reasons. First, thought experiments rarely produce agreement, because people who do not like the conclusion of a thought experiment will just produce another thought experiment that generates different intuitions. Second, thought experiments are inherently circular, in that people make them up to support the views they already hold. Third, thought experiments assume that we have some kind of intuitive capacity to grasp the truth, which is not even true in mathematics. Intuition is actually a process of unconscious parallel constraint satisfaction. So rationalist foundationalism based on axioms or thought experiments is an abject failure.

Similarly, empiricist justification fails when it tries to take sense experience as indubitable and then to derive the rest of knowledge from it. First, experimental work in cognitive psychology shows that sensory experience is not indubitable but rather requires enormous amounts of inference carried out in multiple brain areas in ways that are subject to interpretations and biases. There are more than 100 visual illusions that show how perception can miss the mark. Hence sense experience does not provide the analog of axioms from which the rest of knowledge could be gathered. Second, much important knowledge consists of theories that invoke entities such as electrons and genes that are not reducible to sense experience. Either we abandon theoretical sciences as knowledge or, more plausibly, abandon empiricism as the foundation for knowledge.

Failures of foundationalism engender skepticism, suggesting that there is no knowledge at all. Such skepticism is implausible because billions of people have found their way around the world, many living long and happy lives. Moreover, technologies such as electric lights, computers, and effective drugs such as antibiotics show that scientific theories do have some grip on the world, as argued in chapter 4. Nevertheless, forms of skepticism are currently popular, from the individual relativism that suggests that truth is just a matter of personal feelings to the social relativism that suggests truth is just a matter of cultural variation and interpersonal power relations.

Coherence Justifications Succeed

Fortunately, coherentism provides a powerful alternative to both foundationalism and skepticism. Coherence is not just a vague metaphorical idea about how things fit or hang together but can be made precise by understanding coherence as a cognitive process in which the brain satisfies multiple constraints. There are various kinds of coherence, including those oriented toward action and morality as discussed in chapter 6. For epistemic justification, the most relevant kind is explanatory coherence, where people select mental representations that make sense. Sensory and experimental observations get some priority without being foundations, and hypotheses that explain observations and other hypotheses can be accepted if they are part of the best overall explanatory account.

To take a simple example, suppose that you are expecting to meet a friend for coffee but the friend fails to show up. The surprising occurrence naturally leads you to generate explanatory hypotheses, for example that your friend is sick, forgetful, or in a car accident. You can gather additional evidence by going back through emails or text messages to see what your friend had said about the potential meeting and also contact other people you both know in order to try to find out what your friend is doing.

In your search for the best explanation of why your friend did not show up, you can also consider possible causes for why the different explanations might be true, for example by thinking that your friend has been overwhelmed with work, which could explain the forgetfulness. You will want to avoid overcomplicated hypotheses such as that your friend did not show up because of abduction by aliens as part of a world invasion, and you may also use analogies such as remembering that the last time your friend failed to show up it was because of a broken-down car.

The constraints that operate in picking out the most coherent explanation include positive, explanatory ones between hypotheses and evidence, for example between the supposition that your friend is sick and the failure to show up. These constraints rely on causal connections, for example from being too sick to get out of bed to being unable to make it to the coffee shop. Negative constraints are between conflicting hypotheses, for example sickness versus forgetfulness. It might seem mysterious how the mind can simultaneously consider all these constraints to come up with a strongly coherent overall judgment, but the brain is not the kind of serial process that we find in deductive proofs that proceed step by step. Instead, the brain has billions of neurons working together simultaneously, so parallel calculation of coherence can be an efficient and effective process. It can also be fairly accurate when overall coherence is based on usually reliable inputs such as the senses, memory, and testimony. Such practices are not perfect but are better at accomplishing the epistemic goals of truth, explanation, and human welfare than available alternatives.

Figure 3.1 depicts the simple coherence problem of figuring out why a friend did not show up. The dotted line indicates a negative constraint between the hypotheses of sickness and forgetting, which are not strictly contradictory but provide competing hypotheses. The solid lines indicate positive constraints based on explanations of evidence such as that the friend did not show up and also that there was no regretful email. In addition, there are explanations of hypotheses by deeper ones: the friend forgot because of overwork or came down sick because of a flu going around.

Many important justifications in science also depend on inference to the best explanation based on explanatory coherence. For example, Darwin's theory of evolution is justified because it provides a better explanation of a vast array of evidence about species than alternative explanations such as divine creation. Darwin used evolution to explain many observations such as the geographical distribution of species and the existence of vestigial organs. He also explained why evolution occurs by the hypothesis of natural selection, a mechanism in which the parts are individual organisms that interact with the environment and other organisms to compete for survival and reproduction. The hypothesis of evolution coheres with the evidence explained and with the hypothesis that evolution results from natural selection. Greater coherence developed over time when the theory of genetics provided a mechanism for how inheritance works and when DNA provided a mechanism for genetics.

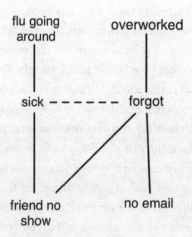

FIGURE 3.1 An explanation choice as a coherence problem, with positive constraints shown by solid lines and negative constraints shown by dotted lines.

Critics of coherence as a method of justification accuse it of vicious circularity, since it seems to justify hypotheses based on evidence and evidence based on hypotheses. There are indeed coherent systems of belief that bear little relation to reality, for example the pseudo-religion of Scientology that maintains that people have infinite capabilities. However, the circularity involved in coherence is more like the mutual support found in a group of friends rather than the illegitimate deductive inference from A to B to A. The nonviciousness of the apparent circularity in coherence justifications is further seen by considering the nature of evidence.

Evidence

Unlike Scientologists, scientists and intelligent people in everyday life take evidence seriously. Useful hypotheses about relativity, natural selection, and the mental states of other people go beyond observation, but that does not mean that observations and hypotheses are the same. The senses are not infallible, for they are subject to hallucinations, emotional influences, and social effects. But they do have a connection to the world by virtue of the generally reliable functioning of our organs such as eyes and ears. Hypotheses can be utterly fanciful, but evidence is not arbitrary as shown by the following features of evidence as practiced in science:

1. Reliability: A source of evidence is reliable if it tends to yield truths rather than falsehoods, for example by systematic observations using instruments such as telescopes and microscopes and by controlled experiments such as those practiced by many scientists.
2. Intersubjectivity: Systematic observations and controlled experiments do not depend on what any one individual says but are intersubjective in that different people can easily make the same observations and experiments.
3. Repeatability: A major source of the intersubjectivity of systematic observations and controlled experiments is that the same person or different persons can achieve similar results at different times.
4. Robustness: Experimental results should be obtainable in different ways such as using different kinds of instruments and methods.
5. Causal correlation with the world: Evidence based on systematic observation or controlled experiments is causally connected with the world about which it is supposed to tell us, for example when telescopes and microscopes provide evidence because reflected light enters the eyes of observers and stimulates their retinas.

Coherence-based justification seems to make knowledge mind-determined, but reliable evidence that meets these five conditions shows how coherence is not just minds spinning stories in the air like fairytales. Evidence in science and everyday life is not guaranteed to be true, because further observation and experimentation may undercut it. Nevertheless, good evidence has enough connection with the world to rescue pure coherence from mere imagination and to discourage the impasse of equally coherent alternative accounts. Evidence intervenes to help avoid the danger that a coherent explanation is just the best of a bad lot.

Because reliable evidence contributes specially to explanatory coherence, my epistemology can be called "reliable coherentism." Reliability is not perfect enough to constitute a foundation, but it suffices to make some inferences to the best explanation fully justified, while subject to future revision. It is a virtue rather than a vice that knowledge on this view turns out to be fallible, because infallible certainty is unreachable. Reliability cannot be judged a priori but is learned from experience, just like all the characteristics of evidence.

Computer simulations show that coherence judgments can efficiently be carried out by neural networks, including the Semantic Pointer Architecture. Unlike rationalist and empiricist foundationalism, my version of coherentism is psychologically and neurologically plausible. Whereas many philosophers have thought that introducing psychology into considerations of knowledge is lethal for objectivity, the opposite turns out to be the case. Only by abandoning specious assumptions about foundations for knowledge can objectivity be restored. The restoration does not come from the adulation of a priori truths or sense experience but rather the inclusion of information from the senses into systematic kinds of evidence that enable coherence to escape from vicious circularity into objective justification.

A final objection to coherence is that it provides no connection with truth, which is discussed in chapter 4. I grant that coherence is no guarantee of truth but argue that it often leads to truth when done well. Coherence has a clear connection with explanation, because hypotheses are supposed to explain evidence in ways discussed in chapter 5. The third aim of knowledge is practical benefit for human welfare, whose achievement in relation to truth is discussed in chapter 4.

Intuition

Philosophical intuitions generated by thought experiments are not evidence because they lack all five of its features: reliability, intersubjectivity, robustness, repeatability, and causal connection to the world. In his bestselling book, *Thinking, Fast and Slow*, Kahneman describes a project to determine when intuitions are reliable. Because of his many experiments that show that people's intuitive judgments about statistical matters are frequently in conflict with logical norms, Kahneman was skeptical about claims that we should trust the intuitions of experts such as firefighters.

Kahneman describes an "adversarial collaboration" with Gary Klein, who has done fascinating work on expertise described in his book *Sources of Power*. Their key question was: When can you trust a self-confident professional who claims to have an intuition? They concluded that their disagreement about the reliability of intuitions was partly due to their experience with different experts. Whereas Klein had studied firefighters, nurses, and other professionals with real expertise, Kahneman had studied clinical psychologists, stock pickers, and political scientists whose forecasts were unreliable. Klein and Kahneman came to agree that the confidence that people have in their intuitions is not a good guide to their validity, because it is easy to achieve high confidence by ignoring failures.

Rather, acquiring a skill for making good intuitive judgments requires both an environment that is sufficiently regular to be predictable and an opportunity to learn these regularities through prolonged practice. Kahneman advocates the rule that intuition cannot be trusted in the absence of stable regularities in the environment. There are, however, domains where real expertise is possible, for example in chess where knowledge of regularities is available and acquired by large amounts of practice and feedback, so that expert intuitions can sometimes be taken seriously.

The implication of these findings for philosophical intuitions is clear: do not trust them. The common philosophical method of making up a story to pump people's intuitions is not based on stable regularities or prolonged practice, just verbal fluency and academic acculturation.

My skeptical conclusion does not mean that thought experiments are useless. Kahneman and Tversky began all their investigations with thought experiments that they used to probe their own intuitions about statistical matters, but they followed good scientific practice in following them up with rigorous real experiments that showed that many other people had the same erroneous intuitions. Great scientists like Galileo and Einstein undoubtedly used thought experiments to advance their research, but acceptance of their ideas depended on the accumulation of experimental results arising from careful observations of interactions with the world.

Philosophers should follow scientists in restricting their use of thought experiments to the generation and clarification of hypotheses and to the demonstration of inconsistencies in rival hypotheses. Real experiments produce evidence that is repeatable and robust in that it can be obtained by different people in different ways through interaction with the world. In contrast, the intuitions produced by thought experiments are not evidence at all.

So where do intuitions come from? An alternative view for which evidence is progressively mounting is that intuitions are neuropsychological reactions generated by unconscious processes operating in parallel. Compare object recognition: If you see an object consisting of wheels, frame, seat, and handlebars, your brain recognizes it as a bicycle with little conscious deliberation, using interactions among multiple brain areas. Similarly, when you hear a story, your brain processes the information in parallel to generate a judgment that can be emotional as well as cognitive, for example that Aristotle's explanation of falling objects is ridiculous or that understanding of language by digital computers is absurd.

I propose that intuitions result from the three primary neural processes that semantic pointers use: encoding representation, neural binding, and interactive competition among representations. Using these, people make sense of a story in a way that generates a reaction in the form of the intuitive sense that a claim is true or false, good or bad. Story comprehension and intuition generation are performed by parallel constraint satisfaction, not by some special faculty for grasping eternal truths. The often-valuable intuitions used by firefighters, nurses, and other experts are based on evidence rather than thought experiments, which makes them very different from the intuitions of philosophers that satisfy constraints that are often derived from motivations and prejudices.

PROBABILITY

Coherence theories of justification are psychologically plausible, are mathematically exact, and fit well with actual cases, but there is a powerful alternative way of doing nonfoundational justification that must be considered. Probabilistic approaches to inference and justification are popular in philosophy and psychology, as well as in some branches of artificial intelligence and linguistics. Hence the prospects of using probability theory for knowledge justification needs to be examined.

Probability and Bayes' Theorem

Probabilities are mathematical quantities that meet the following conditions about a sample space of events such as rolls of dice. The probability of an event is a number between zero and one, for example when the probability of a six-sided die turning up a 1 is 1/6. The probability that at least one of the events in the sample space occurs is equal to one: it is certain that the die will turn up a 1, 2, 3, 4, 5, or 6. If two events are mutually exclusive, then the probability of getting one or the other is just the sum of the probabilities of each: the probability of the die rolling 1 or 2 just equals 1/6 + 1/6 = 1/3. From these simple assumptions, many important theorems can be proven.

For epistemology, the key theorem was derived by an eighteenth-century cleric named Thomas Bayes. Taking the truth of a belief as an event, we can write Bayes' theorem as a way of calculating the probability of a hypothesis given evidence in accord with the following formula: $P(H|E) = P(H) \times P(E|H)/P(E)$. In words, the probability of hypothesis H given the evidence E is equal to the prior probability of the hypothesis times the probability of the evidence given the hypothesis, all divided by the probability of the evidence. The approach to probability and statistics that relies heavily on Bayes' theorem is naturally called "Bayesian."

Many philosophers assume that nondeductive justification should be probabilistic in accord with this theorem, and some psychologists contend that Bayesian reasoning is the core of a whole cognitive architecture for explaining human thinking. While the mathematical basis of probability theory is incontrovertible, its applicability to human psychology and epistemology is problematic. Bayesian ideas are mathematically appealing and practically useful, for example in driverless cars, but that does not mean that they are good for explaining how people think or for laying out how they ought to think when justifying knowledge.

Objections to the Bayes Craze

First, Bayesian architectures fall far short of a unified theory of mind because they do not even mention some of the key ingredients to explaining mental operations. They have nothing to say about many kinds of human thinking, including sensory imagery, analogies, emotions, consciousness, and creativity, all of which can be explained within the Semantic Pointer Architecture as shown in *Brain–Mind*. My coherence view of justification fits well with the account of generation of concepts and beliefs as semantic pointers, because both evaluation and production derive from the operations of neural networks. In human minds, generation and evaluation are coordinated, and creativity operates by recursive binding that is currently lacking in Bayesian models. At best, therefore, probabilistic reasoning is an approximate description of more or less optimal brain processing, not a characterization of the mechanisms that actually make the mind work.

Second, the interpretation of probability is more difficult than Bayesians assume. They usually take probabilities to be subjective degrees of belief, but there is little empirical evidence that peoples' degrees of belief conform to the axioms of probability theory. On the contrary, many studies show that people are not good at thinking probabilistically, and there is even neural evidence that people's judgments about the acceptability of beliefs are more tied to emotions than to the mathematics of probability. For example, when people agree with something they judge to be true, there is activation in parts of the brain (e.g., ventromedial prefrontal cortex) usually involved in positive emotions such as liking. In contrast, judgments that claims are false generate activation in brain areas (e.g., anterior insula) associated with negative emotions or dislike.

While psychologists should be disturbed by these kinds of empirical results, philosophers who are more normatively focused can simply say: even if minds do not work probabilistically, they ought to. The standard argument for saying that people ought to think with probabilities is a proof that if they do not then they can be made to bet in inconsistent ways that would lose them a lot of money. But a better way to avoid losing money in these situations is simply to refuse to bet on situations such as beliefs where betting does not make any sense because no one knows any of the odds. Hence there is no reason to assume that probabilities are degrees of belief in the context of epistemic justification.

An alternative interpretation is that probabilities are frequencies of occurrences in the world, which has the great advantage of making probabilities objective rather than subjectively a matter of beliefs that have no obvious connection with truth. Unfortunately, in many domains it is hard to understand probabilities as frequencies. It is easy to say that the probability of smoking given cancer is a quantity, because there are data that display this frequency in the world. But how do frequencies apply to the claim that smoking causes cancer, based not just on statistical correlations but also on mechanisms of how smoke particles damage cells? This claim is about only one world that we currently inhabit, and no one knows what the frequency of smoking causing cancer would be in some mythical sample space of possible worlds. Sensibly applying probabilities to general scientific theories such as relativity or to philosophical theories such as dualism is even more problematic.

The best way to bridge the gap between subjective beliefs and objective frequencies in the interpretation of probabilities is to introduce ideas about mechanism and simulation. Objective probabilities are not just frequencies but rather properties of setups in the world that have underlying mechanisms that generate the frequencies. For example, dice have structural properties subject to Newtonian laws that generate rolling frequencies. Probabilities are propensities of situations to generate frequencies, but propensities are best understood as dispositions based on underlying mechanisms, as chapter 5 argues in more detail.

If we know a lot about the relevant mechanisms, we can produce computer simulations that look at various operations in various outcomes with varying parameters. This is how forecasters estimate the probability that a candidate will win an election, plugging in various assumptions and running thousands of simulations to see the outcomes. Similarly, the probability of a climate change disaster is calculated by running repeated simulations of computer programs based on understanding of the mechanisms of global warming. This method enables calculations such as the expected probability that a particular candidate will win election but is useless in assessing the probability that smoking causes cancer, or that species evolved by natural selection, or that your friend missed the coffee meeting because of sickness. The coherence approach to justification avoids issues about interpreting probability when mechanisms are unknown.

The third problem with the Bayesian approach is that there are numerous technical difficulties involved in the assessment of probabilities in large systems of beliefs such as the human mind. As a rough guess at how many beliefs each person has, we can start with the estimate that an educated person knows about 30,000 words. Assuming that every word is associated with at least 10 beliefs, that means you have 300,000 beliefs operating in your mind, so that changing your probabilities could involve as many as 300,000 nodes in a Bayesian network. The process of repeatedly updating the probabilities of nodes in such networks is known to be computationally intractable, with numerous simplifications needed to make updating work even in relatively small networks.

Moreover, in the relatively efficient systems that have been built for doing Bayesian updating, the calculations require large numbers of conditional probabilities that nobody knows. In my coffee example, you might be able to take a rough guess about the probability of the person not showing up because of sickness, but for Bayesian updating you also need to know numbers like the probability of not showing up given sickness and not having a car break down.

In sum, Bayesian explanations of justification of knowledge have three main problems. The architecture problem is that they do not accommodate the multimodal and generative aspects of human thought that need to be in an empirically plausible theory of knowledge. The interpretation problem is that they have no reasonable way of interpreting probabilities that fits with the standard axioms and applies to human beliefs in everyday life and in science. The implementation problem is that actual computations of conditional probabilities of the sort required for Bayes theorem appear intractable and require probabilities that are rarely available. Hence probability theory does not provide a satisfactory alternative to coherence-based justification, which fits well within the Semantic Pointer

Architecture, does not require an interpretation of probability, and operates efficiently in large neural networks like those in human brains.

KNOWLEDGE IS SOCIAL

Like cognitive psychology and neuroscience, epistemology has historically concerned individual knowers rather than social groups. Since the 1980s, however, a new field of social epistemology has emerged that treats knowledge as resulting from group activities. This development is appropriate, because knowledge has always had important social dimensions, going back at least to the ancient Athens Academy where Socrates, Plato, and Aristotle developed their ideas in conversation with others. Today, knowledge develops through groups that work together in schools, businesses, and public organizations. Scientific knowledge especially results from group activities, apparent from the fact that almost all publications in science journals such as *Nature* and *Science* have multiple authors. Much of people's knowledge is based on the testimony of others rather than on their own experience.

The semantic pointers approach to cognition and communication is highly relevant to social epistemology. *Mind–Society* (chapter 3) argues that the spread of beliefs and attitudes from one individual to another is a matter of communication of semantic pointers by both verbal and nonverbal means. Such communication is crucial in collaboration, where groups of people such as scientists achieve knowledge that no individual could accomplish alone. Finally, communication is crucial for the development of consensus, when agreement arises concerning what beliefs and values to adopt.

Communication

Brain–Mind shows how concepts, beliefs, goals, values, images, emotions, and analogies are all made out of semantic pointers, patterns of neural firing that are formed by binding a mixture of sensory, motor, verbal, and emotional information. *Mind–Society* argues that communication is therefore much more complicated than merely passing words from the mind of one person into the mind of another, because people do not just think with words. Instead, communication requires the inexact transfer of semantic pointers in one person's head into the head of someone else. Because the patterns of firing in people's heads will never be exactly the same because of their different genetics and learning, transmission can never be more than approximate. But people's brains and environments are sufficiently similar that there can be considerable overlap among the neural patterns in different people.

The complexity of communication explains why education is often so difficult. A teacher may be able to articulate some of what needs to be learned, but teachers have no access to all of the neural structures that constitute their knowledge, so transferring them to their students is hard. Besides words, students need to acquire pictures, nonverbal rules involving motor concepts such as *force*, and emotional values about why the material is worth learning. *Mind–Society* (chapter 12) discusses the implication of semantic pointer theories of cognition and communication for education.

In knowledge communities such as groups of scientists, the same difficulty of communication arises, because one person cannot easily pick up what another person knows. The process of communication is better described as instillation or activation of semantic pointers, rather than as precise transfer of neural representations. Communication of knowledge can sometimes be quite rapid, for example when people read the same journal article or Wikipedia entry. But rapidity depends on a shared set of concepts and beliefs that the new information can be plugged into. These difficulties of communication have implications for collaboration and consensus.

Collaboration

Knowledge in groups is not simply the sum of the knowledge of the individuals, because a successful group can generate concepts, beliefs, and other representations that no individual would arrive at alone. For example, I have a publication on intention with Tobias Schröder and Terry Stewart that no two of us could have produced, because we needed the three-way interaction to develop the key theory and simulations. In line with the analysis of emergence presented in chapter 5, groups have epistemic accomplishments that are emergent through processes of semantic pointer communication.

Collaboration would be much easier if it only required transferring words between the minds of participants. But joint work is often hampered by miscommunication, especially when participants come from different disciplinary backgrounds and hence have different concepts. For example, my first book *Induction* was written in collaboration with a computer scientist, a cognitive psychologist, and a social psychologist. In the preliminary year of discussions, we spent weeks just trying to figure out what each other meant by the word "schema." According to the semantic pointer theory of concepts, schemas combine multiple sorts of information, including sensory-motor and emotional. So collaborators need to be prepared to share nonverbal representations such as pictures and gestures as well as words.

Moreover, a collaboration benefits enormously when the collaborators have values and enthusiasms that require the transfer of emotions, which are much more than emotion words. Chapter 2 described emotions as semantic pointers that bind situations, cognitive appraisals, and physiological perceptions. Face-to-face interactions that include facial expressions, gestures, and body language are all important for communication of epistemic emotions such as interest, wonder, and excitement.

Collaborators need to be able to trust each other concerning the information and the methods that they bring to joint work. Trust is much more than simply an estimate of the probability that someone is reliable. The discussion of romantic relationships in *Mind–Society* (chapter 4) argues that trust is also a matter of emotion, a good feeling that one has about another with respect to some task or goal. Similarly, distrust is not just an expectation of wrongdoing but incorporates negative emotions such as suspicion, dislike, and fear.

Communication of trust is partly based on words and probabilities but is much more based on generation of feelings that a collaborator will be the appropriate kind of person to behave appropriately. Trust is epistemically important, because little information is acquired by individuals on their own, requiring trust in the people and media from which new beliefs are acquired. The semantic pointer theory of trust as emotional communication describes the mental and social mechanisms that make trust function in collaborations and other social situations. When you trust other people, you do not have to enter into a complex coherence computation of the plausibility of what they say but can quickly accept their testimony.

Consensus

Communication sometimes only decreases agreement, for example in the polarization of Democrats and Republicans in the United States. But in science and other groups, social processes can lead to consensus in which the vast majority of participants come to agree about what to believe and what to do. For example, almost all informed scientists now believe that global warming is the result of human production of greenhouse gases.

A simple account of consensus has people exchanging beliefs, evidence, and hypotheses until each arrives at the same conclusion because of individual judgments of explanatory coherence. In practice, consensus is much harder to achieve, because people also differ in their nonverbal representations including emotional attitudes and values. Hence the achievement of consensus requires the approximate transfer or instillation of similar semantic pointers that bind words with images and emotions. This requirement explains why face-to-face meetings are valuable, even in the age of emails, text messages, and video conferences, because they support the full range of nonverbal communication. Face to face, people can pick up on each other's facial expressions, gestures, tone of voice, and body language as a way of detecting and conveying emotional attitudes. Therefore, a full theory of how consensus and polarization develops in scientists, politicians, and other groups needs to take into account the complexities of semantic pointer communication.

Effective social groups serve to counterbalance individual biases in beliefs and values. Because of the neural interconnections of cognition and emotion documented in *Brain–Mind*, chapter 7, individuals have no firewall between beliefs and values. Each of us is prone to motivated inferences that incline us to accept conclusions that fit with our personal goals rather than the evidence. But interacting with other people with different personal goals, especially in groups that value evidence and rational choice, can lead to better thinking overall. Collaborations such as scientific laboratories and political groups can benefit from having emotional communities with similar values.

Discussion of social aspects of group knowledge raises serious metaphysical questions about the reality of groups, group minds, and shared mental states. These questions are addressed in chapter 4.

CONCEPTUAL CHANGE AND THE BRAIN REVOLUTION

A simple description of the growth of knowledge suggests that it is a steady process of building fact on fact, accumulating more and more knowledge. But the history of science shows that growth is much more complicated, requiring deletions as well as additions and substantial kinds of conceptual change. The semantic pointer theory of concepts and the coherence view of justification clarify how science can sometimes be revolutionary, requiring the overthrow of previous ideas. I illustrate the radical nature of conceptual change by showing how it is occurring in the brain revolution, the ongoing realization that understanding of mind requires rejecting and revising conceptions from everyday psychology and earlier psychological theories, in favor of neural theories.

The simplest kind of conceptual change is addition of new concepts, which abound in the brain revolution. When people explain the behavior of themselves and others, they invoke a culturally familiar set of ideas such as beliefs and desires. The brain revolution introduces many alien concepts, including those for neuron, firing pattern, parallel constraint satisfaction, connectome, and semantic pointer. It also requires deletion of concepts that are no longer needed for explanatory purposes, such as soul, immortality, and possibly even free will, discussed in chapter 10. Development of neuroscientific knowledge also requires abandonment of concepts that have been popular in analytic philosophy of mind, such as those for propositional attitudes, supervenience, and massive modularity.

Another important kind of conceptual change is differentiation, where kinds are split into two subordinates. Early cognitive psychology made progress by applying general ideas about computation to the mind, inspired by rapid progress in computer science. But as more is known about neural computing, it starts to look very different from digital computing, for example with respect to parallel processing, multimodal representations, and the importance of emotions. The concept of representation gets similarly differentiated, between unitary representations like words and distributed representations consisting of patterns of firing in millions of neurons. *Brain–Mind* (chapter 8) argues that the concept of consciousness needs differentiation to cover the different degrees that operate in simple animals compared to those that are capable of representations of self and especially of self in society.

Another important kind of conceptual change is coalescence, where concepts that were previously thought to be distinct turn out to be unified. Coalescence occurred in modern physics through the realization that electricity and magnetism are fundamentally the same, with subsequent extension to understanding of light. Similarly, the sharp division between cognition and emotion assumed by folk psychology and most philosophers is breaking down through the realization that both depend on similar neural mechanisms and that the brain does not neatly divide into cognitive areas and emotional areas. Rather, thought depends on extensive interconnections between the so-called limbic system concerned with emotion and the higher cognitive capabilities of the prefrontal cortex.

One of the most radical kinds of conceptual change that occurs in scientific revolutions is reclassification, where concepts jump from one branch of a taxonomy to another. For example, Darwin's theory of evolution radically reclassified humans as animals rather than as specially created beings. The brain revolution compels the reclassification of mental representations such as concepts and beliefs as neural processes rather than as things. Similarly, chapter 8 argues that meanings need to be rethought as processes based on multiple mechanisms rather than as entities. *Brain–Mind* (chapter 12) claims that the self should reclassified as a system of multilevel mechanisms rather than as a spiritual entity.

Finally, the most radical kind of conceptual change, meta-classification, occurs when the whole method of classification is transformed, as when Darwin shifted the method of classifying organisms to take into account evolutionary history, not just similarity. The brain revolution is still too young for clear appreciation of how classification is changing, but there are two notable developments. Mental processes such as emotion are beginning to be classified through deeper understanding of the mechanisms involving interaction of numerous brain areas, rather than through common concepts of folk psychology. Empirical and theoretical advances show how cognition requires emotion and vice versa. The standard of classification should become explanatory adequacy and fit with neural understanding, not everyday language.

Another case of meta-classification is starting to take place in psychiatry, which historically has classified mental illnesses based on symptoms. Modern medicine switched in the nineteenth century from classifying bodily diseases based on symptoms such as fever toward considering causes such as bacterial infections. Similarly, mental illness research is slowly shifting toward classification based on causes rather than on symptoms, hindered by insufficient understanding of how brain mechanisms break down (see *Mind–Society*, chapter 10).

With such revolutionary changes, how can there be continuity in science as well as transformation? The answer is provided by the semantic pointer theory of concepts, because sensory-motor aspects of concepts can continue despite huge theoretical changes in concepts. For example, the concept of fire has undergone major conceptual change from being conceived as an element by the ancient Greeks and Chinese to being considered a process of combination with oxygen thanks to Lavoisier. Despite this reclassification, the sensory aspects of fire such as warmth and bright color continue. Similarly, even though the concept of a person changes dramatically when viewed from the perspective of multilevel mechanisms, there are still continuities based on the look, sound, and feel of the human body.

The semantic pointer theory of concepts that integrates exemplars, typical features, and explanations opens up new way of studying conceptual change. Historians of science could apply dynamic three-analysis to concepts such as *fire*, *atom*, and *mental representation*: for each stage in the development of the concept, I recommend the reader do a three-analysis that captures the relevant exemplars, typical features, and explanations. Identifying differences along these dimensions would provide a new way of analyzing conceptual change. Dynamic three-analysis could also be applied to children's development and social changes, including shifts in emotional values, for example in the conceptual shift from *nigger* to *Negro* to *Black* to *African American*.

SUMMARY AND DISCUSSION

Contrary to Wittgenstein, philosophy does not leave everything as it is but works with science toward dramatic improvements in human lives and understanding. I have used the semantic pointer theory of mind to develop a full account of the structure and growth of knowledge. The bearers of knowledge are no longer abstract propositions but rather patterns of neural firing that constitute mental representations, including concepts, beliefs, nonverbal rules, images, and emotions. This neurocognitive perspective suggests new answers for questions about the generation of candidates for knowledge and their relations to the world via sensory-motor interactions. Epistemology needs to abandon dogmatism, certainty, infallibility, permanence, and foundations to embrace moderate reliability, fallibility, risk, change, and coherence.

Semantic pointers support all three kinds of knowledge: *that* beliefs are true or false, *how* to do things using multimodal rules, and *of* things via sensory-motor experience. Knowledge-how and -of are thoroughly embodied, but knowledge-that can also be transbodied by virtue of the capacity of semantic pointers for recursive binding, which produces combinations that take concepts beyond sensory origins.

The Semantic Pointer Architecture meshes well with coherence-based justification that abandons foundational certainty for fallible attempts to fit diverse elements of knowledge into the best overall explanation. Although coherence does not provide the absolute guarantee that rationalist and empiricist foundations seek, it shows how minds are capable of gaining knowledge about the world through an ongoing attempt to get better and better representations. Along the way, missteps happen in the form of false theories and inadequate concepts that fail to refer to anything in the world, but the generation of new, more explanatory hypotheses helps to overcome these deficits. Conclusive refutation of skepticism and postmodernist relativism is not available, but a sophisticated understanding of mind shows how it connects with the world. The hypothesis that minds inhabit a world that they can learn about has more explanatory coherence than alternatives such as that we are all tricked by a virtual reality.

My emphasis on brain processes in the development of knowledge does not endorse the traditional individualist bias in epistemology. Knowledge has important social dimensions including communication across individuals, collaboration that fosters creative productivity, and processes of consensus and polarization that operate in the social spread of knowledge. These dimensions can be best understood by going beyond the view of communication as passing words from one person to another to seeing it as the approximate transfer and instillation of semantic pointers. Such communication is nonverbal as well as verbal, appreciating the importance of sensory images and bodily actions.

This chapter shows a good fit between answers to philosophical questions about knowledge and mind. The semantic pointer theory of mind supports epistemology by providing a unified account of knowledge-that, knowledge-of, and knowledge-how, and by showing how coherence can operate in the brain using both verbal and nonverbal representations. Conversely, reliable coherentism supports multilevel materialism as the best explanation of current evidence.

The major gap so far is the undefended assumption that there is an external world independent of human thought. This metaphysical question is addressed in the next chapter.

NOTES

Mind–Society (chapter 12) analyzes the vaccine controversy and makes a strong distinction between reasoning and inference. On climate change, see Thagard and Findlay 2011 (reprinted in Thagard 2012c).

Steup 2005 is a brief introduction to epistemology, including the Gettier problem about defining knowledge and the approaches of foundationalism and coherentism. Thagard and Beam 2004 discuss metaphors for knowledge.

Knowledge-how is sometimes called procedural, implicit, or tacit knowledge. Stanley 2011 claims that all knowledge-how is knowledge-that. Knowledge-of is sometimes called knowledge by acquaintance.

On memes, memetics, and evolutionary epistemology, see Dawkins 1976, Campbell 1974, Boyd and Richerson 2005, and Thagard 1988. Dennett 2017 mounts a vigorous defense of memes to explain cultural evolution but underestimates the explanatory power of mental and social mechanisms, as do Caporael, Griesemer, and Wimsatt 2014. *Mind–Society* is one long argument for social cognitivism as an alternative to biological analogies. Of course, brains evolved to support semantic pointers and communication, but social change is much more directly connected to mental representation and transmission than to natural selection.

Another way to move from embodiment to transbodiment is metaphor, which *Brain–Mind* (chapter 10) explains in terms of semantic pointers.

On abductive inference (also called abduction), see Thagard 1988, Magnani 2009, and *Brain–Mind*. Analogy is reviewed in *Brain–Mind*, chapter 6.

The role of emotions in scientific investigation, particularly the discovery of DNA, is discussed in Thagard 2002, reprinted in Thagard 2006. On values see chapter 6, Douglas 2009, and Thagard 2012c, chapter 17.

Further critique of a priori and necessary truths is in Thagard 2010b. Thought experiments are dissected in Thagard 2014e, which is also the source of this chapter's description of evidence. A few paragraphs from Thagard 2014e are reprinted with the permission of MIT Press.

Visual illusions are displayed at http://www.michaelbach.de/ot/.

Social constructivism in science is assessed in Thagard 1999. Postmodernist skeptics about knowledge include Rorty 1979 and Latour 1987.

Coherence is rigorously analyzed in Thagard and Verbeurgt 1998. On explanatory coherence, see Thagard 1989, 1992, 2012c. On the connection of coherence with truth, see Thagard 2012b and 2012c, chapter 6. Experimental reports of coherence effects include Simon, Stenstrom, and Read 2015. Vertolli, Kelly, and Davies 2017 analyze coherence in the visual imagination.

Goldman and Beddor 2015 review reliabilist epistemology.

On intuition, see Kahneman 2011, Klein 1999, and Osbeck and Held 2014.

Bayesian models of cognition are defended by Danks 2014; Griffiths, Kemp, and Tenenbaum 2008; and Piantodosi, Tenenbuam, and Goodman 2016. For critiques see Jones and Love 2011, Marcus and Davis 2013, and Thagard 2000. Poston 2014 contends that Bayesian and coherentist epistemologies are compatible; Olsson 2017 argues to the contrary. The findings about different brain areas for belief and disbelief are in Harris, Sheth, and Cohen 2008.

Hacking 2001 reviews interpretations of probability. Abrams 2012 proposes a mechanistic interpretation. Kahneman and Tversky 2000 report experiments that undermine claims that human thought is probabilistic. My proposal about probability amounts to: $P(A|B) = X$ if and only if there is a mechanism M linking B and A such that a computer simulation of M under assumptions of B yields A approximately X of the time.

My collaboration on intention is Schröder, Stewart, and Thagard 2014. My collaboration on induction is Holland, Holyoak, Nisbett, and Thagard 1986.

On social epistemology, see Goldman and Blanchard 2015 and Lackey 2014. Thagard 2005 discusses testimony from the perspective of explanatory coherence. Thagard 2000 models consensus, and Thagard 1999 discusses group rationality.

Mind–Society describes many applications of motivated inference, based on Kunda 1990 and Thagard 2006.

On conceptual change, see Thagard 1992, 2012c, 2014c. *Mind–Society* contains social examples, including applications to education in chapter 12.

Pessoa 2013 shows how cognition and emotion are integrated in the brain.

Propositional attitudes are obsolete because of the neural interpretations of mental representation in *Brain–Mind*. Supervenience is superfluous because of the account of emergence in chapter 5. Massive modularity is refuted because of the many-to-many relations between cognitive functions and brain areas described in *Brain–Mind*.

Wittgenstein 1968 says that philosophy leaves everything as it is.

PROJECT

Perform a dynamic (historical) three-analysis of some important epistemological concept such as *belief*. Use the techniques in this chapter such as three-analysis and coherence to analyze fake news.

4

Reality

MAKE REALITY GREAT AGAIN

Reality has fallen on hard times. It has been disrespected by politicians who trumpet alternative facts and blur the distinction between fake news and truth. It has been disparaged by postmodernist theorists who think that truth is just a matter of persuasion and consensus. It has been undercut by pragmatist philosophers who think that truth is merely what works, with no consideration that working might result from manipulating reality.

Nevertheless, reality matters. Meeting the challenge of climate change requires dealing with mechanisms of greenhouse gases, energy transfer, and global warming. Vaccinations work to help protect people from serious diseases because viruses, diseases, and autoimmune responses operate in the real world, not only in people's minds and interactions. Humanity could be obliterated by nuclear war because fission and fusion can generate enormous amounts of potentially destructive energy. Such problems presuppose the existence of hydrogen, uranium, energy, climate, viruses, diseases, bodies, and many other entities, processes, and mechanisms.

But what really exists? The question of existence is central to metaphysics, the philosophical theory of reality. Branches of science ask particular questions about the reality of such objects as planets, atoms, cells, and societies. But philosophy poses the most general questions about existence and also considers the normative question of what ought to exist. Do you and should you believe in God, souls, ghosts, trees, rocks, chairs, genes, and/or electrons?

Metaphysics need not be the navel-gazing, angel-counting enterprise that sometimes gives philosophy the reputation of irrelevance. Important practical and scientific pursuits generate inescapable questions about the existence and nature of life, persons, minds, and the universe. We have already encountered metaphysical questions, in chapter 2, which discussed the material reality of mind, and in chapter 3, which assumed that knowledge is about the world that exists independently of minds. Let us now look more carefully at metaphysical questions concerning the nature and interconnections of mind and world, as well as their relations to society.

Some contemporary philosophers and social scientists think that mind and world are just social constructions, and older philosophical views hold that the world is just a mental construction. The fundamental metaphysical question pursued is the relation among mind, world, and society. Understanding their interaction is important for many issues to come concerning ethics, justice, and art.

Some philosophers have opposed metaphysics, dismissing it as meaningless blather. Although reality is often discussed in a priori terms of dubious merit, general questions about reality can be integrated with scientific investigations without being reduced to them. In addition to fundamental questions about existence, including the existence of God and social groups, I address important metaphysical questions concerning truth, space, time, groups, and group minds.

ISSUES AND ALTERNATIVES

There are several available answers to the general question of what exists. Chapter 2 introduced materialism, the view that science provides the best guide to existence through its development of theories about matter and energy. This view assumes the doctrine of scientific realism, according to which science aims to find out how the world is and sometimes succeeds. Natural philosophy cannot simply assume materialism and scientific realism but must argue for them based on overall coherence with philosophical considerations and scientific findings.

The assessment of the coherence of materialism requires comparison and evaluation with respect to alternative views, including dualism, idealism, and social constructivism. Dualism claims that two kinds of thing exist: material objects such as bodies and spiritual entities such as souls, spirits, and gods. More radically, idealism dismisses the idea of matter as distinct from mind and claims that all reality is mind-dependent. In contrast to the materialist conclusion that mind is a relatively recent introduction to the universe, idealism insists that there is no basis for postulating a reality that is not dependent on human minds or the mind of God.

The doctrine that reality is a social construction is a relatively new metaphysical view deriving from social science in the 1960s. There are perfectly sensible ways in which reality is socially constructed. Chapter 3 discussed how knowledge is a social enterprise in which development of knowledge about the world depends on social processes such as communication, collaboration, and consensus. Moreover, it is obvious that social realities such as governments, banks, and sports teams are socially constructed in that they are formed by human interactions. Much less plausible is the stronger claim of social constructivism that all of reality is just social, including physical, chemical, and biological processes. On this view, there is no world independent of its social construction.

These issues about existence connect with questions about truth. The materialist view of the world links with the correspondence theory of truth, according to which truth is a fit between our mental representations of the world and how the world actually is. Idealism rejects this division between mind and world and instead adopts a coherence theory of truth, according to which truth is just a matter of how various thoughts fit together. Although chapter 3 argues for a coherence theory of knowledge, this chapter argues that coherence theories of truth are inferior to correspondence ones. The social constructivist view rejects the independence of mind and world from society, claiming that there is no such relation as truth, merely social processes by which people convince each other of what to take as true.

EXISTENCE

Metaphysics is sometimes described as the study of being, but that is an unduly abstract way of asking about what exists. I argue that plausible candidates for existence include objects, properties, relations, changes, processes, and mechanisms. In contrast, there are no good reasons to believe in the existence of supernatural entities such as souls and gods.

Objects, Properties, and Relations

On the materialist, scientific-realist viewpoint, we have good reason to believe in objects such as trees, properties such as being green, relations such as being part of a forest, changes such as losing leaves, processes such as short-term and long-term growth, and mechanisms such as photosynthesis, which converts the sun's energy into tree development. In accord with chapter 3, the justification for believing in the existence of all of these is inference to the best explanation based on explanatory coherence. Believing in trees and their properties, relations, and mechanisms is part of the comparatively best explanation of available evidence.

Why believe in objects such as trees, planets, and other people? One could adopt the skeptical view that these are all fictions constructed by our minds and social groups. However, idealism and social constructivism have trouble explaining numerous facts about our perceptions of objects. Although people occasionally make mistakes, their observations of trees and other objects are remarkably stable, within and across individuals. People rarely disagree about whether there is actually a tree at the side of the road. Our observations of trees are not based only on the sense of vision, because we can also touch, smell, hear, and taste them. Other people get similar sensory experiences. Moreover, we can act on the world by cutting trees down to become firewood and making them grow faster through fertilization.

Scientists frequently use instruments to measure objects such as the planet Earth, molecules, and biological organisms. Similar considerations operate with many other kinds of natural and manufactured objects, so explanatory coherence with our observations justifies believing that such objects exist. The best explanation of the imperfect but usual reliability, intersubjectivity, repeatability, and robustness of many observations and experiments is that they are caused by interactions with objects in the world.

Despite the existence of objects, you might think that the properties of objects are just in the mind in the form of concepts. Perhaps the greenness, height, and branch structure of the tree are just mental constructions subject to interpretation, for example when we see that leaves seem to be different shades of green depending on the brightness of the sun. But even greenness has an element of mental constancy and intersubjective agreement that belies full subjectivity. Granted, people's perceptions of colors are substantially affected by brain processes, and some brains cannot tell green from red in a common form of colorblindness. But these variations result from well understood processes of light reflecting off the structure of a leaf into the eye to produce perceptions that regularly include green. A major part of the best explanation of why people have experiences of trees is that the trees have properties that make major causal contributions to these experiences.

Relations are just important as properties, because each tree has connections among its branches, leaves, and roots, as well as spatial and temporal relations to other trees, the ground, plants, and animals including humans. People have important relational concepts such as *next-to* and *cut* because brains connect with the world via the sensory-motor processes that help to build up neural representations. Relations are therefore as much in the world as in the mind. The sciences abound with relations central to accepted theories, such as force in physics, atomic bonds in chemistry, genetic inheritance in biology, synapses in neuroscience, buying in economics, power in politics, and attachment in psychology.

The world also contains relations among relations, for example when a dog chasing a cat causes the cat to hide under a bed. Causality is a relation among events, where the events are changes in properties or relations (chapter 5). *Brain–Mind* describes how such iterated relations can be represented in the brain by semantic pointers, neural processes that explain cognition and emotion as results of binding of multimodal representations.

Matter and energy exist because matter is a collection of objects and energy is a property of objects that can be transferred between them in different forms such as heat. Science encourages beliefs about objects, properties, and relations to improve over time. The ancient Greeks and nineteenth-century physicists believed that atoms were the fundamental objects for the rest of matter, but the twentieth century brought evidence for a much more varied set of subatomic particles, including quarks, leptons such as electrons, and bosons such as photons. Properties and relations are in the world but are naturally represented by concepts and beliefs, which are also part of the world as brain processes. Social groups are in the world through interacting brains, so the mind–world–society trichotomy diminishes, because minds and society are part of the real world.

Wittgenstein said that the world consists of facts, not things, reflecting the logic-based prejudice for language over the world. Facts supposedly exist because they are what make sentences true, a quasi-linguistic analog of linguistic representations. But minds in humans and other animals have sensory-motor neural means of representation of the world consisting of objects, properties, and relations. Facts are dispensable, except as an occasionally convenient way of talking about nonlinguistic reality. This dispensability is also desirable because it avoids a metaphysical slum of countless facts that would be needed to correspond to negation sentences (e.g., the fact that Hillary Clinton is not president), universal sentences (e.g., that all crows are black), and mathematical truths (e.g., that there are an infinite number of integers greater than 7). The truth that Hillary Clinton is not president is about objects and relations, not facts.

Changes, Processes, and Mechanisms

What else exists, according to science? An event is what happens to an object or a collection of objects, when one or more go from having a property or relation at one time to having a different property or relation at a later time. For example, a tree changes when it becomes taller, loses its leaves in the fall, and regrows them in the spring. All sciences describe changes: astronomy has planetary motion, physics has falling bodies, chemistry has molecular reactions, biology has evolution, geology has continental drift, psychology has inferences, economics has business cycles, politics has revolutions, anthropology has culture shifts, and so on. Because objects, properties, and relations are not constant, changes and events are real, assuming that time is real.

Processes are series of changes, as when the weather affecting a tree consists of a series of events involving air movement and precipitation. Mechanisms are a special kind of process in which objects that are connected to each other interact in ways that produce regular changes. Chapter 5 describes the importance of mechanistic explanations in science based on the assumption that interactions in the universe cause changes that are approximately regular. Hence we have good reason to believe that mechanisms exist, along with processes and changes. Unlike changes and processes, however, mechanisms are not directly observable, so they raise questions about how science can reasonably go beyond the senses.

Scientific Realism

Consider the theoretical entities that physics introduces, such as atoms, molecules, electrons, quarks, dark matter, genes, viruses and mental representations. Scientific realism claims that these materialist hypotheses are at least sometimes right. Skeptics retort that numerous entities proposed in the past have turned out not to exist, such as Aristotle's aether that was supposed to hold up the stars and the substance phlogiston that chemists thought was given off during combustion. If science turned out to be wrong about these entities, it can also be wrong about electrons and the rest of the lot. We can enjoy the predictive success of science without leaping to the conclusion that its theorized entities and processes exist.

Of course, science could be wrong, in line with chapter 3's insistence on fallibility and uncertainty. But there are large bodies of evidence supporting the existence of many of the entities currently built into scientific theories and no plausible alternative theories that explain the evidence while denying the entities. Science has become far more sophisticated since the days when discredited entities such as aether and phlogiston were proposed, thanks to improvements in instruments, experimental methods, statistical inference, and rigorous evaluation of competing theories. Science not only proposes mechanisms; it often finds underlying mechanisms for these mechanisms, for example when DNA explains genes that explain inheritance in natural selection. Although the history of science contains many early examples of accepted theories later exposed as false, I know of no cases where mechanisms deepened by other empirically supported mechanisms have turned out to be false.

Science has also changed socially with larger communities of competing theorists with interests in overthrowing inadequate views. Competing research teams eagerly generate new hypotheses to compete with existing ones that already have evidential support. In sciences such as physics, chemistry, and biology with hundreds of years of history, there is little risk that the best explanation is just the best of a bad lot. Therefore, scientific realism is defensible, along with the materialist views about the nature of reality that follow from current scientific theories.

Another challenge to scientific realism and materialism comes from the observation that all of our scientific theories resulted from the operations of minds and social groups, so it might seem that reality is mind-dependent and/or socially constructed. But the need of science for mental and social activity does not undercut the inference to the best explanation that science finds out about the world. The following are three kinds of evidence supporting this inference.

First, scientific experiments are remarkably resistant to the desires and efforts of scientists. Many failures and surprises show that experimenters cannot detect whatever they want, because instruments and other forms of observation are strongly constrained by their interactions with the world. If idealism or social constructivism were true, experimental science would be a lot easier than it is because minds could just get together to construct the reality they want.

The obstinacy of experiments is that the world often resists scientists' attempts to get the experimental results that would confirm their expectations. As of 2018, physicists are largely convinced that most of the universe's mass consists of dark matter because of its gravitational effects, but experimental efforts to detect it directly have repeatedly failed. If the world were just socially constructed, scientists should have been able to avoid such failures. In contrast, thought experiments endure no such obstinacy, because you can always make up a story to support your intuitions.

Second, science is unusual compared to other social enterprises such as religion and fashion design in having a remarkable amount of agreement among its practitioners. Although there are many controversies in physics, chemistry, and biology concerning the best theories, there is also remarkable agreement about central theories such as relativity and quantum theory in physics and evolution and genetics in biology. This agreement allows science to be cumulative, building progressively on the results of previous generations, even though theoretical revolutions occasionally occur with substantial conceptual change.

Third, science has been dramatically successful in spawning technology, such as spacecraft, computers, and antiviral drugs. These operate with theoretical hypotheses such as gravity, electrons, and viruses. Without supposing that these hypotheses are approximately true of a world independent of us, we have no way of explaining why technology works as it often does and why there are sometimes technological failures despite the best efforts of minds and groups. As with scientific experimentation, nonmaterialist views have trouble explaining why technology sometimes disappoints. Building technologies requires understanding the world's physical mechanisms that are largely independent of mind and society.

Because scientific realism provides the best explanation of experimentation, agreement, and technology, we are justified in believing that science is often at least approximately true. This justification covers properties, relations, and mechanisms that are theoretical as well as ones that are observed.

Gods, Spirits, and the Supernatural

Scientific realism and mechanistic materialism make no mention of supernatural entities such as souls, gods, and spirits. But their existence could be justified if there are explanatory gaps that science cannot fill, for example if no mechanistic explanation of consciousness is possible. It would be legitimate to infer the existence of soul as the best explanation of phenomena concerning consciousness if scientific explanations are not constructible. However, chapter 2 described the beginnings of scientific theories of consciousness, so postulation of nonphysical entities such as soul is dispensable. The Semantic Pointer Architecture is a crucial part of this argument, because it is the first cognitive theory to cover the full range of mental phenomena, from perception to emotion to creativity, as *Brain–Mind* demonstrates.

The same dispensability applies to other supernatural entities. Before the ancient Greek philosopher Thales originated natural explanations around 600 BC, the standard way that people explained the events in the world was through the intervention of gods. For example, Thales said that eclipses of the sun are predictable events, not divine omens.

Since Thales, science has dramatically increased its scope, ranging over countless phenomena from physics, chemistry, biology, cognitive sciences, and social sciences. Many prescientific groups imbued the world with spirits responsible for the characters and actions of plants, animals, and weather. Now these behaviors fall under sciences such as biology and geology, so hypotheses about spirits are fully dispensable. Today, some people like to say that they are "spiritual but not religious," but they usually mean only that they have a vague sense of purpose and values rather than a belief in spirits.

Hypotheses about gods have many potential explanatory targets. That God created the universe is an ancient explanation of why there is something and not nothing. His intentions explain why the something has the character that it does, for example day and night. Physics and astronomy now provide alternative explanations, for example that the universe originated in the Big Bang that occurred because a previous universe contracted to generate it. Until Darwin's theory of evolution came along in 1859, the standard explanation of the origin of species was divine creation, but today natural selection coupled with genetics and molecular biology shows the superfluous nature of theological explanation of species. Many theologians and some philosophers have seen the origins of morality in divine insistence on principles such as the Ten Commandments, but there are now widely available naturalistic explanations of morality such as the theory developed in chapter 6.

The continuing success of scientific explanations marks theological explanations as inferior, so no inference to the best explanation justifies supposing the existence of God or attendant spirits. Then why do more than 5 billion people continue to believe in the various gods proposed by different religions? *Mind–Society* (chapter 8) answers this question by attention to mental and social mechanisms. Belief in God is appealing to people individually because of the powerful emotional mechanisms of motivated inference and fear-driven inference. In motivated inference, people acquire beliefs that satisfy their goals despite lack of evidence and availability of alternative explanations. People want to believe that God exists to ensure that there is someone looking after them like a benevolent father, that they can survive death by having a soul that can join the company of God, that everything happens for a reason, and that everything will work out in the end. So belief in God satisfies goals of dealing with life's difficulties such as disease, disappointment, and death, irrespective of evidence.

But motivated inference is not good at explaining why people believe in less benign gods, such as the scary deities in the Old Testament, Calvinist Christianity, and some versions of Islam. Why believe in a threatening god? Some people are prone to believe what scares them, for example hypochondriacs who cannot help believing that a small bump is really a tumor and spouses who are jealous despite scant evidence. In all these cases, fear-driven inference operates when emotion shapes attention that pushes belief, independently of evidence. Scary gods are not pleasing to think about, but threats about imminent and eternal punishment can be so emotionally compelling that people acquire beliefs that increase their misery.

These individual, mental mechanisms are only part of the explanation of religious belief, which also has social causes described in *Mind–Society*. Most people acquire their religious beliefs from their parents, through communications that are both cognitive and emotional. Chapter 3 mentions testimony as a legitimate source of some beliefs, and children can usually rely on their parents to tell them truths about the world. So it is unsurprising that children also acquire the religious beliefs of their parents because of emotional attachment, fear of punishment, and influence of religious teachers chosen by the parents. The transmission of religious views operates by a combination of individual motivations and social processes.

Hence psychological and sociological explanations suffice to explain the prevalence of religious belief, which therefore does not need a supernatural explanation that postulates the reality of a god. Natural philosophy therefore has no room for gods, spirits, and souls.

Possible Worlds

Leading philosophers such as Saul Kripke, David Lewis, and Timothy Williamson have tried to use modal logic as a guide to what exists. Modal logic concerns necessity and possibility, feeding into the philosophical aim to achieve truths that are not just contingently true of this world but necessarily true of all possible worlds. Transcendental arguments are supposed to show not just how things are actual but what makes them possible.

Just as epistemology in chapter 3 was doubtful about a priori truths, natural metaphysics is doubtful about necessary truths and possible worlds. The alleged necessity of truths in logic and mathematics has been challenged by the proliferation of alternative logics such as quantum logic and of variants of mathematics such as non-Euclidean geometry. Moreover, the knowability of truths about other possible worlds is rendered suspect by the lack of causal connections between other possible worlds and the one we live in. Chapter 5 argues that causal versions of inference to the best explanation support the truth of what is inferred much more strongly than narrative and deductive explanations. Similarly, no objective probabilities connect what happens in other possible worlds with what happens in ours, so Bayesian reasoning is of little help in justifying belief in other possible worlds. Hence possible worlds and attendant metaphysical doctrines about necessity are best abandoned as vestiges of theology.

Idealism

Materialism and scientific realism support the common-sense view that there is an external world independent of our thinking about it. Although the nature of the world goes far beyond common sense in that it includes exotic, unobservable entities like quarks, black holes, and neurotransmitters. However, there are philosophical traditions that reject this view of the world as independent of mind and society.

Idealism, the claim that the world is fundamentally mental, gains its plausibility from several directions. First, we cannot access the world directly but have to use our minds to represent it. Perception is a complex process that requires many stages of neural processing to go from light hitting the retina to visual images, and countless illusions show that we do not always see the world exactly as it is. When we make generalizations about the world, such as the induction that hawks fly, we inevitably use concepts such as *bird* and *fly*. Because knowledge requires mental representations, and mental representations are not directly and automatically connected to the world, it is tempting to conclude that we should just ignore the world and concentrate on what minds do.

However, we cannot understand what minds do without appreciating the extent to which they are affected by the world. People do not see, hear, touch, taste, and smell whatever they want, because their sensory apparatus is constrained by inputs that come from interaction with external objects, even if the brain has to do a lot of processing to make sense of these inputs. Our thinking is partly embodied in the sense that the kinds of representations we have of the world are much affected by our sensory apparatus, which is shaped by the nature of the world because senses have evolved through millions of years of natural selection. We would not think the way we do unless the world is roughly the way it is, so we should not conclude that what exists is basically mental.

Moreover, the available evidence from astronomy, geology, and other sciences implies that the world existed long before there were any minds to conceptualize it. The universe is approximately 13 billion years old, and the planet Earth has been around for less than 5 billion, with life appearing less than 4 billion years ago. Animals with neurons capable of interpreting the world and generating actions have only been around for about 500 million years. Therefore, based on available evidence, we can conclude that the world existed for billions of years before there were any minds to know it. Because world predated mind, there is no reason to stink that the world is mind-dependent, even though knowledge of the world is mind-dependent because it requires mental representations. This argument may seem question-begging because of use of the presence of neurons to infer the presence of minds, but no other evidence for early minds exists.

The second reason that idealism is appealing is because it restores humans to centrality in the universe, a place that science has progressively challenged. The Copernican revolution dislodged the earth from the center of the universe, the Darwinian revolution dislodged humans from special status different from animals, and the brain revolution is dislodging immortal souls from thought. Idealism is appealing by the motivated inference that a person cannot just be a transient physical object but instead can be the basis of all reality. Some versions of idealism place reality in the mind of God, but this view is also bereft of evidence and only supported by motivated inference.

Social Constructivism

Similar arguments undercut claims that reality is socially constructed. The development of knowledge both in ordinary life and in science is undoubtedly a social process where we rely on testimony and social interactions such as communication and collaboration to develop our mental representations of the world. But the much stronger claim that there is no world independent of social construction falls afoul of the arguments given in defense of scientific realism. Although it often takes a village to conduct scientific experiments, the village does not get to choose the results that it wants.

Daniel Moynihan said that you have a right to your own opinions but not to your own facts. Radical social constructivism rejects facts independent of social processes, but it ignores how much science depends on the world being stable enough to produce reliable evidence. No matter how much you might want to find a cure for cancer, you cannot socially construct a cure unless molecular biology cooperates. No matter how much you want a technology that will transport you instantly to the planet Mars, you cannot socially construct a space ship unless the physical world cooperates. Social interactions strongly affect the way in which experiments get done and their results get interpreted, but they do so with major contributions from a world independent of society.

Exaggerated claims about the social construction of reality are motivated by the goal of placing social sciences and humanities at the center of discourse. These fields are indeed important for understanding the world, for example in explaining prejudice and other phenomena examined in *Mind–Society*. Community entities like banks and governments are socially constructed, but it requires remarkable arrogance to suppose that the same goes for atoms and galaxies.

Consider the case of Ludwig Fleck, a Polish physician who wrote about syphilis in his 1935 book about the social genesis of a scientific fact. He emphasized the social processes that went into the development of early knowledge about syphilis, influencing Thomas Kuhn's *Structure of Scientific Revolutions,* which claimed that the world changes when a paradigm shifts. But by 1943, penicillin was recognized as a cure for syphilis, which only makes sense because this drug stops the division and growth of the bacteria that cause syphilis. Truths about the causes and cure of syphilis are truths about biology, not society. Hence experimentation and theorizing about syphilis were dependent on the world, not just social interactions.

In sum, philosophical reflection on science gives us reason to believe in the reality of objects, properties, relations, changes, events, processes, and mechanisms, independent of mind and society. We need minds and societies to gain such knowledge, but they are strongly affected by the external world while gaining knowledge about it.

TRUTH

Now we can address the ancient philosophical question: what is truth? Truth is not just a social construction, because the arguments for scientific realism against idealism show there is a world that truth needs to be about. The correspondence theory of truth claims that a representation is true if and only if the world is as the representation describes it. The nature of the claim is most obvious when the representation is sentence-like, for example when "Trees are green" is true if and only if trees are green.

More complicated relations between representations and the world operate nonverbally. For example, a picture of a tree with leaves that are a particular shade of green may only be a rough approximation to the actual tree's slightly different shade. Similarly, a recording of the call of a crow may only come close to how crows usually sound but is nevertheless valuable because of the degree of correspondence to reality.

Focusing on truth as a relation between sentence and the world makes it sound misleadingly precise, because words are mentally represented by concepts that are not sharply defined. My concepts of *tree* and *green* are only an approximation of trees and their color in the world. Correspondence does not have to be a binary yes or no relation, so that the match between mental representations and parts of the world can be rough but real. Accordingly, chapter 3 describes knowledge as reliably acquired and approximately accurate representations of reality.

The major alternative to a correspondence theory of truth is based on coherence. It is important to distinguish a coherence theory of truth from the coherence theory of knowledge defended in chapter 3. Justifying a mental representation requires showing that it is part of the best overall coherent account of the world, but showing coherence is very different from saying that the world is just coherence. Coherence is a criterion for truth, not truth itself.

Identifying truth with coherence fails because there are opposing views that are each internally coherent even though they do not agree with each other. For example, the Mormon religion and Scientology are complex worldviews full of internal coherence, but they are not compatible with each other and with scientific findings. Explanatory coherence settles this by comparing competing theories through giving some degree of priority to evidence that results from systematic interactions with the world through scientific experiments. Experiments and systematic observations provide a grip on the world because the results are partly caused by interactions with objects and their properties and relations.

Moreover, theories that correspond approximately to the world provide the basis for modifying the world through technology. The reliable coherentism in chapter 3 accommodates the existence of an external world that interacts with people through sensory inputs and practical outputs. The coherence theory of truth cannot make sense of such inputs and outputs that depend on interactions with the world.

These effects also undercut skeptical interpretations of truth that dismiss it as a mere social construction. In both science and everyday life, people do not get to build the worlds they want because they are constrained by physical, chemical, and biological reality. For example, it would be wonderful to create a world in which everybody enjoys wealth and longevity, but physical limitations such as the amount of resources on the planet and biological limitations such as cell death make it difficult to construct what people want. Chapter 7 on justice discusses further the social processes that establish just societies but also takes into account the nature of the world that our representations and actions are required to manage.

In sum, the best theory of truth invokes approximate correspondence using mental representations that include sensory and motor images as well as sentences. The brain can use them to stand for our world that includes objects, properties, relations, changes, events, processes, and mechanisms. A purely linguistic account of mind is insufficient, but *Brain–Mind* systematically shows how neural representations using semantic pointers can produce perceptions, images, beliefs, rules, and analogies.

Idealism and the coherence theory of truth put truth only in the mind, and social constructivism puts it only in society. Much more plausibly, a correspondence theory of truth tied to the Semantic Pointer Architecture for mind sees truth as a complex relation between mind and world, with social influences but not determinants of mental representations.

In many sciences, explanatorily powerful truths concern mechanisms, which are systems of connected parts whose interactions produce regular changes, where representations of the parts, interactions, and changes can be visual and motor as well as verbal. On this understanding, one mechanistic theory is more approximately true than another if it correctly identifies more parts, more properties of the parts, more relational connections of the parts, more dynamic relational interactions of the parts, and more resulting changes. Approximation is not just a matter of counting truths and falsehoods but rather of capturing more aspects of what the mechanisms are supposed to be about.

SPACE AND TIME

Events are supposed to occur in time and space, but what are these? A surprising number of philosophers and scientists have argued that time is unreal, and some have even thought that space is unreal. An important part of metaphysics is first to ask whether time and space exist and second to ask what they are. In line with the philosophical method presented in chapter 1, the key to answering these questions is not to become fixated on a few thought experiments that reflect the prejudices of their devisers, but rather to consider alternative views and pick the ones that are most coherent with scientific theories and other philosophical conclusions.

The existence of events and changes presupposes that time and space exist at least in some form. Changes require objects to have properties and relations that are different from previous properties and relations. Hence change entails that there are different times at which the different properties and relations occur. Many of the relations are spatial, requiring objects to be in different places and arrangements. Hence denying time and space amounts to denying change, at odds with the entire history of scientific observations and experiments. Can assumptions about changes in events and relations be explained away as illusions? Are space and time just in the mind and society rather than in the world?

Space and Time in the Mind

How do human minds represent space and time? We have abundant words for expressing spatial and temporal relations, such as "besides," "top," "bottom," "moving," "before," and "after." But language only captures some of our ability to think about and operate in space and time, which is embodied in the sense that it depends on sensory-motor representations. Operations such as pushing things, walking, and getting up can be done without any verbal awareness, so we need to understand how neural representations that are not verbal carry out these spatial-temporal. The independence of space and time from linguistic representations is obvious from the operations of nonhuman animals that can accomplish tasks and learn how to do them better without the assistance of language.

Extensive studies have been made on how animals such as rats and pigeons represent time and space. For example, rats can learn to respond to stimuli at different time intervals, thanks to special neurons that keep track of time, and macaque monkeys have neurons that time stamp events. In mice, midbrain dopamine neurons control behaviors that require judgments of the duration of time. For spatial navigation, rats have space cells that fire when an animal is near a familiar location and grid cells that provide an overall map of various locations. Hence mental representation of time and space involve the firing of special arrangements of neurons, not linguistic entities. *Brain–Mind* (chapter 2) describes how visual and other kinds of imagery require binding of representations of sensory features with representations of space and time.

People and other animals also have biological clocks that track external changes such as a light and regulate bodily processes such as sleep. Unlike some philosophers and physicists, biologists have no trouble seeing time as real. Apparent puzzles about how to deal with verbal complexities about time and space can be avoided by realizing that biological organisms represent them primarily by means that are neural and embodied rather than linguistic.

The most plausible explanation of these neural mechanisms is that they are tracking time and space in ways that enable organisms to function better in the world with respect to the evolutionary goals of surviving and reproducing. Finding food, avoiding predators, and mating require attention to objects' locations and changes dependent on time and space. There is no guarantee, however, that the basic representations of time and space provided by biological evolution are the best way of understanding the world, for they only need to be good enough to ensure the fitness of the organisms that have them. Euclidean geometry is fine for everyday life, but relativity theory implies that Riemannian geometry is more apt for describing the universe. Perhaps science and philosophy can do better than innate neural representation of space and time, but they should at least appreciate the benefits of the sensory-motor images that evolution has developed.

Space and Time in the World

Various philosophers and scientists have argued that time is an illusion. Parmenides claimed that reality is eternal and timeless because change requires that something both is and is not. John McTaggart argued that problems about past, present, and future make time incoherent. Kurt Gödel thought that relativity theory showed time is unreal. Julian Barbour argued that reality is timeless because duration is indefinable and redundant. None of these abstract arguments is convincing, although I do see the appeal: if time is unreal, why worry about income tax deadlines?

The physicist Lee Smolin addresses the reality of time in two stages. He first considers a series of reasons why physicists have thought that time is unreal and gives his own reasons why he thinks time does exist. The first reason he gives for doubting time is the Newtonian view that nothing happens except the law-determined rearrangement of particles, so the future is determined by the past. But determinism is not a problem for the reality of time, because Newton's laws of motion all refer to time via concepts such as velocity and acceleration.

Smolin's second reason concerns Einstein's relativity theory, which takes time as just another dimension of space. But time is very different from spatial dimensions, which are reversible: if you can go forward, you can go back; if you can go up, you can go down; if you can go left, you can go right. But there are no known cases where anyone has ever gone backward in time rather than forward. That is why I find time travel implausible, not because there are any good reasons for judging it impossible but just because there is no evidence for it being actual.

Smolin's attempt to rescue time from physics rests on some dubious assumptions. He endorses Leibniz's principle of sufficient reason that requires there to be an answer for all questions about the universe, including why physical constants such as the gravitational constant have the values they do. But Leibniz's principle belongs to theology rather than science or naturalistic metaphysics, for it may well be that some things are just true by chance.

Smolin tries to explain the values of physical constants and the form that laws currently take by a hypothesis of cosmological natural selection, which advocates that universes reproduce by the creation of new ones inside black holes. The postulation of countless universes makes this one of the least parsimonious hypotheses in the history of science. In contrast, Steinhardt and Turok's book *Endless Universe* presents a much simpler account of how our current universe began around 14 billion years ago as the latest in an eternal series of cycles of expansion and collapse. This cyclic theory clearly assumes that time is real, without Smolin's dubious claims about sufficient reason and cosmological selection.

But what is the nature of the time that exists? Smolin makes a good case for the view of Leibniz and Mach that time is a system of relations among events, not an absolute container for events. Oddly, however, he concludes that space is an illusion, rather than an analogous system of relations among objects. I think that both space and time are real, in the relational sense, because that supposition is part of the best explanation of a vast range of physical, biological, and psychological evidence.

Another scientific complication comes from the radically new ideas about space and time developed by Einstein in his special and general theories of relativity. Instead of the common-sense idea of three dimensions of space and one dimensions of time, Einstein advocated four-dimensional space-time. Relativity theory has enormous experimental support, so that the separation of time and space found in the brains of animals such as rats with their different systems of neurons may not be the best way to think about space and time. We can anticipate that understanding of space and time will develop through better theories in physics and biology, but there is no reason to deny their existence.

More positively, it is clear that every science works with time as real. I listed earlier the changes that various sciences describe, and they all deal directly with time. Many of the fundamental constants of physics are stated in terms of seconds: the gravitational constant, Planck's constant, the speed of light, and others. Seconds are a unit of time and so make no sense if time is unreal. Similarly, sciences such as astrology, biology, and economics deal with changes over years, another unit of time that would be bogus if time is an illusion. Abandoning time would be abandoning science and life in general.

Space and Time in Society

Another reason for denying the reality of space and time would be viewing them as social constructions, the result of cultural variations in how people think and talk about their worlds. There are indeed major variations in language and cognition concerning the concepts of space and time different in cultures. For example, some Australian aboriginal languages require every sentence to indicate the spatial orientation of the speaker, so that members of this culture have a much better ongoing spatial representation than English speakers who rarely pay attention to whether they are facing north, south, east, or west. Western societies with capitalist economies are tightly oriented to exact time of day, whereas more agrarian cultures are more relaxed about time. In part, therefore, representations of time and space are socially constructed.

Groups of humans vary greatly in language and culture, but they also have much in common. All humans are descended from mere thousands of people in Africa around 100,000 years ago, with similar genetics, physiology, senses, and neural systems. This common descent limits variation in spatial-temporal concepts that are embodied as well as linguistic, such as *up, down, before,* and *after*. The range of social construction is limited by the structures of the bodies and brains that all people share.

Moreover, people from all cultures are capable of learning the methods and conclusions of science, gaining reconceptualizations of time and space that are empirically warranted, such as appreciation of space-time. Science provides a way of overcoming cultural biases to reach common understanding of physical reality. So far, that understanding includes appreciation of the reality of space and time, even if it must be reinterpreted as relational rather than absolute and has coalesced into space-time.

Therefore, cultural variations do not support denial of the reality of space and time as entirely socially constructed. However, along with changes in scientific conceptions, they point to the need to suppose that theories of space and time are an ongoing project, to be revised based on the basis of neurobiology as well as physics.

GROUPS AND SOCIETY

My approach to mind and society, social cognitivism, considers groups from small ones like marriages to large ones like countries and organizations. But are these groups real, or are they just convenient fictions? In social sciences such as economics and politics, many theorists propound methodological individualism, which says that only individuals exist and have explanatory power. An even more contentious issue is whether there are group minds with shared mental properties such as collective beliefs, desires, intentions, and emotions. Historians and social sciences speak casually about what whole countries think, want, and do. Moreover, some philosophers claim that there are social truths that are not reducible to truths about individuals. Such questions about the existence of groups, group minds, and social facts are important metaphysical matters for natural philosophy to resolve.

Groups Exist

It is justified to believe that groups exist because they contribute to explanations that are better than purely individualist ones. First, the existence of groups explains why people believe there are groups, for example when people speak of their marriages and countries. Such talk might be mistaken but does not seem to depend on truth-distorting motivations such as those that foster religion.

Second, the existence of groups is an important part of explaining the patterns of interaction that affect formation of people's beliefs, desires, and resulting actions. People who are married interact in ways different from people who are not part of such a group, as do members of teams, countries and other sorts of organizations.

Third and most important, groups have emergent properties that are not just the sum of the properties of their individuals. Universities can grant degrees and countries can declare war, with properties that do not reduce to the properties of individuals. Even the president of a university or a country is not entitled to grant degrees or declare war, actions that require interactions of many people. Chapter 5 defends emergence as an important element of explanations, and *Mind–Society* provides numerous examples of emergence in social systems.

For these reasons, groups such as marriages, institutions, and other organizations are legitimate theoretical entities that we can infer exist because they are part of the best explanation of observed phenomena concerning the actions of individual people. Similarly, according to the view of the self in *Brain–Mind* (chapter 12), people are not just their bodies but actually whole systems that integrate mechanisms at four levels: molecular, neural, psychological, and social. As such, people are theoretical entities too, not directly observable as bodies.

But groups seem to be different from theoretical entities such as atoms and genes, justified by explanations that are causal and mechanistic. One might worry that organizations such as universities do not actually have causal effects, unlike the parts of the university such as people and buildings. How can something as amorphous as a university, marriage, or nation cause anybody to do anything?

To answer this question, consider the three-analysis of causality presented in the next chapter, Table 5.2, which assigns these features as typical for causal relations: sensory-motor-sensory patterns, regularities, manipulations, statistical dependencies, and causal networks. Group relations lack the sensory-motor-sensory patterns that characterize simple kinds of causality such as kicking a football, but typical features are not defining necessary conditions so group causality is not ruled out.

There are clearly regularities about groups, for example that countries with many universities have more educated populations. These regularities support manipulations, for example when adding more universities improves the educational level of a country, as happened in China in the twenty-first century. Changing a group can change the behaviors of individuals in it, for example when declaring a group such as a political party illegal forces the individuals in it to conceal their activities. Groups also affect statistical dependencies, as when the probability of a country having educated people given that it has universities is higher than if it does not. For these reasons, we can conclude that groups from marriages to international organizations have causal effects and therefore can be judged to exist based on inference to the best causal explanation.

Group Minds Do Not Exist

Unlike groups, a different conclusion is appropriate for group minds and collective mental states. Many people in ordinary life and in social sciences are inclined to talk about shared properties such as team attitude, national will, organizational beliefs, and institutional knowledge. It might seem that postulation of these group minds and states is justified by inference to the best explanation because they have causal effects in the same way that groups do, but there are several reasons why this justification fails.

First, the common belief that there are group minds is not best explained by the existence of group minds but rather by general ignorance of how minds work. Chapter 2 and *Brain–Mind* show how brain mechanisms involving semantic pointers provide a unified explanation of all mental phenomena, from perception to creativity and consciousness. Brains think using semantic pointers, but groups as a whole do not have these brain mechanisms, so groups do not have minds. They lack specific kinds of mental states such as perceptions, concepts, beliefs, and emotions, all of which are well explained within the Semantic Pointer Architecture. Social cognitivism recognizes that explaining human thoughts and behaviors also requires social mechanisms, but neural mechanisms are a crucial part of the story. Alleged group minds lack such mechanisms, so they are not minds.

Second, postulation of group minds and collective mentality adds nothing to the explanation of observable behaviors beyond the combination of mental and social mechanisms that *Mind–Society* uses to explain many phenomena across the social sciences and professions. Social mechanisms have people as parts that are connected to each other through cognitive and emotional bonds, producing communicative interactions. Communication is not just verbal exchange, because of the importance of nonverbal interactions such as facial expressions, gestures, and body language.

The alliance of (a) thinking mechanisms using semantic pointers and (b) communication by transfer, instillation, and elicitation of semantic pointers provides a broad range of explanations across social psychology, sociology, anthropology, economics, politics, and history. When a sophisticated account of social interactions combines such mental and social mechanisms, the postulation of group minds and mental states become superfluous. Once you specify mechanisms at both levels, group minds go away, unlike groups that still contribute to causal explanations.

Third, group minds and mental states have no emergent properties that anyone has identified. They are just the sum of the mental states of the individuals that make up the group. Team spirit is merely the sum of the attitudes of the members of the team, not something that can be manipulated in order to produce different effects. There are no ways of manipulating group minds that change individual minds. Group minds also lack the other typical features of causal relations that I reported for groups, regularities, and statistical dependencies because there are no reliable ways of identifying them. In contrast, groups are identifiable by means of names and legal arrangements.

Mind–Society (chapter 9) argues that talk of group minds and mentality is at best a figure of speech. Such descriptions are inaccurate but may be figuratively useful when (a) there are mental representations held by the most powerful members of the group, (b) these representations are influenced by how members think of themselves as members of that group, and (c) the representations result from communicative interactions with other members of the group. In general, however, it is better not to resort to such metaphors but rather to provide a detailed explanatory account of the mental and social mechanisms that lead to actions. *Mind–Society* repeatedly shows how to provide such explanations for many social phenomena without invoking group minds and mentality. For all of these reasons, we can conclude that group minds and mental states do not exist.

Social Facts

Some sociologists and philosophers have vociferously defended the existence of social facts, but I have already expressed skepticism about whether any facts exist at all. But if we take facts as just a loose way of talking about real objects, properties, relations, changes, processes, and mechanisms, the question about social facts becomes shorthand for asking whether there are true representations of social objects that are more than true representations of nonsocial ones.

Because groups exist and have identifiable causal roles, we can conclude loosely that there are social facts concerning entities such as marriages, teams, universities, and countries. Methodological individualism claims that there are no social facts, only facts about individuals, but its explanatory inadequacy is well documented in *Mind–Society*. Social facts are more than individual facts because true representations of society require concepts and hypotheses about groups.

However, it would be a mistake to suppose with some social scientists and philosophers that social facts are entirely independent of facts about individual minds. Influential sociologists such as Durkheim have wanted to insist on the autonomy of sociology from psychology because of fear that sociology might simply be reduced to psychology. The fear is legitimate given the hegemonic ambition of rational choice theory to reduce all social explanations to the occurrence of calculations of expected utility in individuals. *Mind–Society* argues that such explanations fail even for individual decision making and therefore are no threat to explaining away group activities.

The best way to understand social phenomena is to view them neither as reducible to psychological ones nor as independent. Instead, what happens in groups from marriages to countries is much better understood by identifying mental mechanisms of cognition and emotion that arise from the operations of human brains and also the communicative interactions that take place when people exchange semantic pointers. Social mechanisms do not reduce to mental mechanisms, nor vice versa. Social theory requires understanding how social communication and individual thinking interact to produce human phenomena. For example, *Mind–Society* (chapter 4) describes how good and bad marriages result from a combination of the cognitive-emotional neural processes in individuals and from how they interact with each other. The developing success or failure of a marriage emerges from all of these mechanisms, in a form of multilevel emergence discussed in chapter 5. Calling this important development a social fact obscures the fascinating complexity of mind–group interactions.

SUMMARY AND DISCUSSION

Following a naturalistic approach to metaphysics, I argue that materialism and scientific realism are much more plausible than their major alternatives: idealism and social constructivism. My philosophical method is to use inference to the best explanation of evidence rather than thought experiments and a priori speculation. Natural philosophy legitimately accepts the existence of objects, properties, relations, changes, events, processes, mechanisms, groups, space, and time. All these concepts and hypotheses are subject to revision as science and philosophy generate more evidence and alternatives. However, skepticism is appropriate concerning the existence of other entities such as souls, gods, spirits, facts, and group minds.

In response to a claim that a particular kind of entity such as trees and gods exist, there are various choices available. If evidence and inference to the best explanation support the existence of the entity, then we are justified in concluding that it exists. Alternatively, we may conclude that the entity is merely fictional, having no existence in reality but just part of some discourse. Then the concepts and other representations that refer to it are erroneous. However, some fictions might be metaphorically useful in some communicative contexts, excusing loose talk about social facts and group minds.

The following is a reliable procedure for determining whether X exists without pure reason or social dogma:

1. Construct hypotheses about X.
2. Collect evidence relevant to X using generally reliable procedures such as perception, memory, testimony, instruments, and experiments.
3. Use arguments to assemble hypotheses and evidence, including alternative hypotheses.
4. Accept the hypotheses most coherent with the evidence.
5. Believe in the entities that the hypotheses are about.

Following this existence procedure does not guarantee that its conclusions are correct, as metaphysics is even more fallible than science. But the procedure is far less arbitrary than other ways of doing metaphysics, such as consulting religious texts, using unconstrained thought experiments, or making ideologically motivated speculations.

The inferences in this chapter about the nature of reality fit well with conclusions about mind in chapter 2 and with the ones about knowledge in chapter 3. The support is mutual coherence rather than deductive implication; the order of chapters is not a sign of logical priority. The triple of multilevel materialism as an account of mind, scientific realism as an approach to reality, and reliable coherentism as a theory of knowledge is highly coherent. None of these is indubitable, but they collectively exceed alternatives such as the package of dualism about mind, idealism about reality, and foundationalism about knowledge. My triple shows how philosophy hangs together not just internally but also with a broad range of facts and theories drawn from science.

More specifically, multilevel materialism supports scientific realism because the mind's sensory-motor interactions with the world help explain the obstinacy of experiments against mind and society, undercutting idealism and social constructivism as explanations of scientific activity. The appreciation of nonverbal representations that comes with the Semantic Pointer Architecture supports a correspondence theory of truth not restricted to language. Conversely, scientific realism supports multilevel materialism because it helps to justify using cognitive, emotional, neural, and social mechanisms to explain the mind.

Similarly, this chapter and chapter 3 are mutually coherent. Explanatory coherence justifies the inference to the best explanation that science achieves approximate truth, at the same time that scientific realism points to a world that knowledge can be about. Such support would be dreadfully circular if the goal were logical proof from axioms but accomplishes the more achievable goal of overall coherence with evidence, which is all that philosophy and the rest of knowledge can reasonably tackle.

Some philosophers are skeptical that science can ever reveal fundamental aspects of reality such as life, mind, meaning and value. Life is already well understood within biology through attention to mechanisms such as evolution, inheritance, and metabolism. Chapter 2 and *Brain–Mind* show how mind is also coming within the scope of science thanks to experimental and theoretical advances such as the Semantic Pointer Architecture. Meaning and value are still contentious, but chapters 6 to 8 bring them within the scope of new theories of mind and society. First, however, we need to address important problems about explanation, causality, and emergence that connect metaphysics and epistemology.

NOTES

Postmodernist rejections of truth include Latour and Woolgar 1986 and Rorty 1979. James 1948 advocates a pragmatist theory of truth less sophisticated than Charles Peirce's view of truth as an ideal point of convergence.

An overview of metaphysics is provided by Le Poidevin, Simons, McGonigal, and Cameron 2009. Ross, Ladyman, and Kincaid 2013 contains discussions of scientific metaphysics.

Guyer and Horstman 2015 review the history of idealism.

On social construction, see Berger and Luckmann 1966 and Hacking 1999.

Wittgenstein 1968 supports facts over things. Armstrong 2004 argues for facts (states of affairs). Ladyman and Ross 2007 argue for structure as more basic than objects, but how can you have a relation without the objects it relates? There is no reason why physics should be taken as the only science relevant to determining what exists. Sometimes physics is wrong, for example when William Thomson said that Darwinian evolution contradicted what cooling calculations showed about the age of the earth, in ignorance of later discoveries about relativity.

Lewis 1986 and Williamson 2013 pursue modal metaphysics based on possible worlds. The interpretation of quantum theory that proposes multiple universes is open to the same objection against inferring entities incapable of causally interacting with our world.

Scientific realism is defended further by Thagard 1988, 2012c, Psillos 1999, and Devitt 2011.

The nonexistence of God is argued for more thoroughly in Thagard 2010b. Holt 2012 considers various answers to why the world exists.

Fleck 1979 (original 1935) discusses the social construction of syphilis, influencing Kuhn 1970.

On the use of diagrams to depict mechanisms, see Bechtel 2008 and Sheredos, Burnston, Abrahamsen, and Bechtel 2013.

Smolin 2013 and Unger and Smolin 2014 discuss the reality of time. Steinhardt and Turok 2007 defend the endless universe.

On how neurons keep track of space and time, see Moser, Kropff, and Moser 2008; Eichengaum 2014; Goel and Bonomano 2014; and Soares, Atallah, and Paton 2016.

On cultural aspects of space and time, see Boroditsky 2011 and Núñez and Cooperrider 2013. In human cognition, space and time are like color and emotion, with considerable cultural variation on top of a common core derived from physiology.

On social facts and collective representations, see Durkheim 1982, Gilbert 1992, and Thagard 2010a. I avoid the term "collective intentionality" because of ambiguity: psychologists use "intentionality" to cover intentions, whereas philosophers use it more broadly to cover any mental state that is about the world.

PROJECT

Do a three-analysis of the concept *real*. Explain the incoherence of the concept of alternative facts. Apply my existence procedure to controversial topics such as extrasensory perception and dark matter.

5

Explanation

KNOWLEDGE MEETS REALITY

What do you find puzzling? Perhaps you wonder why the sky is blue, why cancer is common, and why some voters support evil candidates. Explanations provide relief from puzzlement by fitting events into familiar patterns.

The topic of explanation is important for both knowledge and reality. Chapter 3 argues that the most reliable way to go beyond sensory knowledge is inference to the best explanation but leaves open what explanation is. If anything counts as an explanation, then the best explanation may turn out to be some wild storytelling whose merits are difficult to assess. Explanatory coherence requires accepting hypotheses that fit best with overall evidence taking into account alternative hypotheses, but there may be important differences in the quality of the particular explanations that each hypothesis furnishes for each piece of evidence. Therefore, assessing what explains what is crucial for developing natural epistemology.

The nature of explanation is also relevant to questions about reality. One prominent form of explanation consists of providing causes for what it is observed, but what are causes? The nature of causality and whether it is in the world or merely in the mind are central problems in metaphysics, along with the question whether everything has a cause, which has implications for free will and morality. Explanation and causality must be figured out together as part of the construction of a unified epistemology and metaphysics. This chapter argues that causality is in both the world and in the mind, with attendant social processes, all of which are part of the natural world. Not all explanations are causal, but causal explanation is critical for connecting mind and world.

Explanation also has social and emotional dimensions and is fundamental to questions about reduction and emergence. Despite the materialist metaphysics defended in chapter 4, not all knowledge can be explanatorily reduced to physics. I defend a view of emergence that recognizes properties of wholes that are not properties of their parts and that allows the properties at higher levels to have causal effects on properties at lower levels. In keeping with the view of mind defended in chapter 2, emergence operates at multiple levels, including molecular, neural, psychological, and social.

Issues about knowledge and reality are interconnected. Epistemology affects metaphysics because views of reality are constrained by the methods considered appropriate for developing knowledge. For example, the method of reliable coherentism defended in chapter 3 supports scientific realism, whereas the method of intuitive faith can be used to support theology. Correlatively, metaphysics affects epistemology because the appropriate view of knowledge is constrained by what is taken to be real. For example, if minds are just matter and energy, then there are no souls to gain access to knowledge of the supernatural.

Other philosophical issues besides explanation concern interconnections of knowledge and reality. Chapter 10 provides preliminary discussions of problems about mathematics and free will that are simultaneously epistemological and metaphysical.

ISSUES AND ALTERNATIVES

What are explanations? Are they in the mind, the world, society, or all three? One skeptical view is that they are merely pragmatic conveniences, ways of alleviating puzzlement but not adequate to establish knowledge in the ways assumed by inference to the best explanation and explanatory coherence. Why not just use probability theory? I argued in chapter 3 that probabilities are of limited use in establishing the most important kinds of human knowledge, which require inference to the best explanation.

Hence it is crucial to specify what explanations are. Philosophers have variously proposed that they are answers to questions, logical deductions, causal networks, or narratives. I do not think that these are exclusive alternatives and will consider a variety of patterns of explanations, with varying degrees of value for making inferences about the world.

Another major issue about explanation concerns the analysis and merits of reduction and emergence. Science sometimes seems to operate by reducing the more complex to the simple, for example biology to chemistry and chemistry to physics. Full-fledged reductionism would take the social sciences down to psychology, then down to biology and all the way to physics. Alternatively, antireductionism insists that higher levels such as sociology are independent of anything lower. I argue for a view that is neither reductionist nor antireductionist but instead uses multilevel emergence to indicate how explanations at different levels are interconnected.

STYLES OF EXPLANATION

Most generally, to explain an observation or generalization is just to make sense of it by fitting it into a pattern. There are several kinds of pattern that might work, including stories, deductions, and causal models. Before showing how these patterns work, we can pin explanation down better by providing a three-analysis. I also consider another useful pattern that eliminates the need for explanation in some cases.

Three-Analysis of Explanation

Dictionary definitions say that to explain something is to make it plain, clear, or understandable, but a richer account of the concept of explanation is provided by Table 5.1. Scientific exemplars of explanation include Newton on the motions of planets and falling objects, Lavoisier on the causes of combustion, and Darwin on the origin of species. Medical exemplars include Pasteur's germ explanation of

TABLE 5.1

Three-Analysis of the Concept *Explanation*

Exemplars	Newton on motion, Lavoisier on combustion, Darwin on evolution, Pasteur on disease, Einstein on the photoelectric effect
Typical features	Puzzling phenomenon to be explained
	Explanatory pattern
	Resulting understanding and satisfaction
Explanations	Explains: pursuit of patterns, satisfying understanding
	Explained by: mental and social mechanisms

disease and clinical occurrences such as explaining why someone has a cough because of a viral infection and why someone is severely depressed because of bipolar disorder. Explanations also abound in everyday life, for example why an automobile will not start or why someone is in a bad mood.

The typical features of explanations are something to be explained, something that does the explaining, and the psychological effects of the explanation. What is explained can be particular or general, ranging from specific occurrence such as the economic crash of 2008 to more general phenomena such as the repeated occurrences of crises. The explaining is done by patterns such as narratives, mathematical deductions, and causal models.

What are patterns? I have already used this idea extensively in talking about patterns of neural firing. As usual, no tight definition is possible, but familiar examples of patterns include spatial ones such as tiles and temporal ones such as musical rhythms. Typical features of patterns are consistent and repeating characteristics that match a phenomenon or problem. Use of patterns explains how people make identifications of new situations and obtain a sense of familiarity. For example, you can explain a friend's being late for dinner as part of a general pattern of tardiness.

Explanations have psychological effects, making people feel that they understand something. This feeling of understanding is in part emotional, ranging from quiet satisfaction to excitement that something important and previously unintelligible fits in with other things that are now known. One author has compared explanations to orgasms, although I think that this analogy only works for extraordinarily novel explanations such as scientific breakthroughs and rather mediocre orgasms. I discuss the emotional and social aspects of explanations later, but first review the most important patterns used in providing explanations. For each one, I discuss its relevance to questions about knowledge and reality.

What does the concept of explanation explain? This question is recursive but not circular, because its aim is to capture the cognitive role of explanation, not to define explanation in terms of explanation. The concept *explanation* is cognitively valuable because it helps say why people put so much effort into pursuing patterns that make sense of the world. I argued in chapter 3 that explanation is one of the three values that drive the pursuit of knowledge, besides truth and practical usefulness. People have a desire to understand that is best satisfied by seeking explanations of different kinds. The occasional achievement of such satisfaction is also part of what the concept *explanation* explains, as when people feel good to learn that the sky is blue because molecules in the air scatter more blue light from the sun than other colors with longer wavelengths.

What explains explanation? Much of twentieth-century philosophy took this as a question to be answered by formal logic, but deduction does a poor job of capturing most kinds of explanation, as I will shortly argue. My approach is to explain explanation as a mental process involving representations and processes.

Narrative Explanation

The traditional way to explain events is to tell a story that answers the question of why they happened. Narrative explanations of why the universe exists include the Christian story that God created heaven and Earth, along with light, land, plants, animals, and humans. Many other cultures have creation stories, such as the Chinese myth that formless chaos coalesced into a cosmic egg that hatched a giant whose death transformed his body into the parts of the earth and sky. Narrative explanations are also used in everyday life, for example when someone explains being in a bad mood because of a series of ugly events.

Natural science also uses narrative explanations, for example in the cosmological chronicle that the Big Bang led to formation of stars that include our solar system and planet Earth. Chemistry describes the formation of elements by nucleosynthesis, leading from the simplest elements such as hydrogen and helium to heavier ones such as magnesium and iron. Biological evolution provides a rough account of the development of plants, corals, jellyfish, fish, reptiles, birds, and mammals, including human beings. Medical personnel use narrative explanations when they describe how infection by the Ebola virus can lead to gross symptoms such as expulsion of bodily fluids.

Narrative explanations abound in the social sciences. Economics describes how China has become much more prosperous since 1979 through economic reforms that allowed commercial production of clothing, computers, and automobiles. Politics describes how Western democracy developed from revolts against feudalism such as the Magna Carta to the establishment of Parliament to the eventual practice of universal voting. Sociology explains the development of capitalist society in Europe as a result of religious changes from Catholicism to the Protestant Reformation, leading to a work ethic that supported successful industry. Linguistics describes the evolution of languages, and anthropology portrays the development of various religions.

In line with my three-analysis of explanation in Table 5.1, these scientific and everyday examples of narrative explanation illustrate the typical features of phenomena to be explained, explanatory pattern, and resulting feeling of understanding. The pattern here is just a story in the form of a sequence of events, without much detail about what makes the sequence coherent.

Narrative explanations have numerous strengths that make them appealing. They are easy to understand and use the pattern of storytelling, which is culturally ubiquitous. Narratives recognize historical development of how the presence comes from the past. They are emotionally satisfying, following a cadence from wonder to resolution. Sometimes they specify the connections between events, or at least assume that such connections exist.

Narrative explanations also have numerous weaknesses. Exactly what are the connections between events in the story? Are these connections necessary, merely probable, or somehow causal? It is often unclear whether the connections that are supposed to exist between events are based on evidence (following the standards laid out in chapter 3) or whether they are merely myth or wishful thinking. A major problem occurs when there are alternative narratives, for example the Christian versus the ancient Chinese story about the origins of the universe. What is the basis for preferring one narrative over another? The emotional satisfaction that narrative explanations provide may be an illusory feeling independent of scientific goals such as truth and rigor.

Despite these weaknesses, it would be folly to abandon narrative explanation altogether, because historical patterns sometimes do illuminate phenomena. Instead, we should seek to improve narrative explanations by the following means.

First, we can attempt to establish causal connections between events by identifying underlying physical, psychological, and social mechanisms. Mechanisms provide causal links by specifying parts whose connections and interactions produce changes. For example, Darwin had only a crude story about how inheritance passes from parents to offspring, but genetics eventually provided a mechanism that was deepened further by the discovery of DNA.

Second, the problem of alternative narratives can be approached by evaluating the causal coherence of each alternative with respect to a wide range of evidence. By this standard, the Big Bang narrative wins out over the Christian and Chinese stories concerning the origins of the universe because it draws on evidence-supported physical mechanisms such as gravity and cosmic inflation.

Third, people can resolve to accept a narrative not just because it feels good but because it is inferable as the best explanation of all the evidence, compared to alternative explanations. We can watch out for pitfalls of story acceptance such as motivated and fear-driven inference described in chapter 3. If these three improvements are kept in mind, then narrative explanations provide a link between hypotheses and the world, thereby providing their users with stronger claims about the world.

Deductive Explanation

In twentieth-century philosophy of science, the dominant view of explanation used a logical pattern where specific or general phenomena were deduced from general laws. Why do you have a liver? Because you are a human, and all humans have livers. Similarly, physicists can mathematically deduce why there are eclipses by using laws of planetary motion and light propagation to calculate the effects of the moon coming between the earth and the sun. Generally, deductive explanations have the following structure:

Law 1, law 2, . . .   Example: Thrown objects fall down.
Condition 1, condition 2, . . . Example: You threw the ball up.

What is to be explained.   Example: The ball fell down.

The dotted line indicates that the conclusion follows deductively from the laws and conditions.

This pattern of explanation has many uses, for example in applying Newton's laws such as that force equals mass times acceleration to predict the motion of projectiles. Chemistry uses equations to show how transformations occur, for example when methane plus oxygen leads to carbon dioxide and water. Biology uses mathematical laws in population genetics, and medicine is sometimes able to use statistical laws about infections of viruses.

In the social sciences, there also uses of deductive explanations. Economists sometimes develop mathematical models to predict the effects of economic policies, and political scientists sometimes use game theory to analyze interpersonal behavior. Some psychologists use Bayesian models to explain people's inferences and learning, and some computational neuroscientists explain brain operations as dynamic systems.

The strengths of deductive explanation are substantial. Deduction is logically and mathematically rigorous, and, unlike narrative explanation, provides a tight connection between the explainers and what is explained. Because descriptions of phenomena must be deduced from laws, the connection is not just loose storytelling. When explanations are deductive, the practice of inference to the best explanation is clear, accepting theories that mathematically imply the most descriptions of phenomena. Finally, deductive explanation appealingly connects explanation with prediction, because if you can deduce a phenomenon you can also predict it. Prediction satisfies both the value of truth, if the predictions are accurate, and the value of human benefit, because knowing what is going to happen is often highly useful.

But deductive explanation also has numerous weaknesses that undercut it as a universal pattern of explanation. There are myriad counterexamples to claims that all deductions are explanatory. For example, you could try to explain that a man did not become pregnant because he took birth control pills and all men who take birth control pills do not get pregnant, but this deduction ignores the relevant biological mechanisms. Deduction is not enough to capture causal relevance. The problem is that deduction is a purely syntactic notion, inadequate to capture the semantic (meaning-related) and pragmatic (purpose-related) aspects of explanations.

Statistical explanations using probabilities also have relevance problems, because correlation is not causality. There may be a statistical association between eating ice cream and drowning, since days when more ice cream is eaten also have more drownings, but the actual cause is that hot days lead to more ice cream consumption *and* more drownings.

Moreover, there are numerous explanations in fields in which universal laws are rare, such as biology, medicine, and history. Darwin and Pasteur brilliantly explained evolution and disease, respectively, without stating precise laws or providing mathematical deductions of the important phenomena they explained.

Despite these limitations, deductive patterns of explanation are sometimes valuable, and there are ways to improve their use. We can restrict them to domains like physics where laws are available rather than overextending them into other domains where mathematics is harder to apply. We can also supplement mathematical laws with appreciation of causal relations based on mechanisms. Sometimes it may be possible to mingle deductive explanations with narrative ones that provide the needed historical perspective, for example in biological evolution and cosmology. The narrative of the universe developing after the Big Bang is supported by mathematical deductions using the theory of general relativity.

Mechanistic Explanation

Brain–Mind and *Mind–Society* provide many explanations of cognitive and social phenomena that describe mechanisms that are responsible for human abilities such as perception and emotion and for social occurrences such as economic crashes and religious rituals. These explanations conform to the following pattern:

Mechanistic Explanation Schema:
  Explanation target:
    Why do events occur in an object or group of objects?

Explanatory pattern:
- The object or group has underlying mechanisms, with connected parts that interact.
- The interactions of the parts cause regular changes.
- These regular changes cause the events in the object or group.

Here "regular" need not mean universal as required by the deductive account of explanation but can include more approximate recurring patterns. The meaning of causality is complex and is discussed later. Mechanisms always operate in environments with which they have causal interactions.

As an example, consider how a bicycle moves. Bicycles consist of numerous parts, such as the frame, wheels, gears, chain, and pedals. These parts are connected with each other, for example with the pedal bolted to the gear, which is bound to the wheels. By virtue of the connections, the parts can interact, for example when a person pushes down the pedal that moves the chain and then moves the gears. These interactions cause regular changes, the turning of the wheels. When the wheels turn and engage the road, the bicycle moves.

Mechanistic explanations have been enormously important in the history of science going back to the ancient Greek understanding of simple machines such as the lever, pulley, and screw. Democritus and other Greek philosophers introduced the idea of atoms as the indivisible parts that provide mechanisms to explain the operation of all phenomena, even mind. In 1805, John Dalton introduced a more sophisticated atomic theory to provide a mechanistic explanation for chemical phenomena such as the properties of gases. A century later, the discovery that atoms are divisible introduced new mechanisms involving subatomic particles such as electrons. Modern quantum theory complicates atomic theory by viewing atoms as having wave-like characteristics, like all particles, but still provides parts, connections, interactions, and statistically regular changes that amount to mechanisms.

Similarly, Darwin was able to explain evolution by the mechanism of natural selection, which shows how changes in species can result from interactions of organisms that compete to survive and reproduce. This explanation was deepened once genetics and DNA provided mechanisms for inheritance. The germ theory of disease supplanted the humoral imbalance theory because it provided an evidence-backed mechanistic explanation of how infectious disease results from the interactions of cells and bacteria. Modern medicine supplies mechanistic explanations of many other kinds of genetic, nutritional, metabolic, autoimmune, and cancerous diseases.

Are mechanisms in the world or in the mind? They are clearly intended to be in the world because their parts, connections, interactions, and changes are all physical occurrences. But an explanation cannot simply point to the world and has to use mental representations to describe the relevant mechanism. A mechanism is in the world, but a mechanistic explanation is a mind-generated description of how the mechanism produces the events to be explained. This description can take various forms, including verbal, mathematical, and imagistic, because pictures and sometimes even body movements can be part of the explanation. For example, how a pulley lifts a load is explained by a diagram that shows how imagined pulling on a rope enables someone to simulate the lifting of the load, using multimodal rules of the sort discussed at length in *Brain–Mind*.

Cognitive science abounds with mechanistic explanations in which mental processes are explained by identifying interacting parts. Computational explanations describe parts as mental representations akin to computational data structures and connections as mental procedures akin to algorithms; then interactions consist of applying the procedures to the representations. In neuroscience, brain operations are explained by parts that are neurons connected by synapses that undergo interactions based on excitation and inhibition. Computational neuroscience combines both of these kinds of explanation, because neural processes perform special computations involving the firing of neurons resulting from summation of excitation and inhibition to perform the process of parallel constraint satisfaction.

Mind–Society shows systematically how social changes can be explained by social mechanisms, where the parts are people, the connections are personal relationships such as friendships, and the interactions consists of verbal and nonverbal communication. These interactions cause changes in individual behaviors, social groups, and institutional operations. Numerous phenomena in social sciences and in professions can be explained by the combination of mental and social mechanisms, ranging from economic crashes to effective leadership.

When detailed and backed by evidence, mechanistic explanations can be very satisfying in providing causal answers to questions. Inference to the best explanation allows for evaluation of competing mechanisms, not only with respect to how much is explained but also with respect to evidence for the parts, connections, and interactions that are responsible for resulting changes. An explanation of an occurrence gets better as it fills in details about the operations of the mechanisms, using independent evidence. For example, the claim that smoking causes cancer was initially based on strong statistical evidence but has become even more convincing thanks to the discovery of mechanisms by which carcinogens in smoke damage DNA in cells. Providing a mechanism by which A causes B goes powerfully beyond merely asserting some connection between A and B.

Nevertheless, it would be a mistake to elevate mechanisms to a general theory of explanation, because narratives and deductions do sometimes explain in the absence of known mechanisms. There are some deductive explanations that are mathematically impressive without invoking parts and interactions. For example, quantum theory is mathematically sophisticated and predictively powerful without giving mechanisms for odd phenomena such as quantum entanglement.

Similarly, economic models using differential equations and the abstractions of Bayesian psychology provide deductive but nonmechanistic explanations. I think that all of these mathematical explanations would be better if they were able to draw on underlying mechanisms, but they need to be taken seriously even without a mechanistic underpinning. Mechanistic and deductive explanations can be combined when enough is known about the operations of mechanisms that they can be described with sufficient mathematical rigor to produce deductions or statistical inferences.

In narrative explanations in history, mechanisms are typically unknown, although prospects are good for a new approach to social and political history based on cognitive and social mechanisms. *Mind–Society* advocates the use of mechanistic-narrative explanations in social science, where connections in stories are filled out by evidentially plausible accounts of the mental processes and communications of people. For example, chapter 9 of *Mind–Society* explains the start of the First World War by describing the thoughts, emotions, and interactions of the European leaders whose decisions led to disaster.

Eliminative Explanation and Premature Elimination

An unusual style of explanation operates by eliminating what is to be explained, explaining it away so that no further explanation is needed. Suppose somebody is puzzled about why people are immortal. One way to explain this would be to supernaturally postulate the existence of a soul that can live forever because it is not subject to the scientific laws that predict decay of natural systems. Given current scientific knowledge, a better way to deal with the question of why people are immortal is to notice that there is no evidence that anyone lives forever. Then explanation consists of eliminating the concept of immortality from scientific inquiry. Similarly, miracles are best explained away as delusions resulting from socially reinforced motivated inferences.

How does elimination work as science progresses? Initially, the apparently best explanation of phenomena is given by a theory that uses a set of concepts, for example when life is explained by divine creation. Subsequently, a better theory is developed that provides a better explanation of a range of evidence, using other concepts, as with Darwin's theory of evolution of natural selection. Then rational acceptance of the later theory involves abandonment of the earlier concepts such as creation, which are eliminated rather than explained. Chapter 4 argues that such major conceptual changes do not undermine scientific realism.

The history of science has important examples of elimination. Physics no longer needs to find explanations for the Aristotelian aether that holds up the stars, or the luminiferous ether that provides a medium for light waves, or Descartes's vortices that keep planets moving around the sun. Newton and Einstein made these concepts superfluous, so they were appropriately eliminated. Similarly, the astrology question of how stars and planets influence people's personalities and daily events is best answered by saying that there is no such influence.

Chemical advances have eliminated the alchemical idea of the magical transmutation of metals by magic. Also gone are these putative elements: phlogiston, a principle of combustion superseded by oxygen; and caloric, a fluid responsible for heat before heat was understood as a process of molecules in motion. Similarly, biology once operated with concepts like divine creation and vital force, but these became explanatorily useless thanks to the theory of evolution and the identification of diverse mechanisms responsible for life. In medicine, increases in biological understanding of the causes of diseases have eliminated the ancient Greek idea of imbalance of humors (blood, phlegm, black bile, yellow bile) as responsible for illness. Similarly, modern medicine can do without ideas from ancient Chinese medicine such as *qi* and *ying/yang*. Psychiatry has the beginnings of an understanding of mental illness that already eliminates older religious views about demonic possession.

Cognitive science offers a variety of good examples of successful eliminations such as soul, immortality, and demonic possession. It also shows the dispensability of the philosophical doctrine of propositional attitudes, which are abstract relations between abstract entities, the self, and the meanings of sentences, which can be explained mechanistically.

But psychology and philosophy have also witnessed egregious cases of premature elimination not warranted by scientific evidence, as in the attempts of behaviorists to abolish the mind altogether and just describe behavior. Various philosophers have proposed eliminating the self, qualitative conscious experiences, emotions, concepts, and even the whole idea of mental representation. Science has also occasionally witnessed attempts at premature elimination, as in the neglect of atoms in the many centuries between the ancient Greeks and Dalton.

Behaviorism failed because it could not explain even simple patterns of behavior such as the foraging and eating behaviors of rats, let alone the linguistic abilities of humans. The concept of mind survives because cognitive science has developed a rich set of computational and neural mechanisms that can explain such behaviors. *Brain–Mind* provides mechanistic accounts of concepts, emotions, consciousness, and the self and a broad range of mental representations that shows how these concepts can remain a legitimate part of science and philosophy. There are other concepts in the philosophy of mind whose sustainability is still contentious, for example free will discussed in chapter 10.

These cautionary tales of premature elimination should not obscure successful cases where new theories render previous hypotheses and concepts obsolete, illustrating the useful pattern of eliminative explanation. I will discuss reduction, where phenomena at one level are explained by lower level mechanisms. It is important to appreciate that reduction is not elimination. For example, recognizing that tables are complexes of molecules, atoms, and subatomic particles does not eliminate tables but provides more reason to believe that they are real. Similarly, neural explanations of mental representations and processes supports the reality of the mind rather than eliminating it.

EMOTIONAL AND SOCIAL ASPECTS OF EXPLANATION

I have been describing explanation as a cognitive process in which a person produces a narrative, deduction, mechanism, or elimination that makes sense of phenomena. But explanation is also an emotional and social process.

My three-analysis of explanation included the emotional satisfaction provided by understanding, but what kind of emotion is this? The quest for explanation is often initiated by negative emotions such as doubt, puzzlement, perplexity, and dissatisfaction with existing explanations that can even amount to annoyance. When an acceptable explanation removes these negative emotions, the result is a positive feeling. But the explanatory quest can also be initiated by more positive emotions such as wonder, curiosity, and interest, which generate cognitive goals whose satisfaction is also a positive emotion resulting from the accomplishment of understanding. The search for explanation is also sometimes prompted by surprise, which can be positive or negative depending on whether what was expected was bad or good.

Some explanations make the unfamiliar seem more familiar, such as when mathematical principles, stories, or mechanisms that have already helped explain other phenomena are extended to new ones. The result increases emotional coherence that generates pleasure when things make sense together. Even eliminative explanation can be emotionally satisfying when it gets rid of phenomena that are annoyingly puzzling or even threatening, such as the possibility of eternal damnation that evaporates with the elimination of vindictive gods.

Because of the traditional division between cognition and emotion, allowing the incursion of emotion into explanation might seem to throw rationality out the window. On the contrary, emotion according to the semantic pointer theory has a substantial dimension of cognitive appraisal that is not only compatible with rationality but enhances it when people are strongly motivated to pursue emotional values such as truth, explanation, and human benefit. For example, it was utterly rational for Darwin to be excited about his theory of evolution by natural selection and to communicate some of this excitement to his friends: the theory really did accomplish what he thought it did in providing a better explanation of a broad range of evidence.

Explanation is a process of individual cognition and emotion, but it is often also a social process. Explanation is sometimes what you do for yourself, but the point of explanation is frequently to have one person explain things to others. In science, it may take a team of collaborators to come up with an explanation that they then communicate to a large audience of other scientists and the public.

The semantic pointer theory of communication shows how explanation works as a social process. The person who developed an explanation has mental representations that provide an explanation for him or her, but the social task is transferring an approximation of these representations to a broader audience. Even if transfer is on paper, the communicator can still use diagrams to enhance understanding of the representations to be transferred. In face-to-face communications, the explanation one person has acquired can be transferred by means of words and diagrams but also by gestures and emotional expressions that indicate enthusiasm. *Mind–Society* (chapter 12) provides additional discussion of how explanation in education is an emotional and social process.

As chapter 3 argues, allowing the incursion of social processes into knowledge and explanation does not abandon objectivity and rationality. The objectivity of explanations comes partly from their connections with the world, most strongly in mechanistic explanation where the parts, connections, interactions, and changes are ascribable to the world based on evidence and explanatory coherence. Noticing the social processes of explanation does not make explanation inherently contrary to reason, as some sociologists have insinuated. *Mind–Society* (chapter 11) describes better and worse forms of communication with respect to conveying truth and inciting good legal actions. Some social processes such as mob panics are indeed undesirable, but others can accomplish social goals by informative communication and participatory democracy that are conducive to truth, morality, and justice. Some social groups like scientific communities can enhance rationality by generating more hypotheses and evidence and reducing the effects of individual biases based on personal motivations and idiosyncrasies.

CAUSALITY

Chapters 3 and 4 describe the important role of causal explanation in scientific and ordinary discourse. For example, mechanistic explanation is explicitly causal in my formulation and implicitly causal when parts are said to produce or generate changes. There are conflicting theories of causality in philosophy, ranging from attempted elimination to postulation of special metaphysical powers. Causality is not definable, as is clear from dictionary definitions that use vague expressions such as "give rise to" and "make happen." I provide a three-analysis and then discuss the extent to which causality is part of the mind, the world, and society.

Three-Analysis of Causality

The analysis in Table 5.2 provides exemplars, typical features, and explanations for causality. People may not be able to give good definitions of causality, but they have no trouble drawing on personal experience to provide numerous examples. When you push on a door, it usually opens, and when you pull on a wagon, it

TABLE 5.2

Three-Analysis of the Concept *Cause*

Exemplars	Pushes, pulls, motions, collisions, actions, diseases
Typical features	Temporally ordered events, with causes before effects
	Sensory-motor-sensory patterns
	Regularities expressed by rules
	Manipulations and interventions
	Statistical dependencies, with causes increasing the probabilities of effects
	Causal networks of influence
Explanations	Explains: why events happen, why interventions work
	Explained by: underlying mechanisms

usually moves toward you. When you throw a ball, it flies through the air, and when two cars collide, they change direction and undergo damage. Scientists investigate more esoteric examples of causality, such as how gravity makes objects fall and planets rotate around the sun. Some of these examples are statistical rather than universal: smoking causes cancer, even though most smokers do not get lung cancer.

The original typical feature of people's conception of causality, described in *Brain–Mind* (chapter 3), is sensory-motor rather than verbal. Infants a few months old can already notice unusual causal patterns such as when one object hits another that does not move. Through either innateness or early learning, children expect a pattern in which perception is followed by action followed by perception. For example, a baby sees a rattle, bats at it with an arm, and then sees the rattle move and make a noise. Thanks to these patterns, infants already have the beginnings of a concept of causality linked to perceptual exemplars, long before they learn anything about linguistically described regularities or statistical associations.

The temporal patterns recognizable by infants can be captured by multimodal rules of the sort used in chapter 3 to explain knowing-how: <rattle> → <arm-hits-rattle> → <rattle-moves>. In such rules, the brackets indicate nonverbal representations such as the visual perception of the rattle and the motor action of moving the arm against the rattle. Built into these representations is the assumption of a temporal ordering in which the cause preceded the effect, where time is tracked by the neural mechanisms described in chapter 4. Sensory-motor-sensory patterns such as *rattle–hit–move* are typical features of early and everyday representations of causality, not necessary conditions, allowing for more advanced causal statements about nonobservable causal relations such as the effects of forces.

Once language becomes available, sensory-motor representations can be enhanced with verbal rules that are generalizations from experience or taught by caregivers. These have an *if–then* form, as in: *If you punch your brother, then he will cry*. These rules illustrate the importance of manipulations, suggesting that causality works when people's interventions in the world leads to something happening. In contrast to philosophical disputes about the advantages of regularity versus manipulation conceptions of causality, I see them together as typical features of causal situations.

Characterizing causality purely in terms of manipulation does not work, because there are many circumstances in which manipulations are impossible. Presumably, causal relations operate in galaxies on the other side of the universe that people could never manipulate, so causality should not be so tightly tied to human action, even if children start with some kind of primitive sensory-motor understanding of manipulation.

To try to maintain a manipulation account of causality, philosophers may resort to counterfactuals: A causes B means that if you were to manipulate A, then it would change B. But the truth conditions of counterfactuals are problematic, since they concern circumstances that are not true of this world. Logicians have developed semantics for counterfactuals by considering possible worlds, but chapter 4 argued that these are metaphysical extravagances.

A more cautious account of counterfactuals is that they are neither true nor false but can be judged as plausible or implausible by considering causal networks in which we can tweak some of the variables. For example, it is plausible that if I had jumped off my roof, then I would have broken a leg, because of basic mechanisms about gravity and bone breakage. Such mechanisms depend on causal relations rather than explaining them, so causality illuminates counterfactuals rather than vice versa. In the system of ideas about causes, mechanisms, manipulation, and counterfactuals, none is the conceptual basis for the others. Rather, they are interconnected in a nonvicious circle that is legitimate because of its huge success in describing and explaining the world.

Another popular philosophical interpretation of causality draws on probability theory: A causes B if the probability of B given A is greater than the probability of B alone, in symbols $P(B|A) > P(B)$. For example, smoking causing cancer nonuniversally in that the probability of cancer given smoking is greater than the probability of cancer without smoking. Probability increase serves much better as a typical feature of causality than as a general account because of the difference between causality and correlation. For example, the probability of windy weather given rapidly rotating windmills is higher than the probability of wind alone, but it does not follow that windmills cause wind rather than vice versa.

My account of evidence in chapter 3 emphasized causal correlations, which are correlations that result from causal connections. Manipulation is helpful for figuring out which correlations are causal when manipulating A changes B, suggesting that the correlation between A and B is not accidental and that A happens before B. In addition, there are many cases such as familiar pushes and pulls where people are comfortable with causality without knowing anything about probabilities.

Nevertheless, for trained people such as scientists, probability increase can become an important typical feature of the concept of causality. Education in science and statistics can provide people with a rich understanding of concepts relevant to causality such as conditional probability, regression, dependent variables, and controlled experiments. Probabilistic relations can be organized in causal networks, for example those used in explanations of diseases resulting from multiple genetic and environmental factors. This more sophisticated understanding of causality builds on top of the more basic concepts available to people in everyday life, retaining but enhancing typical features of sensory-motor patterns, regularity, and manipulation. Epidemiology, the statistical study of diseases, employs multiple criteria for identifying causal relationships, such as effect size, reproducibility of findings, and specificity of disease to a particular population and location.

Despite the skepticism of a few philosophers and scientists, the concept of causality abounds in science and everyday life because it has many explanatory uses. The word "cause" and related ones such as "effect" and "influence" appear frequently in scientific journals. That there are causal relations explains why events happen and why interventions sometimes work. For example, people contract pneumonia because bacteria or viruses cause inflammation in their lungs, and antibiotics sometimes cure pneumonia because they can kill bacteria, although not viruses. A more difficult question is what explains causality, which requires understanding causality as not just in the mind but also in the world where mechanisms operate.

Causality in the Mind

The three-analysis of causality clarifies how it is both in the mind and in the world. Relevant mental representations range from sensory-motor-sensory couplings for infants with rattles to mathematically sophisticated probabilistic networks for climate change modelers. Relations between causes and effects can for different purposes be captured by multimodal rules that incorporate sensory and motor manipulations and by verbal rules that express generalizations and mathematical equations. Mental processes of explanation, prediction, and action draw on these diverse representations.

The sensory-motor character of causality is not adequately captured just by particular experiences, because causal relations are more general. The baby can learn the nonverbal rule that a shaken rattle makes a sound. The mental representation of causes begins with sensory-motor images and rules, only later developing into verbal representations such as generalizations and probabilities.

For example, the relation that fire causes smoke can be nonverbal in the form of a multimodal rule <*fire*> → <*smoke*>, where fire is represented by sensations of color and heat and smoke is represented by sensations of color and smell. Later, there can also be a motor manipulation, for example when striking a match produces fire that generates smoke. The operation of multimodal rules allows abductive inference that generates causal hypotheses to operate nonverbally with images, for example when you see smoke and explain it by visually imagining someone lighting a match.

Once language is available, multimodal rules can be translated into words as with "fire causes smoke." Eventually, with statistical sophistication, this rule can be represented as $P(smoke \mid fire) > P(smoke)$. Nonhuman animals are adept at causal interventions in the world, for example to obtain food, but never acquire the verbal and mathematical representations of causality that people manage.

Causality in the World

Do mental representations of causality correspond to anything in the world? Chapter 4 argued that there are spatial relations in the world such as a thing being beside another or on top of it, and there are temporal relations such as two events happening at the same time or one after the other. But are there more complex relations where one event (a change in properties or relations of objects) causes another event to happen? It might just be that processes are series of events that are not causally related to each other. The universe might just be one damn thing after another.

Evidence against this view of universal coincidence comes from all of the sciences, physics to sociology. People can successfully intervene in the world, from a baby moving a rattle to social experiments such as using guaranteed minimal income to alleviate poverty. Familiar exemplars of causality exhibit typical features such as sensory-motor patterns, regularities, manipulations, and increased probabilities. If these features were just a matter of imagination rather than correspondence to the world, then their intersubjectivity, practical usefulness, and openness to experimental testing would be inexplicable except for desperate speculation such as divine control. So people do have good reason to think that causality is happening in the world, not just in their thoughts about it. Except for weird quantum occurrences, a process is not just a sequence of events but rather a series of causally interconnected events that sometimes enable people to manipulate the world.

Mental representations such as rules and pictures correspond to mechanisms in the world in different ways. In verbal descriptions such as "gravitational force attracts the sun and the earth toward each other," there is representation of the objects—the sun and the earth—and the relation between them. Mathematical equations such as $F = MA$ use variables to represent the properties and relations of the objects that change: force, mass, and acceleration. In biology, pictures and diagrams of organs, tissues, cells, and molecules capture the parts in the mechanism and conventions such as arrows show motions and changes, for example the neurotransmitters flowing from one neuron to another. In this way, mechanisms can be understood to be both in the mind and in the world, because mental and published representations of their components correspond to objects, relations, and changes.

Causal mechanisms provide the best way of understanding dispositions, which are tendencies of the world to change in predictable ways, for example solubility and honesty. Table salt is soluble because it dissolves when placed in water, and people are honest because they tend to behave legally and truthfully. Dispositions encompass abilities, capacities, capabilities, propensities, competences, and powers, but what are they? They do not reduce to linguistic analysis as *if–then* statements, because they support counterfactual inferences, for example that if the salt had been placed in water, it would have dissolved. Like counterfactuals, analysis of dispositions in terms of possible worlds tells us nothing about this world.

Dispositions result from underlying mechanisms that cause the system to behave in specific ways under specific circumstances. For example, the solubility of salt is based on the mechanism by which sodium and chlorine ions are separated when sodium chloride is placed in water whose molecules attract the chlorine. The honesty of people is based on the mental mechanisms of beliefs and actions that come into play when people interpret a situation. Chapter 6 identifies human needs as dispositions based on underlying mechanisms. This understanding of dispositions as based on underlying mechanisms is important for reconciling ethical ideas about needs and capabilities discussed in chapter 7. Social power results from emotional relations that cause people to behave in ways constrained by the actions of others, as described in *Mind–Society*, chapter 6.

In sum, causality is a mind-detectable temporal pattern in the world resulting from mechanisms. Saying this does not try to define causality in terms of mechanisms, which would be circular, because my characterization of mechanism described them as interactions that produce regular changes. Here *produce* is already a causal concept. Moreover, *manipulation* is also a causal concept that is effective only because of underlying mechanisms that are in turn caused. If the philosophical goal were to have an axiomatic theory of causality based on taking some terms as primitive, the kind of circularity involved in these concepts would be vicious. But interdefinability is standard in ordinary language, as is evident in any dictionary. Therefore, the interconnections among causality, mechanisms, manipulability, dispositions, counterfactuals, and explanation provide a virtuous rather than vicious circularity based on mutual dependency providing conceptual coherence.

Because causality is in the world as well as in minds, causal explanations are particularly valuable for inferences to the best explanation leading to truths about the world. The sensory-motor-sensory patterns that get babies started on understanding causality lead naturally to explanations in terms of human agency, and by the age of four most children are adept at explaining the actions of other people by ascribing beliefs and desires. Historically, most cultures have adopted agency-based explanations of the natural world, for example in the plethora of ancient Greek gods assigned to the planets, oceans, and so on. Native American Algonquin groups similarly attributed the behaviors of the weather and the environment to Manitou spirits. Thales was one of the first natural philosophers to realize that the world contains causal patterns that are independent of agency.

When narrative explanations are divorced from mechanism-backed causality, as in religious stories, the leap from story to truth is highly risky. Mechanisms reduce the risk, because they lessen the gap between mental representation and reality, as shown by the efficacy of experiment and technology. Narratives without mechanisms are easily matched by alternative stories, but dueling mechanisms can be more rigorously evaluated by asking for the evidence concerning the existence and behavior of their parts, connections, interactions, and resulting changes. As in inference in general, there are no guarantees of truth, just better prospects of getting things right in the long run. Similarly, deductive explanations that use mathematics to describe causal mechanisms are more relevant to the world than merely predictive formalisms.

Causality in Society

That causality explains through correspondences between mind and world eliminates idealistic, skeptical, and social constructivist interpretations of it. Nevertheless, society is relevant to causality in various ways. First, children and adults learn about causality from other people in the form of rules and techniques for getting at the causal structure of the world. I believe that smoking causes cancer based on reports from other people, not because of any manipulations that I carried out myself.

Second, society shapes what causal relations about the world will be learned. For example, much is now known about the cellular mechanisms underlying cancer, which has been extensively studied because cancer is a leading cause of human death.

Third, much causal investigation is targeted at understanding and producing changes in societies, for example to improve economies and social welfare. *Mind–Society* provides causal explanations of many social phenomena by considering the interactions of mental and social mechanisms. Chapter 7 raises questions about what can be done to bring about just societies in which people generally flourish. So there is a great need for more causal knowledge about the social world, for the sake of truth, explanation, and human benefits.

What Causes What?

Do greenhouse gases cause global warming? Does the Zika virus cause babies to be born with small brains (microcephaly)? Asking causal questions is important for understanding why things happen and also for suggesting solutions to problems. Removing a cause can eliminate an undesired effect such as disease and suffering. Introducing a cause can help produce a desired effect such as health and flourishing.

Causal thinking is natural for people, even in infants, but people do not always do it well. We are prone to the fallacy of false cause, thinking that an event that follows another was caused by it. For example, if you take vitamin C and your cold goes away, you might think that the vitamin caused the cure, even though colds go away by themselves. You may also be motivated to think that you have found a cure for colds if it makes you worry less about getting sick again.

Before concluding that A causes B, you need to consider alternative explanations such as that B has a different cause, or that A and B are both caused by something else. We cannot directly observe causal relations but need to infer that they exist as part of the best explanation of systematic observations.

For example, a recent US report on climate change concludes "it is extremely likely that human influence has been the dominant cause of the observed warming since the mid-20th century. For the warming over the last century, there is no convincing alternative explanation supported by the extent of the observational evidence." This report considers many kinds of evidence for the conclusion that human activity has been causing global warming while criticizing alternative explanations such as solar activity and natural fluctuations. The appropriate conclusion is that people need to reduce the cause (greenhouse gases) in order to prevent the disastrous results of the effect (global warming).

Similarly, scientists have come to the conclusion that the recent disturbing increase in microcephaly in newborns is caused by their mothers becoming infected by the Zika virus through mosquito bites. This hypothesis is the best explanation of the statistical association between the introduction of the virus in various countries and the occurrence of microcephaly, taking into account growing information about how the virus affects brain cells. There is also a plausible connection with global warming, which allows the anopheles mosquito that transmits the virus to operate in new places such as Florida.

On the other hand, the conclusion that childhood vaccinations cause autism, for example, was based on unsystematic observations that children are often diagnosed with autism after getting vaccinated and on a study published in the *British Medical Journal* that was found to be fraudulent. Many other studies have found no association between vaccines and autism. The causes of autism are diverse and largely unknown, but there are no grounds to infer that vaccines cause autism.

Judgments about cause and effect are valuable for understanding events and intervening to promote human welfare. But they need to be based on inference to the best explanation that takes into account the full range of evidence and alternative hypotheses.

REDUCTION AND EMERGENCE

The specter of reductionism is raised by my explanations of mental processes by brain mechanisms and by my explanations of social processes by communication of semantic pointers. Reductionism is the view that higher levels of explanation can be reduced to lower levels, for example sociology to psychology to biology to chemistry to physics. Despite the centrality of the brain in all these explanations, my account of mind and society is not reductionist, as can be seen from a careful exploration of the nature of reduction and how it fails because of the importance of emergence.

The standard way of thinking about reduction in philosophy assumes the deductive style of explanation. Suppose that we could express all of psychology in a set of precise, mathematical laws and all of biology in a different set of laws. Then we could reduce psychology to biology by showing that all of the laws of psychology can be deduced from the laws of biology along with principles that translate psychological terms into biological ones. Assuming that psychological explanations are deductive, then the deduction of psychological laws from biological ones would mean, by virtue of the transitivity of deduction, that all psychological phenomena could be explained biologically. At the extreme, one might even think that sociology could similarly be deductively reduced to psychology, and biology to chemistry and then physics, so that transitivity of deduction would then ensure that sociology is reduced to physics! Reductionism wins.

This pretty picture has glaring flaws. First, outside of physics and chemistry, there are few mathematical laws available to accomplish such deductive explanations. Instead of universal laws, explanations in biology, psychology, and the social sciences employ mechanisms, with descriptions of parts and interactions that are only sometimes mathematical. Hence we need a different account of reduction than the deductive one. The mechanistic style of explanation suggests the formulation that a higher level science is reduced to a lower one if higher mechanisms can be mechanistically explained by lower mechanisms.

For example, psychology reduces to biology if psychological mechanisms can be mechanistically explained by biological ones. Specifically, psychological mechanisms concerning mental representations and computational processes would be explained by biological mechanisms concerning neurons and their interactions. *Brain–Mind* shows that much of this translation is already happening through the explanations of numerous psychological phenomena by the neural operation of semantic pointers. But we should not expect it to be complete because of the other two flaws in the deductive account of reduction concerning revision and emergence.

The second flaw in the reductionist picture is that deductive reduction sometimes fails because scientific advances at the lower level lead to revisions in the theoretical understanding at the higher level. Higher level laws or mechanisms are not reduced to lower level ones because science finds out they were wrong. Everyday psychological theories often need to be corrected rather than reduced to scientific psychology, and similarly scientific psychology sometimes needs to abandon ideas that are recognized as crude once neural mechanisms are appreciated. For example, everyday views of memory were revised and enhanced by psychological theories that include important new distinctions such as short term versus long term, episodic versus semantic, and declarative versus procedural. Experiments have determined that memories are not stored exactly like snapshots but rather undergo interpretation and modification.

Similarly, psychological theories become enhanced and revised once neural mechanisms for memory are identified, particularly formation of synaptic connections and transfer of temporary memories in the hippocampus into permanent storage in the cortex. Computational concepts like *buffer* that were important in early theorizing about memory fade away.

Third, and most important, both deductive and mechanistic accounts of reduction fall afoul of the importance of emergent properties and relations that apply at higher levels but not at lower ones. Emergence is sometimes rather mysterious, as in the vague remark that the whole is greater than the sum of its parts. A better account is that *emergent properties are possessed by the whole but not by the parts and are not just aggregates of the properties of the parts because they result from interactions of the parts*.

This kind of emergence occurs even in simple physical systems, for example when water molecules have properties such as being liquid at room temperature that are not properties of their parts, which are hydrogen and oxygen atoms that are gases at room temperature. Emergence also occurs when the motion of molecules in a lake decreases to the point that the water freezes, making it solid rather than capable of flowing. The solidity of ice is not a property of any of its molecules.

Emergence abounds in biological systems, for example when whole organisms such as the human body are capable of functions that no single organ, tissue, or cell could accomplish. Chapter 2 describes how mayonnaise has emergent properties such as being thick and creamy that are not found in its ingredients of oil, vinegar, and egg yolk.

Because of such emergence, higher levels such as psychology and sociology cannot simply be reduced to lower levels such as biology and physics, even though biological mechanisms such as neural firing are enormously relevant to understanding what happens in minds in groups. The best intellectual strategy therefore is to be neither reductive nor antireductive. Complete reduction fails because we need properties that are emergent at higher levels for explaining phenomena that are important to us.

For example, despite the semantic pointer theories of concepts and emotions as neural processes, we still need concepts and emotions to explain people's political behavior such as voting. Even if concepts are basically patterns of neural firing, we need to describe how people apply their concepts and connect them with their beliefs, goals, and emotions. In considering why people voted as they did, mental representation rather than neural firing is the level of mechanism most directly relevant to their choice of action. Concepts and emotions also have emergent properties such as consciousness that are relevant to higher level explanations, for example when voting results in part from people's awareness that they have strong feelings for concepts such as *democracy*.

Similarly, as chapter 4 argued, we cannot dispense with groups as part of our mechanistic explanations of social change, because groups have emergent properties that figure into the explanations of the activities of the individuals in the groups. That universities have the group property of granting degrees helps to explain the individual behaviors of students and administrators.

The fourth reason why reductionism fails is that emergence is even more complicated than the standard picture of one level having emergent properties with respect to a lower level. My account of the self in *Brain–Mind* (chapter 12) showed the importance of multilevel emergence, where the self in humans emerges from interactions of four different levels of mechanisms: molecular, neural, psychological, and social. Despite philosophical objections, downward causation is rampant is human lives. For example, the social interaction of insulting people can cause molecular changes such as raising their cortisol levels, whereas giving a compliment will likely increase their dopamine levels. The social has causal effects on the molecular. What you are as a person depends on your molecular history—genetic and epigenetic—as well as on the learning you have experienced from interactions with other people.

For these four reasons, we should reject the reductive chain from social sciences to psychology to biology to chemistry to physics. However, it is equally implausible to trumpet the autonomy of higher levels and insist that sociology can ignore psychology, which can ignore biology, which can ignore chemistry and physics. Biology has benefitted enormously from grasp of underlying chemical mechanisms, and psychology is increasingly gaining from the appreciation of neural mechanisms at both cellular and molecular levels.

Sociology and other social sciences have been more recalcitrant toward lower level explanations, but my accounts of numerous social sciences and professions in *Mind–Society* provide support for the relevance of psychology and neuroscience. This relevance refutes the strong antireductionist view of autonomy for higher level sciences. But it does not support reductionism either, because multilevel emergence requires social properties and relations to be identified as causally efficacious in the whole system. The best strategy is to marshal all the explanatory resources that enable us to make sense of important phenomena taking into account all the relevant mechanisms and their interactions.

To conclude my discussion of reduction, I make some clarifications about levels, supervenience, and strong versus weak emergence. What distinguishes different levels? There are several interconnected ways of identifying levels, including size, speed, part–whole composition, and function. The objects at higher levels are usually larger than lower ones, as in the difference in size that orders universities, human bodies, organs, tissues, cells, proteins, atoms, protons, and quarks. Correlatively, the basic operations of objects of different sizes operate at different time scales, from the years it takes to earn a university degree to the nanoseconds that measure the motions of subatomic particles. Differences in size enable part–whole relations that help to distinguish levels, as quarks are part of protons, which are part of atoms, and so on. Finally, levels can also be distinguished by the operation of different mechanisms, for example the molecular interactions of genes and proteins and the social interactions of different people. These rough indications of levels are not necessary or sufficient conditions, as there are systems in physics such as galaxies that are very large and unfold over billions of years.

In contemporary philosophy, levels are often discussed in terms of supervenience, where one level supervenes on another if there cannot be changes at the higher level without there being changes at the lower level. For example, psychology supervenes on neuroscience if necessarily for every mental change there are neural changes. This relationship is both too strong and too weak to be useful. It is too strong because it insists on there being a necessary connection between higher and lower levels, but chapters 3 and 4 cast doubt on whether there are any necessary truths at all. Most damningly, supervenience is too weak because it only says that there is some relation between higher and lower levels without saying what it is. A much more productive strategy is to identify the mechanisms that interrelate different levels as well as their emergent properties, abandoning the concept of supervenience.

Another dispensable conceptual tool of contemporary philosophy is a distinction between strong and weak emergence based on whether or not truths at a higher level are deducible from or necessitated by truths at a lower level. Because deducibility and necessity are irrelevant to characterizing the relations between levels, it is better just to address questions about mechanisms and emergence, abandoning the strong/weak distinction.

SUMMARY AND DISCUSSION

Explanation concerns both knowledge and reality and therefore operates at the intersection of epistemology and metaphysics. There is no unitary style of explanation, because different varieties employ different patterns. The most accessible pattern is narrative explanation in which people tell a story concerning how events come about. This form is weak because of the difficulties of finding strong connections between events in the story and in evaluating radically competing stories. Explanation by elimination occurs when a phenomenon is explained away as being nonexistent.

Explanation by deduction is powerful in mathematical fields such as physics but is rarely possible in more qualitative fields such as biology, psychology, and sociology. In these fields, the most useful approach is mechanistic, where events are explained by showing how they causally result from regular changes in systems of connected and interacting parts. Mechanistic explanation can valuably supplement narrative explanations by providing the causal connections between events in the story. Similarly, blends of mechanistic and deductive explanation can flourish when enough is known to give precise mathematical descriptions of the operations of the parts and interactions in the mechanism, as happens in theoretical neuroscience.

Like explanation, causality cannot be captured by a simple analysis in terms of universal regularities, manipulations, probabilities, or causal networks. Rather, a three-analysis marks all of these as typical features of causal relations, on top of familiar exemplars such as pushes and pulls. From infancy, these features build on a nonlinguistic feature of causation also operating in nonhuman animals in the form of sensory-motor-sensory patterns.

Like explanation, causality crosses epistemology and metaphysics, because causal relations are fundamental parts of the world that need to be known by means of human acquisition of representations of the relations. These representations are not just verbal or mathematical but can also be multimodal in the form of pictures, sounds, touches, and bodily movements. Inferences about causal relations require consideration of evidence and alternative hypotheses.

I primarily discussed explanation as a cognitive process, but it is also emotional and social. The pursuit of explanation requires emotional motivations such as curiosity and doubt and generates satisfaction when successful. Explanations are part of a broader social process where what has been figured out by one person can be transmitted to another through multimodal communication covering words, pictures, physical motions, and sometimes feelings.

This talk of mechanisms and causality might seem outrageously reductive, but careful consideration of reduction and emergence implies otherwise. Emergent properties that belong to wholes but not their parts are important at all levels of science. Even though mental representations result from neural firings, we still need concepts, rules, and analogies to make sense of human activities. Similarly, social explanations do not dispense with groups such as nations and institutions just because semantic pointers contribute enormously to understanding the operations of individuals in groups.

This chapter has shown important connections among philosophical issues about knowledge and reality, in keeping with reliable coherentism, scientific realism, and multilevel materialism. A system of philosophy needs an epistemology and metaphysics that are coherent with each other as well as with relevant scientific findings. The rest of the book extends social cognitivism to normative questions that require naturalistic theories of morality, justice, meaning, and beauty.

NOTES

Woodward 2014 reviews scientific explanation. Velleman 2003 discusses narrative explanation. Hempel 1965 is the standard source on deductive explanation. On mechanistic explanation, see Thagard 1999, 2012c; Bechtel 2008; Craver 2007; Craver and Darden 2013; and Glennan 2017. Ramsey 2013 reviews eliminative materialism. On styles of social explanation, see *Mind–Society*, chapter 1. *Brain–Mind* (chapter 1) and *Mind–Society* (chapter 1) provide a three-analysis of the concept of mechanism.

Gopnik 1998 compares explanations to orgasms.

Schaffer 2016 and Pearl 2000 review causality. Woodward 2004 explains causality by manipulation. Hitchcock 2010 surveys probabilistic causality.

Baillergeon, Kotovsky, and Needham 1995 present evidence that infants have a preverbal understanding of causality, also discussed in Thagard 2012c. Ladyman and Ross 2007 report that eight years of the journal *Science* averaged about 90 articles per month containing the word "cause."

Medical causality concerning diseases is discussed by Thagard 1999 and Broadbent 2013. Russo and Williamson 2007 argue that both probabilistic and mechanistic considerations are important for causal reasoning in medicine.

Pearl 2000 explicates counterfactuals in terms of causal mechanisms. He uses probabilistic networks to compute counterfactual claims, but explanatory coherence networks (Thagard, 1989) can perform the same role without probabilities.

Quine 1960 explicates dispositions as resulting from subvisible structures and mechanisms.

The quoted climate change report is at https://science2017.globalchange.gov. Zika virus causation is analyzed by Rasmussen, Jamieson, Honein, and Petersen 2016. Dammann, Poston, and Thagard in press use explanatory coherence to model causal reasoning concerning the Zika virus and other cases in epidemiology.

For reviews of reduction, see van Riel and van Gulick 2014 and McCauley 2007. Bechtel 2017 examines downward causation. McCauley 2009 (emphasizing time) and Craver 2015 analyze levels.

Bunge 2003, McClelland 2010, and Wimsatt 2007 provide accounts of emergence that fit with scientific practice. *Mind–Society* (chapter 3) provides further discussion of emergence, including a three-analysis.

PROJECT

Compare and contrast narrative and mechanistic explanations in medicine and other fields. Use the three-analysis of causality to explicate determination of causality in fields like epidemiology. Do three-analyses of the concepts *disposition* and *capability*.

6

Morality

RIGHT AND WRONG

During the 2016 American election, news reports revealed that Donald Trump had bragged about groping women. Even he and his supporters admitted that groping was wrong, but what makes it wrong? Ethics is the branch of philosophy concerned with morality and aimed at figuring out the difference between right and wrong. Doing so requires an ethical theory that can justify evaluations of the rightness and wrongness of different actions and can answer additional important questions such as why people should be moral at all.

Traditionally, such questions have been answered by religion, with moral codes laid down in sacred scriptures such as the Bible and Quran. Chapter 4 argued, however, that there is no good evidence for the existence of God. Even if such evidence should arise, religious ethics still has the unsolvable problem that there is no apparent basis for choosing one religious code over another. Another possible source of morality might be ethical theories arrived at by a priori reasoning, but chapter 3 argued against this route to knowledge. No one has ever succeeded in using pure reason to establish an ethical truth.

One response to the failure of religious and a priori ethics is the skeptical view that right and wrong are illusions. Cultural relativism claims that, because different societies have different standards of right and wrong, there is no objective way of choosing between them. Relativism can even go down to individuals, proclaiming that right and wrong are just a matter of personal preference.

It might seem that natural philosophy tied to scientific facts has to descend into skepticism and relativism, because there is no way to derive judgments about what ought to be from empirical matters. Failure of derivation holds if "derive" means "logically deduce," but there are many other legitimate kinds of inference besides logical deduction. I argue that ethical objectivity is not only compatible with natural philosophy but depends on it, through an account of morality that coheres with semantic pointers (chapter 2), reliable coherentism (chapter 3), and multilevel materialism (chapter 4).

I begin by laying out some of the fundamental issues that a theory of morality must address and outline some of the traditional ways of approaching them. After discussing values as both psychological and potentially objective, I develop an ethical theory, needs-based consequentialism. On this view, actions are right or wrong depending on the extent to which they lead to the satisfaction of human needs. Needs are far more objective than wants because they are tied to underlying psychological and physiological mechanisms in all human beings. Good ethical theory can take needs as the basic values that go into ethical judgments, which are inherently emotional but can nevertheless be objective when value judgments are based on needs. Applying values to assessing the morality of actions requires balancing different needs against each other in a process of emotional coherence.

As in epistemology and metaphysics, philosophical method in ethics has often relied on thought experiments, for example asking what people would do if they had to push someone in front of a train in order to save five others. I think that such thought experiments are misleading because they oversimplify real-world problems by neglecting important psychological and social factors. They tend to glorify ethical intuitions, which are as subjective and biased as the intuitions about knowledge criticized in chapter 3. Instead, my aim is to elucidate the moral difference between right and wrong based on a combination of coherence with evidence-based scientific theories and well-supported philosophical theories. The result should help to make it clear that naturalism in philosophy is the best route to ethical objectivity, not an obstacle to it.

ISSUES AND ALTERNATIVES

Sciences such as evolutionary biology, social psychology, and cultural anthropology discover phenomena relevant to understanding morality, but they cannot by themselves develop an ethical theory that is capable of providing useful answers to questions such as the following:

What are ethical values and how do they affect human thinking?
Can values be objective rather than merely subjective?
Are ethical judgments about right and wrong true or false, or are they merely expressions of personal opinion?
Are ethical judgments cognitive, emotional, or both?
Does the objectivity of moral judgments derive from principles about rights and duties or from the consequences of actions with respect to human happiness or other values?
Can rights be based on needs, and what makes needs objective?
When moral actions depend on conflicting principles and needs, how can people achieve ethical and emotional coherence?
Why should anybody be moral?

To answer such questions, philosophers have developed a range of ethical theories that provide alternatives to the religious ethical doctrines familiar to the billions of adherents of faiths such as Christianity, Islam, and Hinduism. Secular approaches to ethics fall into two main camps, deontology and consequentialism. Both deontology and consequentialism take ethical judgments to be objectively true or false but find a different basis for this objectivity.

Deontology bases morality on principles concerning rights and duties, where people are assigned duties such as telling the truth, treating other people well, and respecting other people's rights to autonomy and equality. From this perspective, Trump's action of groping women is wrong because it violates principles of autonomy and beneficence.

In contrast, consequentialism looks at the consequences of actions with respect to human goals such as happiness. The most famous version of consequentialism is utilitarianism, which says that actions are right if they promote the greatest good for the greatest number of people, where good is understood as the achievement of pleasure and the avoidance of pain. Utilitarianism tries to turn morality into calculation, where an action can be judged to be right or wrong depending on the extent to which it promotes pleasure and avoids pain. For example, Donald Trump's action of groping women can be judged wrong because of the distress it causes the victims.

There are important overlaps between deontology and consequentialism, because the duty of being good to people rather than harmful is applied by estimating the consequences of actions. I argue in chapter 7 that the best way of identifying human rights is by considering human needs, which are also a more solid basis for identifying morally relevant consequences than just pleasure and pain. Both deontology and consequentialism were developed in the eighteenth and nineteenth centuries independent of scientific understanding of how human minds work. I now develop my own version of consequentialism by starting with a theory of values tied to the semantic pointer theory of emotions.

VALUES

Values are ideas, things, and goals that matter, but what matters? Nihilists say that nothing matters, that there are no objective values. Egoists say that the only thing that matters is yourself, whereas altruism says that other people also matter. Anthropocentrism insists that only human beings matter, but the broader view of biocentrism says that all living things are also important, right down to the lowliest insect. According to theism, nothing matters more than God.

How does mattering happen in human minds? The many people who objected to Donald Trump's claims about groping women had an emotional response based on values that include respecting women, avoiding harm, and upholding human rights such as autonomy and privacy. If values were abstract ideals in Plato's heaven, their status would make it mysterious how they can have such strong effects on human thinking. People who feel that their values are threatened often become angry and inclined to act against those they think are making the threats. On the other hand, when proposed actions fit with people's values, they feel happy and inclined to act accordingly. We need a psychologically plausible theory of values to explain their effects on both judgments and actions.

Three-Analysis of Value

Definitions of value that report on how the word "value" is used in everyday life make vague allusions to worth, merit, and importance. The three-analysis in Table 6.1 gives a broader and deeper understanding of the concept of value. In Western society, standard examples of values include freedom and happiness, but other cultures attach more values to other concepts such as conformity, religion, authority, and social hierarchy. For each concept associated with positive emotions, there are correlative concepts with opposite negative emotions, such as coercion, misery, death, slavery, dictatorship, disease, and indolence.

Typically, values are associated with concepts, which are mental representations corresponding roughly to single words such as "equality." By extension, values can also be associated with goals, as in the value that everyone should have enough to eat. These associations are emotionally positive, indicating concepts or situations

TABLE 6.1

Three-Analysis of the Concept *Value*

Exemplars	Freedom, happiness, life, equality, democracy, health, work
Typical features	Concept, goal, or belief
	Positive emotion
	Influences judgments
	Leads to action
	Varying degrees of importance
Explanations	Explains: judgments, behaviors
	Explained by: mental and social mechanisms

that people like and want to accomplish. People have good feelings about what they value, influencing their judgments and actions, but some values are more important than others.

These connections among concepts, emotions, judgments, and actions make the concept of values important for explaining human thinking and behavior. Moral judgments that applaud or condemn people's actions of themselves and others always depend on values. Ethical theories vary in their central values: the word of God in religious ethics; principles, rights, and duties in deontological ethics; and outcomes with respect to happiness and other goals in consequentialist ethics. All ethical judgments are value judgments, and chapter 3 described how scientific inferences can also be affected by values.

The final component of my three-analysis of the concept of value is what explains it, about which there is much more disagreement. Religious views claim that values result from God's communications to people about what really matters. Deontology tries to make judgments of value a matter of pure reason, as in Kant's categorical imperative that demands acting in ways that can be made universal. Utilitarianism tries to make values more empirical, relying on the observation that people value pleasure and the avoidance of pain. My explanation employs cognitive science to consider values as operating in the mind through neural mechanisms that support social interactions.

Values as Semantic Pointers

A plausible neuropsychological theory of values arises by combining the semantic pointer theories of concepts and emotions mentioned in chapter 2 and developed in much more detail in *Brain–Mind* and allied articles. Recall that semantic pointers are neural processes that explain cognition and emotion as results of binding of multimodal representations.

Values are associated with concepts such as *freedom* and *justice*. Within the Semantic Pointer Architecture, concepts are patterns of neural firing that integrate sensory, motor, and verbal information. Similarly, emotions are neural patterns that integrate physiological perception with appraisal of the goal-relevance of a situation. Integration is performed in both cases by the binding operation of convolution, which converts two or more patterns of firing into a unified one that retains much of the information from the contributing patterns. Binding is recursive, so that more and more complicated patterns can be built up using bindings of bindings of bindings. Accordingly, to make values, concepts are bound with emotions, as when feminists attach a strong positive emotion to the concept *women* and misogynists attach a negative emotion to it. The resulting values are semantic pointers that combine a concept and an emotion.

As emotional concepts, values can contribute both to judgments and actions. Reports of women being groped generate emotional reactions such as outrage and dismay, with correlative judgments that the behavior is wrong. These emotions also encourage actions such as verbal condemnation, in line with the neural theory of intention and action in *Brain–Mind* (chapter 9). Emotions include evaluations that are bound into intentions that compete to cause actions.

On common views of rationality, recognition of the neural and emotional aspects of values seems like a threat to objectivity. There are philosophical views, emotivism and expressivism, that take ethical conclusions to be just a matter of the expression of emotions, implying that there is no objective, rational basis for ethics. If emotions were just a physiological response with respect to such variables as heart rate, breathing rate, skin conductance, and cortisol levels, then emotions would indeed eradicate rationality and objectivity from value judgments. An involuntary response such as sweaty palms cannot be the basis for rationally concluding that something really matters.

But the semantic pointer theory of emotions includes cognitive appraisal as a key component of emotions that is integrated with physiological perception. This integration is required to explain why physiologically similar emotions such as fear and anger have different effects on both conscious experience and action. Unlike physiology, cognitive appraisal can be judged according to how well it takes into account information and goals that ought to contribute to the evaluation, in line with consequentialist theories of ethics.

Value Maps

Moral judgments and decisions are rarely made on the basis of a single value. For example, the choice to support a political party requires a good fit between your values and those of the party, which come in systems of interconnected concepts. Such systems are conveniently displayed by a technique of cognitive-affective maps, value maps for short. *Mind–Society* uses value maps to exhibit systems of values in many important social occurrences, including romantic relationships, prejudice, political ideologies, economic cycles, and religion.

In such maps, ovals are used to indicate positive values, that is, concepts toward which people have favorable emotions. Hexagons are used to indicate negative values, concepts where people have disapproving emotions. When color is available, ovals are green, like "go" traffic lights, and hexagons are red, like stop signs. Solid lines indicate that concepts are mutually supportive, while dotted lines indicate that concepts are emotionally incompatible: if you like one, then you probably will not like the other.

Consider for example the American presidential election of 2016, where victory by Donald Trump confounded the pollsters and the pundits. Why did so many people, especially Whites without a college education, support Donald Trump? Voting might be understood as a purely practical enterprise in which people use self-interest to produce a cold, cognitive calculation of which candidate accomplishes their personal goals. But it is much more psychologically realistic to recognize that people's voting decisions are both ethical and emotional. People vote for candidates thought to deserve support and for whom they feel positive rather than negative emotions.

Figure 6.1 provides a value map for Trump supporters who liked the promises he made in his many campaign speeches. It shows the mental representation *Trump*

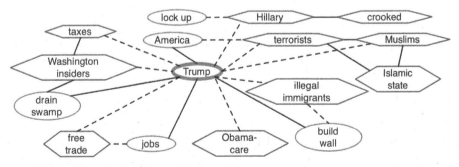

FIGURE 6.1 Value map for Donald Trump supporters. Ovals indicate positive values and hexagons indicate negative ones. Solid lines show emotional support (positive constraints), while dotted lines show incompatibility (negative constraints).

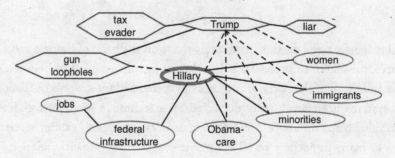

FIGURE 6.2 Value map for Hillary Clinton supporters. Ovals indicate positive values and hexagons indicate negative ones. Solid lines show emotional support (positive constraints), while dotted lines show incompatibility (negative constraints).

gaining emotional support from many directions. The Trump oval gains positive support because it is linked to favorable emotions via the concepts *America* and *jobs*, but it also gets support because it is incompatible with negative values such as *terrorist* and *Obamacare*.

In contrast, supporters of Hillary Clinton had different values portrayed in Figure 6.2. The Hillary representation gains emotional support from its links to other positive values such as *women* and *jobs* and because it is incompatible with negative values such as *gun* and *Trump*. Value maps show how an emotional preference and an ethical judgment result from complex configurations of values.

An undecided voter in the 2016 election might have had sets of values and beliefs that involve a combination of the extreme positions shown in Figures 6.1 and 6.2. Then the ultimate decision of which candidate is right to vote for depends on balancing competing values using the coherence mechanisms to be described.

MORAL EMOTIONS

Value maps are useful for depicting the considerations that go into a decision, but they are inadequate for displaying the full emotional range of ethical judgments. They only show degrees of positive and negative emotions: strongly negative, negative, neutral, positive, and strongly positive. They fail to show any of the specific emotions that contribute to moral reactions, but writers since Adam Smith have recognized a variety of moral sentiments, ranging from negatives such as shame to positives such as pride. Donald Trump gained support in part by tapping into many emotions such as pride in America, sadness about job loss, anger about immigrants, fear of terrorists, disgust toward Hillary Clinton, and hope for a better future.

According to the semantic pointer theory of emotions, different feelings come from different patterns of neural firing resulting from combinations of physiological perception and cognitive appraisal. Different emotions have different physiological signatures, although these are not so finely tuned as to discriminate between similar emotion pairs such as anger and sadness, shame and guilt, and gratitude and pride. Finer discriminations require recognizing different cognitive appraisals, with different goals being satisfied or not in different contexts generating different emotions. For example, the cognitive difference between fear and anger is that anger is directed at someone who is thwarting one's goals.

Some of the negative emotions associated with moral judgments that something is wrong are guilt, shame, embarrassment, disgust, anger, and contempt. Unlike values, which are associated with general concepts, these emotions are usually associated with actions and situations, although they can also concern general behaviors. For example, bestiality (sex with nonhumans) is generally associated with disgust.

The most prominent emotions associated with moral judgments that something is right are approval, satisfaction, pride, generosity, and gratitude. All of these involve feeling good, through a combination of physiology and appraisal. When a situation fits with your values, you feel approval and satisfaction that can take on different forms depending on the nature of the action. If you perform the good action yourself, then you feel pride, but if the action was performed by someone else in a way that benefits you, then you feel gratitude. Forgiveness usually comes with positive emotions such as relief, following negative emotions such as anger associated with blame.

Sympathy and compassion are moral emotions in which you feel sad because of the misfortunes of others. They depend on the more general emotional attitude toward others of caring, showing interest and concern. Sympathy can arise from the process of empathy in which you put yourself in somebody else's shoes and thereby generate some approximation to their emotions, although empathy is more complicated in ways to be described.

If emotions were just raw feelings based on uncontrollable physiological responses, then there would be no point to evaluating them morally. But consider the following statement from Canadian politician, Jack Layton, written shortly before his death: "My friends, love is better than anger. Hope is better than fear. Optimism is better than despair. So let us be loving, hopeful and optimistic. And we'll change the world." We can easily expand Layton's comparisons with respect to other moral emotions: caring is better than contempt, compassion is better than pity, trust is better than suspicion, and usually forgiveness is better than anger.

To make such judgments, we need to be able to evaluate the appropriateness of emotions, which depends on the appropriateness of the goals that they satisfy or fail to satisfy, which in turn depends on the values underlying such appraisal. What makes some values better than others and thereby makes some emotions morally appropriate?

OBJECTIVE VALUES AND RATIONAL EMOTIONS

Sadly, the world abounds with despicable values and irrational emotions. There are egotists, narcissists, and psychopaths who value only themselves, placing their own wealth and comfort far ahead of the misfortunes of the downtrodden. Some are even angry at attempts to use taxes to diminish their wealth even slightly. Values and emotions, such as the Nazi ones described in *Mind–Society* (chapter 6), undoubtedly contributed to some of the most horrendous events in human history, including war, genocide, and slavery.

It is therefore tempting to try to eliminate values and emotions from human rational deliberation and to base morality on pure reason or religious doctrine. Unfortunately, these attempts fail, opening the abyss of relativism through the prospect that ethics is only a matter of opinion.

Such relativism fits well with contemporary postmodernism that denies all forms of truth and objectivity, but the tolerance that it pretends to offer is illusory. Saying that anything goes appears to be an attractive alternative compared to religious dogmatism and political oppression, but it prevents critique of other kinds of dogmatists. For example, many leading American conservatives including Donald Trump are fans of Ayn Rand, who espouses a crude philosophy that she calls "objectivism." This view extols an extreme version of individualism, where ethics is largely a matter of following self-interest and fighting incursions on individual freedom, including ignoring demands of racial and other minorities to eliminate discrimination. Ethical relativism provides no way of saying why this view is any worse or better than religious extremism or altruistic benevolence.

We need a way to establish some values as objective and some emotions as rational, without attempting to pull ethical rabbits out of the hat of pure reason. Instead, values need to be based on scientific knowledge concerning human minds and their operations in society. If there are objective values, then emotions are rational when their cognitive appraisals are based on them rather than on mere wants, whims, and interests. Objective values provide a basis for critique of egotism, racism, and sexism, expunging the zeal, anger, and hatred that go with them as irrational emotions. More positively, objective values would provide the basis

NEEDS

Needs are crucially different from wants and whims, which subjectively depend on accidents of culture and psychological history. Someone may say "I need an iPhone" when all they can legitimately say is "I want an iPhone." Vital needs are those that people require to live as human beings, distinguishing objective values from capricious ones. Emotions can be viewed as rational when cognitive appraisal is performed to evaluate situations based on their relevance to vital needs.

Demonstrating that needs can provide the basis for objective values and rational emotions requires several steps.

1. Distinguish between instrumental needs and vital needs, where only the latter are the basis for objective values.
2. Specify what vital needs are, which is best done not with a definition but by a three-analysis.
3. Provide an account of how needs result from biological and psychological mechanisms.
4. Show why people should care about the needs of other people besides themselves.
5. Lay out a psychologically plausible and rationally effective method for reconciling conflicting needs, as people unavoidably try to balance out different needs in their lives.

Vital versus Instrumental Needs

For some people, saying that they need something is just another way of saying that they really want it. This equivalence is only true for instrumental needs, which are wants subordinate to other wants. For example, you may think you need a fancy car because you want to impress other people, where having a fancy car accomplishes that social goal.

Vital needs are much narrower, covering goals that are not merely subordinate to others but are required for you to pursue any goals. Contrast your need for water with a desire for a new dishwasher. Dishwashers are conveniences that save people time and effort, but you can operate as a human being without one. In contrast, a few days without water usually leads to death, the end of pursuit of any goals. Hence water is clearly a vital need, whereas a dishwasher is only an instrumental need, better called a want. Vital needs are enduring, objective, and universal, whereas wants are transient, subjective, and variable across individuals and cultures. Important wants are those that are instrumental to vital needs, for example if you want a pump to provide a reliable source of water.

Three-Analysis of Vital Needs

The best way to characterize vital needs is to give a three-analysis as sketched in Table 6.2. The most obvious exemplars of vital needs are the biological ones for oxygen, water, food, shelter, sleep, and health care. Failure to satisfy these needs can lead to death, as it does for lack of oxygen in less than 10 minutes. People will starve to death in a month or two without food, and poor-quality food can lead to nutritional diseases such as scurvy that enormously diminish capacity to function as a human being even if they do not immediately cause death. Lack of shelter in a Canadian winter or in a cyclone can also lead to severe bodily harm. The absence of health care can be immediately lethal in the case of heart attacks or strokes and more slowly lethal in cases of cancer, diabetes, or other diseases. Sleep deprivation only indirectly leads to death, for example by automobile accidents, but can cause intense reduction in the capacity to pursue goals. So these six biological needs of oxygen, water, food, shelter, sleep, and health care clearly qualify as vital. I argue that there are also psychological needs that should be counted as vital, but they are less familiar so I do not yet list them as exemplars.

The typical features of vital needs include the person who has the need and relations rather than simple properties. The biological needs require people to possess entities crucial for their survival, including water, food, and housing, and the psychological needs with respect to other people that are discussed in the next section are also relational. A need is not just a state of human being, because relations are

TABLE 6.2

Three-Analysis of the Concept *Vital Need*

Exemplars	Oxygen, water, food, shelter, sleep, health care
Typical features	Person, relation to world and other humans, desires, crucial for life, absence leading to great harm
Explanations	Explains: why people die or suffer intensely
	Explained by: biological and mental mechanisms

actually ongoing processes operating in human minds and bodies through mechanisms to be described.

Typically, people have conscious desires with respect to their needs, but consciousness is not crucial because people sometimes have needs of which they are not aware. For example, a severely dehydrated person in a coma has a desperate need for water but no desire. Vital needs are crucial for life, although it would be too narrow to insist that "crucial for life" refers only to forestalling imminent death. Sometimes, failure to satisfy a need produces great harm to a person, with life greatly limited rather than fully eliminated.

The concept of need helps to explain variations in the quality of people's lives, across individuals and across societies. Failure to satisfy strict needs such as oxygen and water causes rapid death. When people's other biological and psychological needs are not satisfied, they usually have shorter lives and are less satisfied with them. Vital needs are thus more explanatorily powerful than instrumental wants that can drift across time. Satisfaction of minor wants can generate pleasure and happiness in ways that depend on personal and cultural histories, but vital needs tap into universal human nature. Hence needs-based consequentialism is normatively superior to utilitarianism.

Normative concepts play more than an explanatory role because they also generate prescriptions as well as descriptions. One of the most important prescriptive roles for the concept of vital needs is providing a basis for understanding human rights. Chapter 7 uses vital needs to explain rights and to provide guidance concerning just societies.

Finally, what explains needs? Needs are dispositions that support conditionals and counterfactuals such as: If you had not had water last week, you would have died. Chapter 5 argued that the basis for dispositions and counterfactuals is underlying mechanisms, which clearly applies to needs like water. If you become severely dehydrated, your metabolic processes such as sodium levels are disrupted, leading to failure of organs including the heart. Mechanisms can similarly be specified for why people need oxygen, food, sleep, shelter, and health care. Less familiar mechanisms explain human psychological needs.

Psychological Needs

Merely keeping people alive is not sufficient for enabling them to operate as human beings: bacteria in a dish may be growing and reproducing, but they are not capable of human operations. What are the psychological factors that are required for human performance and therefore can qualify as vital needs?

The most famous psychological theory of needs is Abraham Maslow's hierarchy, depicted in Figure 6.3. In this pyramid, the most fundamental needs are physiological ones that need to be satisfied before other needs can be taken care of, including safety, love, belonging, esteem, and self-actualization. These needs are important, but no research has ever established that they are as hierarchical as Maslow suggested. Even in Nazi concentration camps where people were in constant threat of annihilation by starvation and brutality, inmates still attempted to self-actualize by pursuing music, art, and literature. For the sake of esteem and self-actualization, people risk their safety by mountain climbing, hang gliding, and car racing.

A nonhierarchical and empirically supported theory of human needs is proposed by psychologists Richard Ryan and Edward Deci. They identify three fundamental psychological needs: relatedness, autonomy, and competence. Relatedness is the requirement to feel closeness with others through attachments and feelings of security and intimacy with others we care for and who care for us. Other psychologists call this need belongingness or affiliation. There is substantial evidence that desire for interpersonal attachment is a fundamental human motivation and that people suffer grievously when they lack such attachments. Loneliness, the feeling of personal isolation, increases rates of lethal disease such as heart failure and cancer.

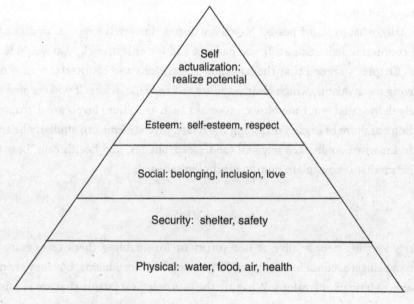

FIGURE 6.3 Maslow's hierarchy of needs as a pyramid.

Autonomy is the need to feel that activities are self-chosen in ways that enable people to organize and regulate their own behavior, following their own interests and values. When people lack autonomy, they are more likely to become depressed, with accompanying medical problems that range all the way to suicide. The desire that some people feel for power is in part a need for autonomy. Autonomy in this psychological sense does not require absolute free will or complete independence from other people.

Competence is the need to engage in challenges that display mastery of the physical and social worlds in ways that make people feel effective in their activities. As with autonomy and relatedness, success in satisfying the need for competence is associated with people's judgment of well-being, whereas people who cannot satisfy them end up with poor motivation and performance. The need for competence often manifests as a need for achievement.

All three of these needs are relational rather than mere properties of individuals like weight. Relatedness obviously concerns relations to other humans and for some people also to nonhuman animals. Autonomy is largely also a relation to other people who are not coercively restricting a person's freedom. The sense of achievement that comes through satisfying the need for competence depends on doing things in the world and with other people, so it is also relational.

Failure to satisfy these three psychological needs will not need lead to death as rapidly as deprivation of water and shelter but Ryan, Deci, and their collaborators report much evidence that people thrive in their presence and suffer in their absence. Hence, they deserve to be counted as vital needs that enable people to develop and function in healthy ways. The relative priority of psychological needs may vary across cultures, with some emphasizing relatedness more than competence and autonomy, a difference found in Asian versus Western societies.

If Ryan's and Deci's theory of psychological needs continues to have evidential support, we should add relatedness, autonomy, and competence to the list of exemplars of vital needs. The three explain much of human well-being beyond basic biological needs but can also be viewed as biological when brain mechanisms are specified that show their origins.

Mechanisms for Psychological Needs

We saw that biological needs are easily explained as resulting from underlying bodily mechanisms. The mechanisms for psychological needs are less well understood, but knowledge is growing at molecular, neural, psychological, and social levels, especially for relatedness. Investigations of neuroendocrine stress mechanisms suggest that chronic social isolation increases the activation of the hypothalamic pituitary adrenocortical axis. Research on social isolation has implicated brain areas involved in social threat surveillance and anxiety (e.g., amygdala), social reward (e.g., ventral striatum), and attention to self-preservation (e.g., orbitofrontal cortex). People who are lonely as a result of lack of relatedness undergo epigenetic changes with different gene expressions that can lead to medical problems such as strokes and heart attacks.

People suffering from lack of autonomy and competence undergo changes in psychological dimensions such as confidence and self-esteem, also connected to biological changes. Control by others and failures in work cause stress, with increased cortisol levels that contribute to heart and other diseases. To understand how people's perceptions of their situations and other people lead to stress, we require an account of neural representation and inference.

Simple biological needs and effects work without neural mechanisms, because we can explain why people died from thirst merely with respect to cells and organs. However, human actions that are driven by biological needs are mediated by thoughts, intentions, and emotions, all of which can be understood as neural processes involving semantic pointers. If you want to understand why people are willing to migrate across the world in order to satisfy their biological needs, you cannot ignore the thought processes built out of cognitive-emotional representations.

The relevance of neural representations is even clearer with respect to psychological needs. Many people devote time to thinking about their relationships with other people and to thinking about their work, which is often a major source of competence. The attendant plans and emotional reactions to success and failure are best understood as neural processes involving semantic pointers, as *Brain–Mind* documents.

Semantic pointers also help to explain how the satisfaction of needs leads to good human outcomes and why the failure of satisfaction of psychological needs diminishes human lives. For example, the psychological suffering of major depression is partly to be understood through molecular mechanisms involving neurotransmitters such as serotonin, but also relevant are emotional mechanisms that combine cognitive appraisal and physiological perception via semantic pointers. *Mind–Society* (chapter 10) explains mental illnesses such as depression as breakdowns in molecular, neural, psychological, and social mechanisms.

Mechanisms show the crucial difference between vital needs and casual wants. When unimportant wants are not satisfied, people can switch to more attainable goals, such as having a beer when the wine is gone. In contrast, biological needs are based on mechanisms not easily adjusted: people cannot find alternatives to oxygen, water, food, sleep, shelter, and health care. Similarly, most people cannot simply decide to stop caring about connections with other people, because the need for relatedness is built into human brains. This need is crucial for establishing morality by providing people with concern for the needs of others, not just themselves.

Needs are interdependent rather than isolated. For example, meeting the biological need for food makes it easier to pursue the psychological needs of autonomy and competence, whose restriction can lead to shortfalls in the biological need for health. Satisfaction of needs for relatedness and autonomy can enhance satisfaction of the need for competence, when work collaborations that leave room for freedom lead to increased achievements.

Vital needs are a much better measure for morality and justice than pleasure and happiness. Pleasure can be achieved, at least in the short run, by means such as narcotics and trivial television shows that are orthogonal to vital need satisfaction. Happiness works best not as a goal in itself but as a result of satisfying goals, only some of which may be relevant to a fully satisfying life. From the perspective of the semantic pointer theory of emotions, happiness and pleasure come from both physiological changes and evaluations that goals are being accomplished. This combination leaves room for the higher order evaluation of whether the goals accomplished are legitimate, which is best considered by evaluating the extent to which they meet vital needs. The human capacity for emotions about other emotions enables people to be happy or unhappy about being happy, depending on the source of happiness.

THE NEEDS OF OTHERS

You want to ensure fulfillment of your own vital needs, both biological and psychological, that allow you to satisfy a much broader range of instrumental wants. Perhaps you take pleasure in reading philosophy, but reading is difficult if you are asphyxiating, starving, seriously ill, or deeply depressed because you are lonely. Hence individual rationality requires each person to pay close attention to vital needs. The existence of such needs across societies is enough to undermine cultural relativism as a challenge to all objective morality.

But why should you care about the needs of other people? The view that rationality should focus on self-interest is endemic to contemporary economics and political science, with disastrous results both empirically and morally. *Mind–Society* (chapter 7) argues that the inability of mainstream economics to explain both individual behavior and large-scale events such as recessions derives from its obsession with nonemotional individuals, ignoring the emotional groups that drive economic change.

Alternatively, we can draw on neuroscience, psychology, and anthropology to build a case that people do care about others. For the vast majority of people, there are strong interconnections between meeting one's own needs and the needs of other people, some of whom are also concerned with meeting your needs, as my discussion of relatedness suggested. Let us now consider the full range of evidence for this conclusion across multiple fields.

My argument about why you should care about the needs of others is not deductive from axioms to theorems but rather consists of several interlocking strands. First, there is empirical evidence from several fields that typical human beings do care about other people's needs. Second, underlying neural mechanisms explain how and why people care about other people's needs. Third, if you care about other people's needs, then you will be more likely to have your own vital needs met. Fourth, societies in which people tend to care about the needs of others are much better than ones in which people are more selfish.

People Do Care About Others

What is the evidence that people generally do care about the needs of other people rather than being as egoistic as Ayn Rand advocates? A minority of the population does treat other people with callous indifference and complete lack of empathy, care, and conscience. Such people are called psychopaths or sociopaths, or assigned in the terminology of the *Diagnostic and Statistical Manual of Mental Disorders* to have a mental disorder of antisocial personality. This disorder can be diagnosed using a psychopathy checklist that uses a semi-structured interview to assess the extent to which people such as criminals can be assessed as selfish, callous, remorseless, unstable, antisocial, and socially deviant.

As many as 1% of the population can be diagnosed as psychopathic based on the test, but the good news is that 99% are not psychopathic and show genuine concern for others. The percentage of psychopaths among prisoners is much higher, particularly among offenders who killed police officers. But most people are naturally inclined to be caring and altruistic, even as infants.

These psychological findings are fully compatible with evidence from anthropology and evolutionary biology about the development of the human species. Both the anthropology of current human groups and the archaeology of ancient ones show that humans have always been part of a highly social species rather than isolated individuals. Hunter-gatherer groups typically operate in bands of 50 to 150 people, so people rarely are on their own. Operation in groups of this size contributes to the biological goals of survival and reproduction, because such groups can foster both the acquisition of food and the raising of children. Because human infants are so helpless and need years of care before they can operate on their own, human societies have developed social means for gaining food by hunting and gathering, for maintaining protection against predators, and for nurturing and teaching offspring. In order to function effectively in groups of this size, early humans had to develop the ability to appreciate the thoughts and emotions of others, including mechanisms of empathy to be discussed.

The historical record therefore completely contradicts the individualist view that morality arises when individuals realize that they can maximize their own self-interest by making contracts with others about respecting each other's rights. From Thomas Hobbes to John Rawls, the social contract has served as an intellectual device to get from individual self-interest to social morality. But as a historical account it is completely inaccurate because of the long history of social dependence in human societies. And as a thought experiment it is severely flawed because it anchors morality in an assumption of individual self-interest that is at odds with both the psychology and the anthropology of human nature. Social contracts fit better with the pure-reason approach to morality than with the human-nature approach. In contrast, based on psychological and anthropological evidence, we can conclude that people usually do care about the needs of others.

Such caring, however, has important limitations. People are much more prone to care about their immediate social contacts such as families and clan members rather than people in general. Many cultures exhibit xenophobia, manifested in various ways such as racism and war. The expansion of caring about the needs of others to include humanity in general is a relatively recent cultural development. It became formalized only with the development of ideas about human rights in the eighteenth century, eventually codified in the United Nations Declaration of Human Rights in 1948. Sophisticated secular ethical theories such as Kant's deontology and Bentham's utilitarianism came along only in the eighteenth and nineteenth centuries. These imply that moral obligation be directed toward all people, not just personal associates.

Neural Mechanisms for Care

Understanding how and why people care about each other requires attention to mechanisms at molecular, neural, psychological, and social levels. I have already described how human psychological needs result from multilevel mechanisms, and the same biology happily encourages people to meet those needs in others.

Your need for relatedness encourages your actions that help others to satisfy their needs for care.

Molecules important for understanding human caring include oxytocin and dopamine. Oxytocin makes a major contribution to maternal care and also to trust. The role of oxytocin alone would be enough to substantiate a claim that caring about the needs of others has in part a molecular basis that derives from our genetic history. Dopamine is also important for caring about other's needs when people gain pleasure from beneficial acts. Normal people find the suffering of others stressful, indicating a role for the stress hormone cortisol. Other hormones contribute to maternal caring behavior including prolactin, progesterone, and estrogen.

Numerous brain areas contribute to caring for others. The autism researcher Simon Baron-Cohen has identified 10 areas that contribute to an "empathy circuit," including parts of the prefrontal cortex and the amygdala. How does your brain use interconnected neurons in multiple areas to understand and care about other people?

First, you need brain mechanisms to represent others, which you can do by words, vision, sound, touch, and emotions. For example, your representation of the singer Leonard Cohen can combine verbal representations of him being a Canadian folksinger, visual representations of what he looked like young and old, auditory representations of what his music sounds like, and emotional representations of how you feel about him as a person and a singer. These representations require neural firing in brain areas dedicated to language, vision, sound, touch, and emotion. Combining them into a unified representation of the singer requires binding by convolution in convergence zones where neurons receive input from multiple brain areas involving different modalities.

Second, your brain needs representation of the needs of others, which can range from verbal descriptions such as "Leonard needs medicine," to visual images such as Leonard looking sad, to olfactory images such as a baby needing a change of diapers. The recognition of the relational information that someone has a need requires binding a firing pattern for the person with a firing pattern for the particular need. *Brain–Mind* (chapters 2 and 3) describes how semantic pointers can recursively build up such relational complexities operating across several different modalities.

Third, your brain needs to put caring into the package, using areas important for emotions such as the ventromedial prefrontal cortex and the amygdala. Like other emotions, caring emotions such as compassion and sympathy require both physiological changes and cognitive appraisals. Your perception of someone suffering causes physiological changes such as alterations in heartbeat and breathing, which interact with unconscious appraisals of how this fits with your goals. For most people, such goals include other people being happy rather than suffering.

The brain puts all this together using the mechanism of recursive binding. Caring about someone's suffering requires binding the three representations of the person, of the person's need, and of the emotional reaction to the need. Caring results from neural mechanisms but can also be encouraged by cultural developments such as religious and secular moral codes.

Why is the brain so equipped? Babies are more likely to survive the rigors of infancy if they have parents looking after them, so that parents who have the capacity to care for their children are more likely to have their genes passed on to subsequent offspring. Moreover, human societies historically have had relationships that provide a basis for caring for children that involves not just parents but also grandparents and other caretakers. So, caring about others need not be externally imposed for the vast majority of people, because it builds on brain mechanisms that are neatly equipped to care for children.

In sum, people's brains go beyond concern for individual needs thanks to molecules such as oxytocin, brain areas such as the ventromedial prefrontal cortex, and mechanisms such as recursive binding that produce semantic pointers to implement caring about the needs of others. Additional brain mechanisms for empathic understanding of other people are discussed later.

According to neurophilosopher Patricia Churchland, moral norms are shaped by four interlocking brain processes: caring, recognition of other's psychological states, learning social practices, and problem solving in a social context. Hence the origins of morality are both neural and social. Attempts to improve society by making it more responsive to people's needs should consider all of these processes. *Mind–Society* (chapter 5) analyzes social norms as multimodal rules built out of semantic pointers.

Caring People Flourish

Ayn Rand's writings proclaim that selfish people are superior and better off, but evidence indicates otherwise. The oxytocin researcher Paul Zak set out to test the Aristotelian idea that being virtuous makes people happy. In a game where people choose to give money to others, those who had the largest surge in oxytocin were the ones who gave the most money, and also the ones who had greater life satisfaction and resilience to adverse events. Elizabeth Dunn and her colleagues report that people who spend money on others are happier than those who spend it just on themselves.

The final reason for thinking that people do and should care about the needs of others comes from comparisons of countries rather than individuals. Chapter 7's discussion of social justice reports studies that look at international differences in happiness and reported well-being. The countries at the top of such lists include those like Denmark with high degrees of trust, equality, and social concern. In that country, the national school curriculum includes an hour per week for building empathy. A Canadian organization called Roots of Empathy has had much success in reducing bullying and aggression in schools, as described in chapter 7. How does empathy work?

EMPATHY

An intelligent computer could make inferences that a person is suffering by using general verbal rules such as: persons who are crying are suffering, and people who are suffering have unmet needs. Humans, however, can get a better idea of how others are suffering thanks to empathy, the process of feeling something like what the other feels. The word "empathy" is sometimes used more broadly to cover any ability to identify what someone else is thinking or feeling and to respond emotionally, which covers compassion and sympathy. But empathy more specifically requires that the empathizer feel something like what the other person is feeling, understanding the emotion by sharing it.

Brain–Mind (chapter 7) describes three modes of empathy that contribute to interpersonal understanding by different neural mechanisms: analogy, mirroring, and embodied simulation using multimodal rules. The most verbal is analogical, where you consciously put yourself in someone else's shoes and get an approximate feeling about his or her emotional experience. For example, if you have a friend who got fired, you can remember some occasion on which you lost a job, think about how you felt on that occasion, and then attribute that feeling to the person who was fired. This kind of empathy is usually conscious and substantially verbal, a kind of explicit emotional analogy that constructs a mapping between your mental state and that of another person.

Another kind of empathy is much more direct than analogical mapping. Mirror neurons fire in people's brains when they perceive other people's behavior, with the amazing result that the brain areas in which your neurons are firing when you see someone in pain are the same as the brain areas belonging to the person in pain. For example, if you see someone get hit in the stomach with a ball, you may also have a feeling of being hit, because pain areas in your brain are firing in ways similar to what happens when you are hit. Mirror neurons were first discovered in monkeys by single-cell recordings, but brain scans suggest that they also operate in human beings with feelings such as pain perception. If you see somebody crying, you may therefore get neural firings similar to the firings of the person crying, yielding a similar feeling.

A third form of empathy is not as verbal as emotional analogy but not as direct as mirror neurons. You can appreciate people's emotional situations by simulating them using multimodal rules, which have an implicit *if–then* structure where the *if* and *then* parts are nonverbal semantic pointers rather than sentences. For example, you can imagine someone dancing by running a mental simulation in which you move different parts of your body in accord with nonverbal rules. Similarly, you can use multimodal rules such as <*insulted*> → <*hurt*> to imagine how someone who has been insulted might feel, where <*insulted*> and <*hurt*> are semantic pointers rather than words. The semantic pointers in the multimodal rule can be visual, auditory, tactile, motor, or emotional, and using a sequence of them to produce a simulation gives you an extended idea of what someone else is feeling.

So human minds are equipped with brain mechanisms for appreciating the needs of others by recognizing the distress and suffering that comes from serious unmet needs. To care about other people, you do not require an argument such as the social contract that indicates that it is in your self-interest to care about other people, nor do you need a religious doctrine to direct you. Barring pathologies such as psychopathy and narcissism, your brain enables you to care about the needs of others via empathy. A 2016 study found that people with more empathy traits are more likely to behave altruistically in a real-world setting where people stopped to help someone in pain. Empathy works because it activates your own needs-based emotional processes to appreciate the needs of others. It is therefore more psychologically natural and efficient than drawing on explicit ethical theories concerning consequences and rights.

Although empathy is important for bridging the gap between your own needs and the needs of others, it falls well short of providing a full basis for ethics. Paul Bloom has pointed out serious defects in empathy as a full-blown source for morality. People are much better at empathizing with those close to them such as members of their own families and clans, whereas modern ethical theories, both secular and religious, are universal in demanding moral concern for all humans. I advocate needs-based consequentialism as the best way to achieve such universality.

Nevertheless, the neural mechanisms for empathy are an important step in the progression from moral concern with self to concern for everyone. Why be moral? The simplest answer is just because you are a human being, naturally equipped to care for others using the molecular, neural, and psychological mechanisms just described. This answer is too simple for several reasons. It could be that this equipment is ineffective, just part of our biological heritage, which may be antiquated given current cultural needs. Human males are equipped to impregnate many females, but exercising this ability today would be destructive to society. Reason is required to establish the universal ethics found in deontology and consequentialism. Moreover, I have only considered caring for others where the others are human beings, but there may be grounds discussed in chapter 10 for also caring for nonhuman animals and maybe even robots.

The strengths and limitations of empathy as the basis of morality are similar to those of purely emotional approaches in general. One popular view of decision-making is that you should just go with your gut, which assumes that your gut reaction is going to be ethically appropriate. But guts can be dumb, for example around the emotion of disgust. You may find some foods and behaviors disgusting for largely cultural reasons rather than because they involve some objective wrong. For example, many people in Western societies find it disgusting to eat insects, whereas other societies find them to be a useful source of protein. Going with your gut accesses physiology but neglects the appropriateness of a thorough appraisal concerning vital needs.

Morality benefits when the emotional impetus of empathy combines with the moral emotions of shame, guilt, pride, gratitude, and compassion to turn abstract moral reasoning into actions by ourselves and heartfelt evaluations of others. But extensive inference with respect to consequences concerning the needs of all concerned is crucial for justifiable, intersubjective ethical judgments. Empathy without reason is blind, but reason without emotion is empty.

The three modes of empathy, working with different neural mechanisms, contribute to morality in a variety of ways. The contribution of empathy by mirror neurons is minimal, merely highlighting someone in pain as suffering and therefore a likely candidate for moral concern. Empathy by multimodal rule simulation provokes greater involvement with a person witnessed as having ongoing suffering and thereby stimulates a higher degree of moral concern. Both these modes of empathy are local and immediate and therefore incapable of supporting the universality of full moral concern.

In contrast, empathy by analogy has the capability to provide more abstract and universal understanding by virtue of its use of words as well as emotions. For example, you can ask yourself how you would feel if you were starving in a Sudan famine. Combining verbal, visual, and internal representations of hunger, you can generate some emotional understanding of the degree of suffering in starving people. This understanding is not as local and parochial as that afforded by mirror neurons or rule-based simulation, because you can create an image of someone different from yourself but suffering because of the deprivation of the biological need for food. Using words in a form of reverse empathy, this kind of understanding can be instigated by suffering people who ask: how would you feel if you were in my situation?

Unlike neural mirroring and multimodal rule simulation, analogical empathy is not so heavily dependent on immediate sensory stimuli and therefore can focus more on vital needs. Nothing is gained morally by empathizing with some people's arbitrary desires such as owning an airplane or having sex with children. The standard alternatives of using utilitarian calculations and deontological reasoning are hard to execute and lack the direct connection to action brought by emotional judgments based on empathy.

Therefore, although empathy is neither necessary nor sufficient for ethical concern and caring, it is a valuable contributor to enabling people to appreciate the needs of others. Especially in the emotional analogy mode, it helps to expand the circle of morality beyond self and family or clan to universal human needs. Literature and film have played valuable roles in expanding people's empathic concern by portraying characters that people can identify with as analogous to themselves in respect to human needs, for example with *Uncle Tom's Cabin* and *Schindler's List*.

CONFLICTING NEEDS AND ETHICAL COHERENCE

Decisions would be easy if they each concerned only a single need, want, or goal. But decisions such as how to vote are rarely based on a single consideration. As in the value maps for Donald Trump and Hillary Clinton in Figures 6.1 and 6.2, people find candidates attractive for multiple reasons that outweigh negative concerns that they might have about them. For example, some people judged Trump to have valuable qualities such as leadership that outweighed his past treatment of women.

How does the mind/brain reconcile conflicting values and goals? It would be convenient if such conflicts could be overcome numerically, as when cost-benefit analysis boils everything down to a matter of money. The ethical theory of utilitarianism supposes that there can be a calculus of pleasure and pain that allows mathematical calculations of right and wrong, a tradition that continues in the expected utility approach of current economics. But values do not reduce to money, nor to any other common currency, so how can people make decisions with conflicting values and needs?

The most plausible neuropsychological answer is that practical inferences about what to do are made in the same way as knowledge inferences about what to believe, through coherence understood as constraint satisfaction. Chapter 3 describes how the brain can make parallel inferences that accomplish constraint satisfaction in order to determine what to believe given the evidence. Similarly, minds can decide what to do given conflicting goals and possible actions.

The value maps for Trump and Clinton are not just static diagrams but translate immediately into a computational model that uses artificial neurons to accomplish constraint satisfaction. Think of each node as represented by a group of neurons and the solid line indicating mutual support as excitatory links between the neurons in the respective nodes. For example, if liking Trump is captured by firing in a neural group, and liking America is captured by firing in another group, then mutual support is captured by excitatory links between various neurons in the two groups. These links have the effect that liking Trump makes you like America, and liking America makes you like Trump.

Other constraints require satisfaction in voting decisions, including negative ones such as between the Trump and Hillary nodes. The dotted lines in Figures 6.1 and 6.2 indicate negative constraints, for example if you like Trump then you do not like Hillary and vice versa. For neural computation, such negative constraints can be captured by inhibitory links between the neurons representing the relevant nodes. Firing all the neurons in the whole network balances excitation and inhibition until a stable solution is reached that approximately maximizes coherence.

For the typical strong supporter of Trump shown in Figure 6.1, the whole complex of positive and negative constraints provides emotional support for Trump. Undecided voters may get stuck in the process of balancing sets of constraints that do not all fit together, for example in the case of someone who likes Trump's attitude toward terrorists but not toward women. The result can be a virtual tie in firing between Trump and Hillary neurons, or an oscillation in which one group seems preferable for a moment and then the other.

In general, decision making can be understood as a process of parallel constraint satisfaction executed by emotional coherence. Emotions are inputs to the process of decision because of the values represented by the nodes, but they are also outputs because the judgment of what to do includes feelings about the action chosen. Such feelings are crucial for ensuring that the action actually gets carried out, in accord with the theory of action in *Brain–Mind* (chapter 9). Emotional coherence can require reconciling conflicts between specific emotions rather than just positive and negative values. For example, a Trump supporter might feel proud about his exaltation of America but embarrassed about his harassment of women.

Reconciliation depends on readjusting cognitive appraisals as well as concomitant physiological responses.

Ethical judgments are a special case of decision making where the values are not just personal wants but rather moral requirements based on needs. When you have conflicting needs, for example between relatedness and autonomy, you need to balance these by the neural process of parallel constraint satisfaction. It would be reassuring if you could translate such judgments into spreadsheet arithmetic, but the relevant information about the precise implications of different actions for different values is rarely available.

When you make a personal decision or moral judgment, what comes to consciousness is only a small manifestation of the values and constraint satisfaction that contributed. Such conscious experiences are intuitions, and some have thought that moral intuitions express solid ethical truths. But if ethical intuitions are brain processes resulting from emotional coherence, they should not be relied on much, because it is often difficult to figure out what values and unconscious balancing went into them. The operating values may be ones that, on reflection, you do want to be paramount because they represent human needs or some other part of your ethical system. Oppressed people sometimes have needs for autonomy of which they are not aware, because needs, like desires, are sometimes unconscious.

The unconscious operations that generate your moral intuition may be based on concerns that you would judge reflectively to be trivial or even odious. For example, you might be unconsciously racist so that your intuition results in part from values that you would prefer to excise. Ethical intuitions are as suspect as the epistemic ones critiqued in chapter 3.

Accordingly, conscience as an inner voice about right and wrong is not to be trusted much. On some religious views, your conscience derives from communication from God, but divine communication lacks evidence. Much more plausible is that conscience consists of emotional intuitions that become conscious as the result of competition among semantic pointers, in keeping with the neural theory of consciousness presented in chapter 2 and in *Brain–Mind* (chapter 8). A reliable conscience would generate moral judgments supporting actions that contribute to human needs. But moral intuitions and conscience are not to be trusted unless you can unpack them to get some idea of whether they are based on morally legitimate considerations.

In sum, the psychological process of parallel constraint satisfaction, carried out by neural mechanisms, provides a plausible account of how people can make practical decisions and moral judgments that incorporate multiple conflicting values, goals, and needs. Such decisions and judgments can be capricious, but you can approach objectivity by asking three questions. First, are the values contributing to constraint satisfaction legitimate because they capture vital needs, or at least ones that are instrumental to vital needs? Second, are the constraints—the links between actions and needs—based on good evidence according to the strictures of chapter 3? Third, has the process of parallel constraint satisfaction been pursued thoroughly to come as close as possible to maximization?

Consider these three questions in a concrete case. Was it ethical for people to vote for Donald Trump for president? The first factor requires asking whether the values that went into their decisions were based on people's vital needs rather than more subjective wants. The desire for jobs is tied with people's psychological need for competence as well as with biological needs for food. But the desire to live in the greatest country in the world, making American great again as Trump promised, is not tied to vital needs. Failures of empathy can lead people to consider only their own needs, ignoring equally legitimate needs of other people.

The second question requires asking whether people had evidence that Donald Trump would actually accomplish what he promised, including fighting terrorists and stopping illegal immigration. Otherwise, the constraints contributing to the coherence judgment are illegitimate.

The third question requires asking whether voters for Trump did a good job in balancing all the values that they think are relevant to making the decision. Because of the limitations of conscious cognitive processing, people may become obsessed with a small number of values such as making America great, rather than balancing multiple considerations against each other.

In all these ways, by neglecting values, by working with false constraints, and by poorly assessing coherence, people can fail to make good emotional decisions or ethical judgments. Sometimes, the resulting actions can be so horrendous that they qualify as evil.

WHY IS THERE EVIL?

If people are biologically equipped to care about others, why do they so commonly treat each other badly? If brains are set up for caring and empathy, why is the world full of child neglect and abuse, battered spouses, rape, murder, corruption, and war? Would it not be more plausible to suppose the people are inherently evil, so that good behavior is actually the rare accomplishment of cultural developments such as religion? Religions have long worried about the problem of evil, of how there could be so much bad in a world created by a supposedly benign God. The Christian doctrine of original sin supposes that people are inherently evil and need to be redeemed by the church. Even without religious impetus, there is a major problem of explaining why people who are theoretically capable of good are so often prone to fall short of ethical standards.

This question is analogous to the question of how disease arises, considered in *Mind–Society* (chapter 10). Bodies function normally most of the time but can easily malfunction given environmental problems such as viruses or just wearing out through age. Normal functioning depends on the operations of biological mechanisms, and diseases are best understood as breakdowns in mechanisms.

Immoral behavior is not in itself a disease, although it sometimes results from diseases such as brain tumors. I described earlier the neural mechanisms that make caring about the needs of others natural, so that we can ask the same question that *Mind–Society* (chapter 10) asks about diseases. What are the breakdowns in mechanisms that lead people to behave in evil ways? Here evil just means being severely immoral, acting strongly contrary to the needs of people.

Consider an intense form of evil, pedophilia, in which adults take sexual pleasure in children. Most people find children deserving of care, so how does it happen that approximately 1% of males are inclined to have sex with children? Pedophilia qualifies as evil because of the great harm caused to abused children, from immediate distress to long-term mental illness and difficulties with relationships, sometimes leading to suicide. The causes of pedophilia seem to be varied, including brain defects in frontotemporal regions, psychological problems with impulse control, and past history of being abused as a child. No comprehensive account is currently available, but, by analogy to the mental illnesses discussed in *Mind–Society*, pedophilia will likely be another case of multilevel emergence, where breakdowns in a combination of molecular, neural, psychological, and social mechanisms interact to produce evil acts.

More general breakdowns in moral behavior occur in psychopaths, who routinely engage in manipulative, self-serving behaviors with no regard for others. The causes of psychopathy are still up for discussion but are likely as disparate as other aspects of personality, deriving from combinations of genetics, epigenetics, early childhood learning, and learning from later environments. Baron-Cohen describes psychopaths as having zero degrees of empathy because of neural deficits in brain areas such as the ventromedial prefrontal cortex, but he also discusses psychological causes such as inability to recognize fear and social causes such as parental neglect. Molecular causes may operate through epigenetic effects such as the methylation of the gene for oxytocin receptors. Hence psychopathy also seems to be a case of multilevel emergence from breakdowns in interacting mechanisms.

Even non-psychopaths can do evil things because of neural, psychological, and environmental effects. Parallel constraint satisfaction can fail to produce good moral judgments when people forget about relevant values that ought to contribute to processing of emotional coherence. According to Baron-Cohen, empathy deficits can result from stress, fatigue, or alcohol, contributing to neglect of the needs of others. Men with high testosterone or low oxytocin may be more inclined to violent behavior. Environmental stresses such as lack of food, water, and sleep can lead to faulty emotional judgments through limited coherence calculations.

In addition, ordinary people are adept at giving self-serving justifications for why their actions are not immoral, a kind of motivated inference. Guilty feelings can be resisted by concocting stories about how situations are ambiguous, actions have complex consequences, and other people are worse. For example, corrupt politicians may be able to convince themselves that they really are doing their best.

Evil actions can also have social causes. Peoples' mental states are intensely influenced by emotional communication from others, and groups can end up engaging in more risky behaviors than the individuals would normally do on their own. Authority figures giving orders can produce behaviors in people that they later consider inappropriate, as in the famous Milgram experiments where participants agreed to give intense electric shocks to learners.

Thus, even without mental disorders such as pedophilia and psychopathy, evil behaviors can result from breakdowns in molecular mechanisms (alcohol), neural mechanisms (frontotemporal dementia), psychological mechanisms (poor impulse control), and social mechanisms (peer influence). When these mechanisms commonly interact, then evil results from multilevel emergence.

SUMMARY AND DISCUSSION

There are four main candidates for sources of morality: religion, pure reason, culture, and human nature. The ancient Greeks were familiar with all of these, with traditionalists looking to the gods, relativists looking to culture with the saying that man is the measure of all things, Plato looking to pure reason, and Aristotle and the Stoics looking to human nature. Natural philosophy is open to considering the enormous influence of culture but draws back from the relativist conclusion that there are no universal ethical truths. The source of morality lies in human nature, where scientific progress across many fields makes it possible to develop a much deeper understanding of humanity based on multilevel mechanisms.

Ethics does not come from religion, for there are no gods. Nor does it come from pure reason, for there are no a priori truths. Instead, ethics has to come from an understanding of the nature of humans and the world they live in, drawing on the cognitive and social sciences but going beyond them to develop normative conclusions concerning what ought to be done. The connection between evidence and norms is no simple derivation of "ought" from "is" but rather a coherentist application of the normative procedure described in chapter 1. Ethics identifies methods that best satisfy the goals of morality concerning universal satisfaction of human needs. Introducing psychology and neuroscience into moral deliberations does not undermine the objectivity of ethics but rather provides a nonarbitrary basis for it.

Ethical judgments depend on values in ways that would be utterly mysterious if values were abstract entities or behavioral preferences. Instead, the Semantic Pointer Architecture makes it possible to see how values can be processes of neural firing produced by binding concepts with emotional attitudes, which in turn result from binding of physiological states and cognitive appraisals. As semantic pointers that integrate cognition and emotion, values contribute to thought and action biologically, in contrast to values as abstract entities whose mental impact is inexplicable. Identifying values with preferences is also explanatorily inadequate, because preferences are just dispositions to choose, and dispositions need to be fleshed out by underlying mechanisms that are both cognitive and emotional: values as semantic pointers.

Because emotions include a crucial appraisal dimension, they can be judged to be rational or irrational, so that values can be objective or subjective. Immoral judgments and irrational emotions are those in which the underlying appraisal was performed with respect to trifling goals rather than vital needs.

The key to translating human nature into morality lies in needs not just as instrumental wants but as basic requirements for living as a human being. Vital needs include biological ones from oxygen to health care, but there is more to living as a human being than just staying alive. You also need to function satisfactorily in all the major pursuits of people, including love, work, and play. Full human functioning requires satisfaction of psychological needs for relatedness, autonomy, and competence. The cognitive sciences are starting to understand the mechanisms that are the basis of these needs, both for generating them and for accomplishing their satisfaction. We can hope that neuroscience will eventually provide as deep an understanding of the mechanisms for psychological needs as physiology currently provides for the mechanisms for biological needs such as food and water.

Values do not operate in isolation, but as systems that can conveniently be displayed using value maps that show conflicting values. Resolution of these conflicts can be performed naturally by parallel constraint satisfaction, combined with brain mechanisms for emotion and consciousness to yield feelings of emotional coherence, more commonly recognized as intuition or conscience. Values and moral emotions are not simply a matter of feeling pro and con, but can involve specific emotions such as joy, pride, guilt, shame, anger, disgust, indignation, and vengefulness.

Most people are biologically equipped to care about the needs of others as well as themselves, through a combination of molecular, neural, and psychological mechanisms that also underlie their own needs. Empathy is important for enabling people to appreciate the suffering of others by recognizing that their vital needs are not being met. Empathy alone is not the basis for ethics which needs to be more universal in considering the consequences of actions for the needs of all people affected. But empathy helps humans to identify and also to care about the needs of other people, operating in the three modes of neural mirroring, multimodal rule simulation, and emotional analogy. Empathy is therefore an important contributor to judgments of emotional coherence that sometimes work well to accomplish the assessment of needs-based consequences.

Despite empathy and other mechanisms for caring, people sometimes fail to attend to others needs and commit evil actions. In extreme cases such as psychopaths and pedophiles, a complete lack of empathy and responsible emotions makes some people particularly prone to committing evil acts. But evil can also be perpetrated by more ordinary people subject to molecular, neural, psychological, or social disruptions in the making of decisions that impact the needs of others. Evil is thus best viewed as the result of breakdowns in moral mechanisms that are chronic in the case of psychopaths but situational in the case of ordinary wrongdoers.

One major topic omitted in this chapter is the problem of free will. It might seem that a neurobiological approach to ethics implies that free will and moral responsibility are illusions, making the whole discussion of values and moral decisions completely pointless. But the question of free will is difficult to resolve given current understanding of how the brain works, so I save it for more tentative discussion in the final chapter of this book.

Philosophy is supposed to show how things hang together generally, so we should be striving for coherence of ethics with philosophy of mind, epistemology, and metaphysics. The semantic pointer approach to mind provides new ways of understanding important ethical ideas including values, intuition, conscience, and caring about the needs of others. Multilevel materialism explains thought and behavior as resulting from interacting molecular, neural, psychological, and social mechanisms, with applications both to morally good behavior that takes into account the needs of all relevant people and to evil behavior resulting from breakdowns of these mechanisms. Conversely, ethical thinking displays the need to integrate cognition and emotion, as performed by semantic pointers.

Natural philosophy also achieves a good fit between reliable coherentism about knowledge and needs-based consequentialism about ethics. Both positions avoid thought experiments and other bogus routes to a priori truths and instead seek guidance from theories and experiments about minds and brains. Values construed as emotional mental representations operate in both knowledge and morality. Coherence understood as a neural process of parallel constraint satisfaction accomplishes reconciliation of conflicting values, evidence, hypotheses, and actions. The fit between ethics and epistemology is not that they imply each other but rather that they can depend on the same objectivity-generating mental mechanisms.

Similarly, ethics and metaphysics fit together because needs-based consequentialism and multilevel materialism work with similar theories of mind. The demise of dualism about mind undercuts hopes for a priori ethical truths, but the view of human selves as resulting from the interaction of molecular, neural, psychological, and social mechanisms provides a different way of pursuing ethics based on needs. Human nature derives from these mechanisms that determine the vital needs that must be met for all people.

Now we can turn from consideration of ethics as largely the judgments of individuals to much broader social concerns about justice in society and government. In keeping with needs-based consequentialism, I consider how countries can best operate to further the satisfaction of human vital needs.

NOTES

For an overview of philosophical ethics, see http://www.iep.utm.edu/ethics/. Decety and Wheatley 2015 contains articles reviewing the neuroscience of morality. Doris 2010 surveys moral psychology. The online *Stanford Encyclopedia of Philosophy* contains articles on moral naturalism, deontological ethics, and consequentialism.

Much philosophical and psychological work on moral psychology uses hypothetical moral dilemmas, but Francis et al. 2016 employ virtual reality to suggest that actions differ from hypothetical judgments. Bostyn, Sevenhant, and Roets 2018 report that people respond differently to hypothetical and real-life dilemmas. Bernhard et al. 2016 find that differences in moral judgment are associated with variations in the oxytocin receptor gene. Findings such as these further undermine the evidential status of ethical thought experiments.

Holyoak and Powell 2016 defend deontological coherence as a descriptive theory of morality, but Greene 2014 advocates utilitarianism (happiness-based consequentialism) as both descriptive and normative. The other major modern ethical theory is virtue ethics, according to which moral acts derive from good character, but it is both psychologically naïve (lacking a theory of character) and philosophically useless unless you already have a theory of what makes actions good and virtuous. In philosophy, ethics tied to emotion is sometimes called "noncognitivism," in ignorance of the appraisal dimension of emotions.

Schwartz 1992 and Elliott 2017 discuss values. On cognitive-affective maps (value maps), see http://cogsci.uwaterloo.ca/empathica.html.

For reviews of moral emotions, see Haidt 2003; Prinz and Nichols 2010; and Tangney, Stuewig, and Mashek 2007. Haidt 2012 ties moral emotions to six moral foundations: care, fairness, loyalty, authority, sanctity, and liberty. Jack Layton's letter is here: http://www.theglobeandmail.com/news/politics/ottawa-notebook/laytons-last-words-love-is-better-than-anger-hope-is-better-than-fear/article617801/.

Ayn Rand's work is excerpted in Hull and Peikoff 1999. *The Washington Post* reported on the influence of Rand's ideas in the Trump administration: https://www.washingtonpost.com/posteverything/wp/2017/03/03/ayn-rand-is-dead-liberals-are-going-to-miss-her/?utm_term=.14dda0ec740c.

On the ethical significance of needs, see Miller 2013, Reader 2007, and Thagard 2010b. Various authors use different terms for vital needs: basic, true, fundamental, real, essential, intrinsic, genuine, natural, or grave.

Maslow 1987 presents his theory of needs, but Wahba and Bridwell 1976 describe empirical shortcomings. Ryan and Deci 2017 describe the three psychological needs; see also http://selfdeterminationtheory.org. Church et al. 2013 show that these needs operate across cultures.

On the costs of social isolation, see: Cacioppo et al. 2015; Steptoe, Shankar, Demakokos, and Wardle 2013; and Valtorta, Kanaan, Gilbody, Ronzi, and Hanratty 2016. On needs versus capabilities, see chapter 7.

Psychological, anthropological, and biological evidence that people are naturally altruistic includes Bloom 2013; Dunn, Aknin, and Norton 2014; Hrdy 2009; Pfaff 2015; Tomasello and Vaish 2013; and Wilson 2015.

On the neural basis of empathy, see Baron-Cohen 2011, Bernhardt and Singer 2012, and Decety 2014. On oxytocin and neuromolecular processes see Zak 2012 and Churchland 2011. Crockett and Rini 2015 describe how neuromodulators influence moral cognition.

Held 2006 and Slote 2007 defend care and empathy as important for philosophical ethics. Bloom 2016 sees empathy as more of a hindrance than a help to good ethical thinking. On Danish teaching of empathy, see https://qz.com/763289/denmark-has-figured-out-how-to-teach-kids-empathy-and-make-them-happier-adults/.

The Roots of Empathy project is at http://www.rootsofempathy.org. Bethlehem et al. 2017 find that empathy promotes altruism in real-life situations. Zaki and Cikara 2015 describe ways of overcoming empathy failures.

Emotional coherence is modeled in Thagard 2006, and its significance for ethics is discussed in Thagard 2010b. Thagard and Finn 2011 discuss conscience and moral intuitions, criticizing theological and a priori conceptions.

On pedophilia, see Seto 2009 and Joyal, Plante-Beaulieu, and De Chanterac 2014. On psychopathy, see Baron-Cohen 2011; Blair, Mitchell, and Blair 2005; Dadds et al. 2014; Debowska, Boduszek, Hyland, and Goodson 2014; and Remmel and Glenn 2015. Self-serving justifications for immoral behavior are reviewed by Shalvi, Gino, Barkan, and Ayal 2015. Darby, Horn, Cushman, and Fox 2018 identify brain lesions associated with criminal behavior.

PROJECT

Conduct three-analyses of concepts such as *moral* and *sympathy*. Construct contrasting value maps of individualists such as Ayn Rand and altruists such as Gandhi. Describe in more detail the psychological mechanisms underlying the needs for relatedness, competence, and autonomy.

7

Justice

FROM MORALITY TO JUSTICE

Is it wrong that 1% of the world's population owns half of its wealth? If so, what should be done about it? Do governments have any kind of legitimate role in redressing such imbalances? These are questions about social and political justice, which I answer based on the account of needs and morality developed in chapter 6.

I depart from my usual practice of beginning with a three-analysis, because the concept of justice is so controversial that different views disagree even about exemplars. For libertarians, an exemplar of justice is the state leaving people untouched, particularly their property. In contrast, for socialists, an exemplar of justice is the state taking strong means such as taxation and expropriation to ensure equal rights and welfare. Accordingly, I develop a theory first and then use it to guide three-analysis of the concept of justice afterwards.

As in my interpretations of mind, knowledge, reality, and morality, the natural philosophy approach to justice builds on evidence-based accounts of how humans operate in the world, paying particular attention to mental and social mechanisms. This method is very different from how theories of justice have usually been developed.

Some theories of justice are based on religion through principles derived from holy texts. As usual, this approach is not open to natural philosophy, because there is no basis for determining that one text is holier than any other. Equally inappropriate is the attempt to base principles of justice on pure reason, using a priori considerations to determine how people ought to establish their social arrangements and political institutions. Like foundationalism in epistemology, dualism in metaphysics, and deontology in ethics, pure-reason justice founders on the failure to arrive at uncontentious principles that have even a whiff of necessity about them.

Since John Rawls landmark *A Theory of Justice* was published in 1971, a favorite philosophical method for arriving at principles of justice and morality has been reflective equilibrium. This method consists of alternating ethical judgments based on intuitions about cases with consideration of moral principles that might justify those intuitions, adjusting both intuitions and principles until equilibrium is reached among them. Reflective equilibrium has the advantage over religious and a priori methods of being dynamic, using cases to suggest principles that justify intuitions, and developing more and more sophisticated cases and principles. Rawls pumped readers' intuitions by asking them to imagine themselves behind a veil of ignorance about their own place in society, and he arrived at principles such as that inequalities are only justified if they benefit the worst-off members of society.

Reflective equilibrium sounds like scientific method where theory and observation undergo a similar dance, with observations suggesting new theories which suggest new observations, aiming for a general fit among them. The crucial difference is that science is not just pure coherence, because the observations have causal connections with the world that give them a moderate degree of acceptability on their own. This connection is not absolute enough to serve as a foundation, but it still provides external constraints on the theory/observation process. However, the reflective equilibrium method has no such constraints, because the cases can be arbitrarily concocted. Reflective equilibrium is a sophisticated version of the method of thought experiments I criticized in earlier chapters. Hypothetical cases meant to support principles can be self-serving and capricious, with no independent validity, reflecting prejudices more than knowledge.

Another problem with reflective equilibrium is that it can be too easy to reach, with satisfaction in an apparent fit of principles and merely resulting from motivated inferences. It can lead people into a local maximum of coherence remote from global validity. In contrast, the social conventions of science enjoin taking observational and experimental data seriously and considering alternative hypotheses. The scientific acquisition of knowledge does not provide a guarantee of truth, but the consideration of evidence with respect to competing theories gives it a big advantage over reflective equilibrium. Instead of imaginary cases, we can consider mental and social mechanisms relevant to the accomplishment of human needs.

ISSUES AND ALTERNATIVES

Theories of justice can be evaluated according to how well they answer important questions concerning how the world is and how it ought to be. The first question might be "what is justice?," which sounds like a futile call for a definition but will instead be answered by a three-analysis at the end of this chapter.

More concrete questions concern particular kinds of justice: social justice about equality and opportunity; legal justice about system of laws, courts, and judges; and political justice about what sort of government is most conducive to social and legal justice. The key practical question looks for effective ways of moving countries with different political and legal systems toward greater justice.

Justice is often linked with equality concerning freedom, opportunity, wealth, income, rights, happiness, or needs. Why is equality a requirement of justice, and what are the most important of these concerns? If some kind of equality is required for a just society, what are legitimate changes in the political and legal systems that would bring about that kind of equality?

The following is an eloquent description of a just society by Canadian Prime Minister Pierre Trudeau in 1968:

> The Just Society will be one in which the rights of minorities will be safe from the whims of intolerant majorities. The Just Society will be one in which those regions and groups which have not fully shared in the country's affluence will be given a better opportunity. The Just Society will be one where such urban problems as housing and pollution will be attacked through the application of new knowledge and new techniques. The Just Society will be one in which our Indian and Inuit populations will be encouraged to assume the full rights of citizenship through policies which will give them both greater responsibility for their own future and more meaningful equality of opportunity. The Just Society will be a united Canada, united because all of its citizens will be actively involved in the development of a country where equality of opportunity is ensured and individuals are permitted to fulfill themselves in the fashion they judge best.

This statement reflects a social-democratic understanding of how government action ought to increase equality and justice. Social democracy promotes justice within a capitalist economy by means of political interventions that redistribute income. Alternative views provide a starkly different view of justice.

One extreme alternative is a relativist or postmodernist position that denies any objective basis for understanding what justice is or how it can be increased.

There are indeed cultural variations in what is considered to be a just society, but these differences can be surmounted if there is a common human nature that guides choice of morality.

Another alternative to social democracy is the libertarian view that all of Trudeau's concerns are not legitimate parts of the theory of justice, which should be aimed only at protecting individual liberty from intrusions by others. Then the great disparities in wealth in modern societies are not of ethical concern, because morality can only justify a minimal state that protects people from harm by others. Redistributing income or wealth, for example by income taxes, would be illegitimate. Society is not actually unjust but would be unjust if it tried to interfere with people's freedoms.

Libertarianism is usually taken to be right wing, because it defends private property and the untrammeled operation of capitalist societies. But at the other extreme lies a view that also rejects the role of the state in ensuring social justice. Anarchists advocate the complete elimination of the state, along with its defense of capitalism based on rights to private property. Left-wing anarchism seeks justice through the mutual cooperation of people whose behavior will somehow improve once the state and the capitalist system are eliminated.

Anarchist procedures such as consensus-based decision-making have worked for short times among small groups of people but are implausible for managing a complex modern society with millions of people and advanced technologies. Moreover, lack of government would make it hard to deal with the aberrant personalities described in chapter 6, the small percentage of people who are psychopaths.

With similar goals to anarchism but vastly different strategy, socialism demands extreme measures to establish equality by abolition of private property and the whole capitalist system in which trade and industry are controlled by owners for profit. Communism claimed that unjust inequalities are endemic to capitalism and can only be overcome by putting companies and factories in public ownership. The implementation of socialism in the Soviet Union and China revealed serious problems with this plan.

First, inequality and privilege were maintained in socialist systems by giving power and resources to members of political elites with special access to goods and services. Second, maintaining a socialist system in the face of public dissatisfaction required severe social controls, including reductions in freedoms of speech and assembly. Third, socialist systems had difficulties generating the economic growth that occurred in successful capitalist economies, so that overall needs satisfaction was not increased. The Soviet Union dramatically abandoned socialism in 1989, and China has moved increasingly to a capitalist and hybrid economy since the 1990s. The failure of such regimes makes socialism an implausible option for fostering social justice.

Another alternative to social democracy is liberal democracy, which shares with it concern for inequalities but also shares with libertarianism the desire to maintain a high degree of freedom. The tension in liberal democracies between personal freedoms and social welfare can lead electorates to swing back and forth between right-wing and left-wing parties, for example Republicans and Democrats in the United States.

A major philosophical problem for both liberal and social democracy is what justifies focus on the rights of minorities, women, and the poor. In John Rawls' theory of justice, such concern is prompted by a thought experiment that asks people to imagine that they will come into the world without any idea of what their social status will be. Rawls supposed that people would choose principles that give people an equal right to liberty and opportunity, with inequalities justified only to the extent that they benefit the least-advantage members of society.

These principles are well-intentioned, but Rawls' thought experiment has failed to convince conservatives and libertarians who resist calls for redistribution of wealth to aid the least-advantaged members of society. Knowing that they have ended up relatively privileged in wealth, income, and freedom, the well-off find the veil of ignorance unappealing, as often happens when thought experiments oppose personal motivations.

JUST SOCIETIES: NEEDS SUFFICIENCY

I propose a starkly different approach that justifies democratic judgments about inequality based on objective vital needs. As chapter 6 argued, vital needs that are universal to humans are markedly different from wants and preferences, which can vary capriciously among different people. Equality and justice can then focus on what kinds of social, legal, and political systems promote human needs. Then it does not matter whether people have nearly the same wealth and income, as long as they all have sufficient resources to satisfy their vital needs ("needs sufficiency").

Needs and Justice

It is clear how needs sufficiency works with biological needs: to be just, a society has to ensure that all individuals have food, water, shelter, and health care that are adequate to maintain life. Some people may have better food or fancier housing as long as the basic biological needs are met for all members of society. Legal and political systems qualify as just when they work toward the goal of satisfying the vital needs of everyone. The reason to care about the worst-off people is not a thought experiment about an imaginary social contract but rather the empirical finding that minorities, women, and the poor have more unmet vital needs.

Application of the needs approach is more complicated for the psychological needs of relatedness, autonomy, and competence. The centrality of these needs to human existence serves to justify rights that go beyond those required just for satisfaction of biological needs. Freedom of expression, for example, is important for development of all three needs, because without it people are limited in developing relationships with other people, avoiding coercion, and pursuing achievements. Unwarranted incarceration similarly interferes substantially with need satisfaction. Competence and autonomy are severely restricted when people do not have the opportunity to pursue their goals by having accomplishments and achievements in domains appropriate to their talents.

Organizations such as the United Nations make proclamations about myriad human rights, but on what basis? Religious thinkers see rights as god-given, and some philosophers have tried to derive rights from duties that are established by pure reason. For example, if you have a duty not to lie because lying cannot be made universal, then I have a right not to be lied to. But it is just as plausible that duties derive from rights rather than vice versa, for example if your duty not to harm me derives from my right not to be harmed. The American Declaration of Independence proclaimed it self-evident that men are created equal and endowed with unalienable rights to life, liberty, and the pursuit of happiness, but billions of people have doubted such equality and rights.

A more empirically plausible theory of rights derives them instead from human needs. People have human rights to satisfaction of their vital needs, so that other people have a duty minimally not to interfere with people satisfying these needs and maximally to act in ways that contribute to such satisfaction. So, vital needs explain and justify the existence of human rights because they identify the prerequisites of being human.

Is it a problem or an advantage of the needs-sufficiency approach that it does not demand strict equality of wealth and income? It is an advantage, because of the enormous difficulty of changing current societies to produce radical equality in all respects. For reasons of genetics, epigenetics, and learning, people vary in abilities, talents, and luck arising from social and medical circumstances.

One problem with remaining inequality, even when everyone's vital needs are satisfied, is that people often envy others who have more than they do. People's evaluations of their own situation are not absolute but relative to how they think they are doing compared to their peer group. The solution to this problem is psychological rather than ethical, through education to convince people that their lives are fine even if others are somewhat better off. Moreover, the measures required to ensure that everyone's needs are met will require governmental measures plus increased taxation that will reduce inequalities of income and wealth to some extent.

Envy is an emotion in which people make cognitive appraisals tied to their physiological states, judging that it is bad that somebody else is better off than they are. Historically, envy could be combated by the religious view that the present world does not really matter because everyone can have the same fabulous afterlife. Such reassurance is no longer rational, but education can inform people of studies that show that happiness is less affected by wealth than by personal relationships. Comparisons both within and across countries suggest that the effects of income on health and life expectancy tend to level off after people have enough to support vital needs.

Education can also help people to acquire social norms based not on individuals striving to be richer than everyone else but on ensuring that society operates in a way that produces equality with respect to vital need satisfaction. These social norms also have the advantage of discouraging the kinds of consumption that deplete the world's resources and accelerate climate change in the form of global warming, which is increasingly likely to lead to disasters for future generations.

Alternatives: Utilitarian, Flourishing, and Capabilities

This needs-sufficiency theory of justice requires a society to ensure that everyone's vital needs are sufficiently met to allow them to function as humans, ignoring highly variable preferences and wants. This view of justice based on needs is very different from a utilitarian account that says that societies should be evaluated on the extent to which they produce the greatest good for the greatest number, where good is understood as pleasure and the avoidance of pain.

Calculations to maximize utility across society are devilishly complex, since utilities are not easily measurable, but assessment of the extent of satisfaction of biological needs is easy because of the observability of whether people have enough food, water, shelter, and health care. Observation of the satisfiability of psychological needs is more complicated, but it is not hard to identify when people are lonely rather than related, coerced rather than autonomous, and restricted rather than competent.

A standard objection to consequentialism is that it seems to warrant causing a few people pain in order to increase the happiness of the greater number, with resulting violation of human rights. But needs-sufficiency justice blocks such violations when they drop people below minimum levels. Vital needs provide a much better standard for equality and justice than pleasure and happiness. The needs-sufficiency view only requires that everyone's vital needs be met. It also contrasts with libertarian approaches that only emphasize autonomy and freedom from bodily harm, neglecting the equally important psychological needs of relatedness and competence.

Vital needs are also a better standard for justice than Aristotle's standard of flourishing, which is also more than happiness. Aristotle thought that flourishing for human beings is largely a matter of rational activity based on virtue. But he was vague about the sources of virtue and the goals of rational activity. In contrast, the needs-sufficiency theory is based on the underlying psychological and neural mechanisms described in the last chapter, which show more precisely why people suffer when vital needs are not satisfied. Flourishing is indeed desirable but is best understood as the satisfaction of vital needs that result from the biological nature of human beings, including psychological needs for competence, relatedness, and autonomy.

The needs-sufficiency approach to justice is largely compatible with the capability account of Amartya Sen and Martha Nussbaum. They argue that justice is not just a matter of freedom or welfare but rather of a person's abilities to achieve and exercise basic human activities and opportunities. According to Nussbaum, the core capabilities that society should support are life, bodily health and integrity, senses plus imagination and thought, emotions, practical reason, affiliation, relations to other species, play, and control over one's environment.

This list fits fairly well with my account of vital needs, because life and bodily health and integrity are connected with biological needs and the others are connected with psychological needs. Emotions and affiliation support the relatedness need. Imagination and thought, practical reason, and control over one's environment support the needs for autonomy and competence.

Nevertheless, I think that the needs-sufficiency approach is superior to the capabilities approach for two reasons. First, Sen, Nussbaum, and their collaborators give no account of the nature of human capabilities. Capabilities are dispositions to act in particular ways under particular circumstances, but what are dispositions? Chapter 5 argued that dispositions are best understood as based on underlying mechanisms, which are systems of connected parts whose interactions produce regular changes. The capabilities approach says nothing about the underlying psychological and biological mechanisms that dispose people to flourish when their capabilities are exercised and to suffer when their capabilities are thwarted. In contrast, the vital needs account builds on combined physiological and psychological accounts of emotion and human functioning.

Second, because of its grounding in mechanisms, the needs-sufficiency approach provides a way of singling out some capabilities as more important than others. One can challenge, for example, Nussbaum's concern about other species as vital to humans independent of their contribution to human needs such as food. Nonhuman animals could be connected to relatedness in the form of pets and with competence in the form of working on the world, but pets and farming are instrumental wants, not vital needs for most people.

Applications: Inequality, Discrimination, and Future Generations

Let us now return to the question of whether it is wrong that 1% of the world's population owns half of its wealth. From the needs-sufficiency perspective, this inequality would not be wrong if the vital needs of the 99% were satisfied. But we know otherwise: many people live in abject poverty with inadequate food, water, shelter, and health care, leading to short and miserable lives. Moreover, people under those circumstances have difficulty satisfying their psychological needs for autonomy, competence, and relatedness.

In rich countries such as those in North America and Europe, there is enough overall wealth to ensure that all people have their vital needs satisfied including psychological ones that require medical and educational systems that are both high quality and widely accessible. Medical systems are essential for satisfying the vital need for health, and educational systems are essential for satisfying the vital needs for competence and autonomy.

A theory of justice should say what is wrong about discrimination against people on the basis of their sex, race, ethnic origins, disabilities, or sexual orientation. *Mind–Society* (chapter 5) provides a psychological explanation of how discrimination and prejudice work as systems of values driven by motivated and fear-driven inference. But that descriptive account does not address the normative question of why discrimination is bad.

Extreme forms of prejudice and discrimination violate the ability of people to satisfy their biological needs. Slavery puts people at risk of death because of the drastic means required to control and transport people; it also dramatically reduces their autonomy, competence, and relatedness. Milder forms of discrimination can also severely hinder needs satisfaction through reduced life expectancy, poor educational opportunities, and psychological stress resulting from shame and lack of respect.

For example, the indigenous people in Canada, formally known as Indians and Eskimos but now more appropriately called First Nations and Inuit, have suffered centuries of maltreatment. More than 100,000 children were taken away from their parents and forced to attend residential schools, where they were subject to emotional, physical, and sometimes sexual abuse. Today, many indigenous people live in communities with substandard water and limited health care and education. The result is diminished satisfaction of both biological and psychological needs, evident in much higher suicide rates, lower life expectancies, drug use, high infant mortality, and high rates of sexual abuse. It is therefore obvious that the vital needs of such minorities are not being met by Canadian society, which therefore can be judged to be unjust, contrary to the plan of Pierre Trudeau and the pronouncements of his son, current Prime Minister Justin Trudeau.

Similarly, discrimination against women violates basic biological needs when it results in physical assaults, dangerous illegal abortions, and sexual exploitation. Less extreme prejudices violate psychological needs when they limit women's abilities to achieve autonomy and competence. Relatedness can also suffer when coercion by husbands or fathers interferes with the attainment of closeness and care.

The needs-sufficiency theory of justice also provides reasons for caring about future generations. Traditional ethical theories only concern people who are currently alive, considering their happiness, freedoms, and rights. The argument in chapter 6 concerning why you should care about the needs of others generalizes to considering the needs of others in future generations.

Just as your human brain enables you to empathize with other people who are currently alive, so you can imagine the lives of future beings as analogous to yours. This exercise is easiest to do with prospective offspring, especially children and grandchildren, but empathy by verbal and other kinds of analogy can extend to people 100 years from now who are suffering from bad current practices such as pollution. You can put yourself in the shoes of people suffering from floods, starvation, and disease caused by global warming and realize that you do not want your current lifestyle to contribute to misery in billions of future people. Such realization can be blocked by being a psychopath who does not even care about people who are alive now or by motivated inference that climate change must be a hoax because dealing with it would hurt your economic status. Nevertheless, needs-sufficiency justice applies to future generations because future people can be expected for biological reasons to have the same vital needs as those currently alive.

In sum, the needs-sufficiency theory of justice is superior to alternatives because it provides plausible answers to questions about equality, discrimination, and future generations, based on the biological and psychological mechanisms that constitute human nature. Implementing justice requires attention to politics and the nature of just governments.

JUST GOVERNMENTS

Basing social justice on the universal satisfaction of vital needs provides direction for thinking about political and legal justice. What kind of government is most conducive to needs sufficiency? Answering this question should take into account the historical record of various kinds of government but also should be open to the possibility of inventing new forms of government that might be superior for needs satisfaction. The risk, however, is that some new form might have unintended consequences that would undercut its ability satisfy needs. For example, communism was originally concerned with universal need satisfaction, going back to Marx's dictum: to each according to needs, from each according to abilities. But as it unfolded in the hands of Lenin, Stalin, and Mao, communism brutally trampled on the vital needs of millions of people, restricted autonomy, and often eliminated life altogether.

There are four main issues to be addressed for a needs-based account of political justice. First, what general form of government is most conducive to the satisfaction of vital needs? The three main candidates are dictatorship in which a monarch or other leader controls the society, democracy in which people get to choose leadership, and anarchy in which there is no leadership at all.

Second, given the argument to come that democracy is the best approach for the satisfaction of human needs, what particular kind of party ideology best contributes to human needs satisfaction? Today the main alternatives are right-wing conservatism, moderate liberalism, and more aggressive social democracy.

Third, what role should the legal system play within a democracy that aims at promoting human needs? Fourth, going beyond the governments of particular countries, what kinds of international organizations and agreements are desirable for fostering vital need satisfaction all over the world?

General Form of Government

Are there grounds for thinking that democracy is the best overall political system? Representative democracy based on elections has identifiable flaws, such as the inequality and discrimination that still exist in countries with elected governments. Voting can give power to people who are at best concerned with the needs of the majority, ignoring needs of citizens not required for getting them in office. At worst, power can be gained by self-serving special interests who are able to use media to gain votes. New technologies encourage other forms of manipulation, for example the fake stories on Twitter and Facebook that affected the 2016 American election. Democracies can narrowly elect demagogues who make decisions that quickly run against the interests of most people, as with Hitler in Germany in 1933.

Nevertheless, Winston Churchill was right when he said that democracy is the worst form of government except all those other forms that have been tried. Comparison of present governments strongly suggests that democracies are better at satisfying human needs than dictatorships. There is no international survey of vital need satisfaction, but various measures can serve as approximations.

The United Nations Human Development Index takes into account three factors: life expectancy, education, and income. Life expectancy provides a surrogate for satisfaction of biological needs, because people who lack food, water, shelter, and health care die early. The education and income measures serve as rough approximations to the satisfaction of the need for competence, because educated people with well-paying jobs gain more opportunities to accomplish their achievement goals. According to the 2015 Index, the highest scoring countries, in order, were Norway, Australia, Switzerland, Denmark, Netherlands, Germany, Ireland, United States, Canada, and New Zealand. All of these have democratic forms of government where elections choose leaders and parties that can be voted out of office periodically. In contrast, many (but not all) of the lowest scoring countries were dictatorships.

The Human Development Index ignores the need for autonomy, but there are freedom indices that attempt to measure the extent to which people have personal, civic, and economic freedoms. According to the Freedom Index produced by a Canadian think tank, the Fraser Institute, the most free countries in the world in 2016 were, in order, Hong Kong, Switzerland, New Zealand, Ireland, Denmark, Canada, United Kingdom, Australia, Finland, and the Netherlands. As for the Human Development Index, these countries (except Hong Kong) are democracies, and it is not surprising that the least free countries were usually dictatorships.

Neither the Human Development Index nor the Freedom Index provides a measure of satisfaction of the psychological need for relatedness. But this factor is known to be important for personal happiness, so it should contribute to reports of happiness and well-being that have been measured by world surveys. According to the World Happiness Report of 2016, the countries in the world with the happiest people are Denmark, Switzerland, Iceland, Norway, Finland, Canada, Netherlands, New Zealand, Australia, and Sweden. Again, almost all the lowest ranking countries were dictatorships.

Even though these three indices are only rough approximations to vital need satisfaction, they provide evidence that democracies are generally more just than dictatorships. The correlation between democracy and happiness has become weak since the emergence of countries such as Moldova from the former Soviet Union that are more democratic but still miserable. But residual problems such as lack of trust and security show that democracy is not sufficient for needs-sufficiency even if it usually contributes to it.

Five countries appear on all three top-10 lists: Australia, Canada, Denmark, New Zealand, and Switzerland. All of these are far from perfect according to the standard of vital need satisfaction for the entire population. Australia, Canada, and New Zealand have despicable histories of mistreatment of indigenous peoples. All 10 have pockets of poverty that limit the opportunities of children and others to satisfy their psychological needs, while shortening their lifespan. Nevertheless, comparatively, democracies far exceed dictatorships in promoting vital need satisfaction.

The alternative to dictatorship and democracy is anarchism, which tries to completely do away with leadership. This form of administration has occasionally worked on small scales such as in the Occupy Movement of 2011, which practiced highly distributed decision-making and the explicit rejection of leaders. Anarchist movements organized around worker control of industries were influential in Catalonia before being quashed by fascists in 1939.

But no case exists of anarchist conduct of a hugely complex society with systems for public health, medicine, waterworks, roads, education, and safety without formal leadership in government. Moreover, anarchism has no mechanism to keep society from degenerating into a kind of feudalism in which psychopaths manage to control local fiefdoms.

Particular Parties and Ideologies

Choosing democracy over dictatorship and anarchy as government is relatively easy, but a much more challenging question is what kind of leadership within democratic systems is most conducive to satisfaction of vital needs. All of the top countries according to various indices have political systems in which different parties vie for electoral supremacy. The three most common kinds of political parties are conservative, liberal, and social democratic. Most of the countries in the top 10 lists have had each of these in power or close to power at various times. The major exception is the United States, which has a two-party system with the conservative Republicans and the liberal Democrats, although there have been small groups operating within the Democratic Party who show concern with social democracy with respect to eliminating poverty and discrimination against minorities.

In their official policies, the greatest concern with vital need satisfaction is shown by social democratic parties such as the British Labour Party, the German Social Democratic Party, and the Canadian New Democratic Party. All aim to provide social benefits such as universal health care and programs to reduce poverty. Social programs that work to reduce discrepancies in need satisfaction have been introduced by parties in power that either espoused social democratic values or felt threatened by alternative parties that did.

For example, Canada has had universal health care since 1984 because it was introduced by the Liberal Party responding to the success of a comprehensive health plan introduced in the province of Saskatchewan by the social democratic party in power in 1960. All of the five countries that made top 10 in the three indices have had substantial traditions of social democratic parties. Social democracy balances vital needs for health, relationships, and competence with capitalist capability for freedom, economic growth, and innovation. Therefore, both in policy and practice, it seems that social democratic parties do a better job of promoting needs satisfaction.

One could respond that conservative or liberal governments that do a better job of fostering economic growth and personal freedoms are actually better at promoting overall needs satisfaction, even if they do not have the same commitment to satisfying the needs of all people. This response raises questions about which needs should be satisfied for which people, with trade-offs among various needs such as the relative priority of autonomy versus health care. Republicans in the United States argue that people should be able to choose whether they spend their money on health insurance. The trade-offs among needs to be satisfied may be different in wealthy countries such as Switzerland compared to developing countries such as Nigeria. In the short run, a country might be better off aiming more for economic growth than for universal needs satisfaction.

Nevertheless, placing vital needs as central to justice supports the long-term aim of ensuring that all people reach a minimal threshold that provides everyone with food, water, shelter, health care, autonomy, and the opportunity for relatedness and competence. This aim is already achievable by many countries by means of the basic income policy discussed later.

Benjamin Radcliffe has systematically examined how voters' choices determine the quality of life in Organisation of Economic Co-operation and Development countries. He looks at the effects of government policies on self-reported satisfaction and concludes that generous welfare states where the state plays a major role in the economy have greater overall happiness. He recognizes that happiness and well-being as reported in surveys are just convenient stand-ins for satisfaction of human needs.

I have omitted socialist parties from this discussion because of the dismal record of socialist governments in promoting economic growth and personal needs satisfaction. Another alternative would be Green parties that have had influence in some countries such as Germany, but their principles about sustainable growth and climate change could be incorporated into social democratic approaches.

Legal Justice

With social justice oriented to vital need satisfaction, and with political justice oriented to social democracy, we can approach the question of legal justice. What role should the system of laws, courts, and police play in the fostering of needs? A conservative or libertarian view would maintain that the role of the legal system is merely to protect people from having their freedoms interfered with by the actions of others. The right to property, for example, needs to be rigorously enforced by countering crime. Liberals and social democrats favor the activist view that the legal system is a legitimate force for bringing about greater need satisfaction.

The recognition of autonomy through personal freedom is only one of the vital needs for the legal system to promote. When democratic governments introduce measures that reduce social and economic inequalities, the role of the courts and police should be to enforce the measures and to restrict people who are resisting them. Needs-based consequentialism in ethics and the attendant needs-sufficiency approach to justice support an active role by the legal system to ensure that human rights based on needs are not violated.

Conservatives often rail against taxes, government regulation, and legal enforcement because they limit the freedom of citizens who have large incomes and wealth. Such complaints are legitimate only if freedom is the paramount value behind justice, but I have argued that justice needs to be as much concerned with the satisfaction of other vital needs. Taxes and regulations are morally justified when they actually contribute to needs satisfaction by improving health, equal opportunity, and the ability of the downtrodden to satisfy their vital needs. Taxation is a small price to pay for civilization.

International Justice

Justice is not just a matter of what goes on in an individual country, and my opening question concerned the distribution of wealth in the whole world. Without going to war, countries have limited control over the actions of other countries. The human development, freedom, and happiness indices show huge disparities between different countries. The United Nations estimates that more than 10% of the people in the world are suffering from chronic undernourishment.

There all also disparities in the amount of equality of wealth in different countries as measured by the Gini coefficient. A country with complete inequality would have a Gini index of 100, whereas a country with complete equality would have a Gini coefficient of zero. For example, countries like South Africa and Brazil have high Gini index values above 50, whereas countries such as Norway and the Czech Republic have more equal coefficients lower than 30. Inequality in itself has serious social effects, for example in the lessening of empathy among the better-off. Surprisingly, lower class individuals are more engaged with the needs of others than upper-class ones.

The argument in chapter 6 about why you should care about the needs of others does not simply apply to other people in your own country but extends to all people in the entire world. Hence justice requires that people act to reduce inequality and promote needs satisfaction internationally. You have more power over what happens in your own country, so your efforts are probably more sensibly placed there, but there are also ways in which people in one country can be concerned about the meeting the vital needs of people in other countries.

Politically, you can support international organizations dedicated to meeting worldwide needs. Good candidates include the United Nations High Commissioner for Refugees and the World Health Organization. More personally, you can donate money to organizations that work to reduce poverty and disease in the most disadvantaged countries, such as Oxfam and Doctors Without Borders.

Other kinds of international political efforts are required to deal with major problems of inequality in needs satisfaction. Thomas Piketty argues that wealth inequality is steadily increasing around the world in a way that cannot be dealt with just within individual countries, because of the ease with which individuals and corporations can move their wealth across countries in order to dodge taxation. He therefore argues for an international wealth tax that would help reduce cross-nation inequality in wealth and satisfaction of vital needs but that would be unfortunately difficult to enact and enforce.

JUST SOCIAL CHANGE

Mind–Society explains social change as a result of interactions among mental mechanisms in individuals and social mechanisms in groups, covering social changes concerning romantic relationships, prejudice, politics, economics, religion, and international relations. But these explanations ignore the normative question of what social changes ought to take place; in other words, when is social change justified?

The needs-sufficiency theory of justice answers these normative questions by saying that society is morally good when it brings about increased satisfaction of vital needs, moving toward a state where all people can lead the best lives of which they are capable. Oppressive societies led by dictators have historically been abominable at meeting the vital needs of their citizens, so overthrow of such regimes is fully legitimate. Revolutions are justified when they are needed to overcome social control that prevents many people from fulfilling the biological and psychological requirements for a fully human life.

In democracies, however, there are less-extreme means than revolution for making society more just, depending on educating the electorate to support political parties that will bring about changes to increase justice. *Mind–Society* describes how people can change beliefs and attitudes through interacting cognitive-emotional adjustments based on coherence with evidence and values. Providing people with evidence, explanations, and systems of values that fit with their needs is crucial for rational change. Provision requires social interactions with both cognitive and emotional communication that goes beyond the verbal to include pictures, songs, gestures, and group rituals. Forming new groups is a powerful way to bring about the required interactions, along with the resulting rituals such as rallies, meetings, and street protests.

There is no direct way of transforming systems of values and beliefs to move leaders and followers toward acceptance of the importance of vital needs satisfaction. A crucial step is transition from values based on arbitrary personal preferences and absolute personal freedom toward more socially responsible sets of values that systemically work towards increasing justice. Positive values such as vital needs, caring about others, and human dignity battle their opposites such as inequality, neglect, and exploitation. Progressive packages of values can be made more appealing to people, especially children, through literature, songs, plays, and interpersonal interactions including setting a good example.

Just social changes are concerned not only with the needs of people currently alive but also with those of future generations whom we can appreciate through imagination, empathy, and shame if we fail to meet their needs. Full appreciation of future needs requires overcoming natural thinking tendencies, including misunderstandings of probability, emotionally framing short-term losses as more important than long-term gains, motivated inferences about current wants, and overreliance on ill-informed intuitions. The natural human inclination to think that everything will be okay must be countered by values and evidence about the consequences of current actions for future people.

In the short run, social changes designed to increase needs satisfaction require difficult trade-offs for individuals and groups of individuals. Contrary to the needs hierarchy of Maslow, there is no strict ordering that says which needs should be met first in an individual. Moreover, there is no formula that says how to adjudicate among the needs of different people. The point of the needs-sufficiency theory of justice is to ensure that all people have enough needs satisfaction to be human.

For increasing social justice, the relevant normative questions include not only the thought processes of individuals—how they should think—but also the social processes of groups. Ethical decisions in families, workplaces, hospitals, and other organizations depend on interactions among individuals. Because such decisions are inherently emotional as well as cognitive, evaluation of group processes has to pay attention to different kinds of emotional communication. One desirable form of communication is empathy, in which people gain emotional appreciation of the situations and suffering of others.

Mind–Society (chapter 6) describes the negative effects of power relations in affecting people's thoughts and behaviors, not only by overt coercion but also through social norms. Norms in the form of multimodal rules make their subjects acquiesce to the plans and goals of an individual or group through agreements that only seem to be voluntary because they do not rely on coercion or the offering of benefits. Instead, the mixture of positive emotions such as pride and negative emotions such as shame can lead people to accept norms that run contrary to their own interests and the vital needs of others. Then communication to promote just social change has to contribute to reconfiguration of the whole system of beliefs, values, and emotions that allows power to control people against their needs satisfaction.

In bringing about social change that accords with the needs-sufficiency theory of justice, it is important to recognize limitations on what people can accomplish ethically. Unlike the pure-reason approach to ethics of deontology and the calculation approach of utilitarianism, natural philosophy recognizes psychological mechanisms that make it hard for people to be completely universal in their ethical judgments. It is probably impossible for most people to feel the same kind of empathy for strangers that they feel for their loved ones or even for their friends. But ethical education for a just society needs to build beyond this limitation to encourage people to expand their circle of empathy to enable them to appreciate the needs of other people and the social changes required to satisfy them.

A Toronto-based organization called Roots of Empathy is dedicated to increasing empathy in schools. Their method involves bringing babies and mothers into classrooms to invoke interactions and discussions among students, from kindergarten to grade 8. This technique has been so successful in improving classroom behavior that it has spread from Canada into several other countries involving more than 1,500 classrooms and 600,000 children.

The Roots of Empathy method activates all three modes of empathy, using mirror neurons, verbal analogies, and multimodal rule simulation. The children naturally mimic the facial expressions and body language of the babies, putting them in physiological states similar to the baby such as smiling, so that nonverbal noninferential processes can put them in approximately the same emotional state. This mimicry is instantaneous using mirror neurons, but the students can also simulate more complicated behaviors by invoking their own rules that are visual, auditory, tactile, and kinesthetic.

Teachers trained by Roots of Empathy also invoke a kind of empathy that is more deliberate, conscious, and verbal, when the teacher uses words to draw parallels between the baby's situation and the situation of the students. The teacher also uses words to draw parallels among the situations of the various students, thus increasing the empathy that they have with each other. The overall goal is to reduce aggression and bullying by enhancing the ability of students to take each other's perspective.

Society needs more such procedures, in schools and other organizations, to battle the efforts by irresponsible politicians and social media to emphasize differences and the ascendancy of individuals. Lacking the immediacy of having a baby in the classroom, social change oriented to social justice has to rely more on verbal analogies than physical imitation, encouraging people to imagine relations between the suffering of others and their own lives.

Literature has contributed to social change by helping people to appreciate the situations of others apparently different from themselves. Some great books that have expanded readers' circles of morality through empathizing with their characters are *Uncle Tom's Cabin*, *Les Misérables*, *The Grapes of Wrath*, *The Diary of Anne Frank*, and *Roots*. Unfortunately, literature can also be used to shrink morality to raw egoism, as in Ayn Rand's *Atlas Shrugged* and *The Fountainhead*.

Empathy can also make political contributions to democratic decision-making when people try to see the perspectives of others. Michael Morell argues that empathy contributes to openness toward others, reciprocity, tolerance, mutual respect, inclusion, cooperation, and fairness, all of which foster democracy. Fairness is open to several interpretations, as being equal, getting what you deserve, and satisfying needs.

BASIC INCOME

As a concrete application of the needs-sufficiency theory of justice, consider proposals for a basic income, also known as guaranteed annual income, "mincome," and "demogrant." Basic income is a government payment to a person or family to provide a minimum income. It has never been implemented by any nation, but pilot programs are underway in 2018 in Spain, Finland, the Netherlands, Kenya, Scotland, Ontario, Alaska, and California. In 2016, a referendum in Switzerland rejected a proposal to establish a basic income. The aim is to reduce poverty and bureaucracy with a program that is more generous and easier to administer than standard programs such as welfare and disability.

I avoid the practical details of basic income such as what amount should be provided and whether it is best implemented through giving grants, topping up income, or instilling a negative income tax. My concern is with the ethical justification of such a program, of whether it would be legitimate for a state to use tax money extracted from other people to benefit the poor. Such a justification might be based on religious doctrines, a priori rights, or utilitarian principles of maximizing happiness, but all of these are problematic. In contrast, vital needs provide a strong argument for basic income.

Start with basic biological needs for food, shelter, and health care. Basic income targets people whose poverty makes it difficult from them to pay for adequate food and shelter. The poor are often malnourished and have difficulty balancing the costs of housing against expenses for food and medicine, sometimes leading to homelessness. The taxation required to provide a basic income to all people would not have the effect of lowering taxed people below the line required for adequate food and shelter, so the satisfaction of these vital needs is already enough to suggest that basic income can contribute to a just society.

There is also evidence that basic income can contribute to overall health. A pilot experiment in Dauphin, Manitoba, Canada in the 1970s found an 8.5% reduction in hospitalization rate for participants, particularly for accidents and mental health. Even in countries with universal health coverage, basic income can be expected to improve health (and potentially decrease health costs) by improved nourishment, reduced homelessness, and lowered psychological stress. *Mind–Society* (chapter 10) discusses how mental illness arises from social causes as well as from breakdowns in molecular, neural, and psychological mechanisms. So the vital need for health also supports basic income.

Psychological vital needs are also promoted by basic income. Autonomy increases in two ways. First, the provision of enough money to live on gives people more choices about how to lead their lives. Second, basic income programs are intended to be less controlling and intrusive than standard welfare and disability payments that come with much oversight and many strings attached, along with the psychological baggage of social stigma.

Competence is also increased by basic income because it provides people with the ability to choose educational and career paths that fit with their desires and talents, improving achievements. Critics of basic income worry that it will enable people to become lazy slackers, but the Dauphin experiment found that people are generally willing to pursue education and work, in accord with most people's need for competence and accomplishment. Psychological studies have shown that scarcity of resources induced by poverty decreases cognitive functions needed for achieving competence.

When people do work less, it can benefit another vital need for relatedness. By providing no-strings support for children, basic income supports families that are the most common source of social relations. People with adequate income also have more time and psychological ease to promote other kinds of relatedness such as friendships. The Dauphin experiment found a reduction in domestic violence. Through stronger families and friendships, basic income can help to promote relatedness as well as the other vital needs, biological and psychological.

All of these claims about the benefits of basic income are empirically testable, which is why pilot experiments in various countries are crucial. Basic income is ethically warranted if it increases broad satisfaction of vital needs, which can only be determined by examination of consequences rather than thought experiments. The Ontario and other pilots are looking at the effects of various ways of implementing basic income on health, education, careers, and work behavior, with randomized trials to determine the comparative accomplishments of various amounts. These experiments should also be able to determine whether basic income encourages autonomy by being less intrusive and stigmatizing than welfare and whether it encourages competence by strengthening rather than weakening desires to work.

Poverty leads to illness, early death, substance abuse, poor education, and legal problems. Basic income is a plausible social mechanism for alleviating poverty and thereby providing vital needs satisfaction for many people.

SUMMARY AND DISCUSSION

This chapter develops a theory of justice based on evidence concerning social development and mental and social mechanisms, rather than on religious principles or thought experiments. I have not relied on counterfactual judgments about what people would say if they were ideal observers or behind a veil of ignorance. As chapter 5 argued, such counterfactuals have no plausibility unless there are underlying mechanisms that can support them. Counterfactuals without mechanisms are vacuous. Mental processes of intuition and imagination in thought experiments are often affected more by emotion and culture than by objective rationality.

Instead, we can use increasing knowledge about how minds and societies work in order to make evidence-based judgments about justice. Autonomous action does not require full-blown metaphysical free will of the sort rejected in chapter 10, but it does require people to have physical and social circumstances that enable them to figure out how best to run their lives.

The crucial bridge between observations and values in the study of justice is vital needs, to be satisfied if people are going to function as human beings. A just society meets both the biological needs of all its members for water, food, shelter, and health care and the psychological needs for relatedness, competence, and autonomy. Such needs are much more fundamental that individual variations in wants, desires, whims, and preferences.

Human rights are neither god-given nor self-evident but can be identified by what is required for the satisfaction of vital needs. Justice does not require complete equality of wealth, income, or preference satisfaction, as long as people are equal in having their vital needs satisfied. Enough is enough. Inequality is not evil in itself and is not completely eliminable given variations in abilities and luck. But all people should be equal in satisfaction of vital needs sufficient to enable them to lead the lives of which they are capable, and a just society is morally obliged to meet these needs.

The needs-sufficiency view of social justice has strong implications for establishing political and legal justice, including taking into account the needs of future generations. To contribute to social justice, the political system in a country needs to support the population's vital needs. Democracy is the best available system for accomplishing this support, in contrast to the known inadequacy of dictatorship and the untried and infeasible prescriptions of anarchism. Within democracy, different parties offer competing ideologies and strategies, but social democracy has provided the greatest contribution to general needs satisfaction. Justice is not just the concern of individual countries but requires considering international organizations that can promote interconnected needs across borders.

The promotion of social changes that increase justice can be accomplished by attention to mental and social mechanisms such as empathy. One strong candidate for a major change is a basic income that will increase satisfaction of biological needs such as food and psychological needs such as autonomy.

I omitted a three-analysis of justice earlier because the concept seemed too ideologically contestable. The needs-sufficiency theory of justice defended here stabilizes the concept to allow the three-analysis of *justice* in Table 7.1 Complementarily, Table 7.2 provides a three-analysis of *injustice*, which is just as important for influencing actions.

The concepts of justice and injustice together provide a basis for understanding and judging social policies and actions. Whenever past, current, or planned actions

TABLE 7.1

Three-Analysis of the Concept *Justice*

Exemplars	Countries such as Denmark, policies such as universal health care, actions such as promoting women to responsible positions
Typical features	Sufficient satisfaction of vital needs, equal, fair
Explanations	Explains: why some countries are better than others; why some policies are right or wrong
	Explained: by human nature consisting of physiological, mental, and social mechanisms

fit better with the exemplars, typical features, and explanations of injustice than with those of justice, then we can judge them to be immoral.

Vital needs connect the individual morality discussed in chapter 6 with social justice. Just as evaluation of the actions of individuals depends on the consequences of actions with respect to needs satisfaction, the evaluation of social practices and policies as just or unjust depends on consequences for needs.

The theory of justice as needs-sufficiency connects with the theory of mind in chapter 2 because it relies on the same cognitive-emotional mechanisms. It also fits with the coherentist epistemology in chapter 3, relying on evidence rather than pure reason. Finally, by working with real people rather than idealizations, this approach to justice aligns with the multilevel materialism of chapter 4, avoiding any supernatural requirements of justice found in religious theories. Justice requires multilevel emergence from social interactions among people whose minds combine molecular, neural, and psychological mechanisms. By considering how to

TABLE 7.2

Three-Analysis of the Concept *Injustice*

Exemplars	Countries such as Nazi Germany, policies such as slavery, actions such as jailing protesters
Typical features	Interferes with satisfaction of vital needs, unequal, unfair, discriminatory
Explanations	Explains: why some countries are worse than others; why some policies are wrong
	Explained by: breakdowns in physiological, mental, and social mechanisms

increase the general satisfaction of vital needs, the natural philosophy approach to justice applies the normative procedure described in chapter 1, warranting social changes such as the introduction of basic income.

The needs-based approach to morality and justice also serves to provide an answer to the perennial question of the meaning of life. Chapter 8 argues that personal and social meaning come from the satisfaction of vital needs for oneself and others.

NOTES

The top 1% own more than the bottom 99%: https://www.theguardian.com/business/2016/jan/18/richest-62-billionaires-wealthy-half-world-population-combined.

On philosophical theories of justice, see Rawls 1971, Sen 2011, and http://www.iep.utm.edu/justwest/.

Daniels 2016 reviews reflective equilibrium.

The Trudeau quote is from Trudeau 1998, pp. 18–19.

Narveson 2001 defends the libertarian idea. Marshall 2010 gives a history of left-wing anarchism. For value maps of anarchism and other ideologies, see *Mind–Society*, chapter 6.

Orend 2002 connects rights to needs, drawing on Wiggins 1987. See also Thagard 2010b. Doyal and Gough 1991 examine needs in relation to social justice. Heller 1976 discusses Marx on needs.

Arneson 2013 reviews equality including the sufficiency view. Atkinson 2015 and Stiglitz 2013 discuss the costs of and remedies for inequality. Marmot 2004 documents that social status is a major cause of illness. See also the World Inequality Report for 2018: http://wir2018.wid.world. Thagard 2018 provides a cognitive-emotional explanation of conflicting views on equality.

Wilkinson and Pickett 2010 chart the leveling of income on happiness and life expectancy with national income per person over about US$25,000. Stevenson and Wolfers 2013 dispute a satiation point for well-being with increasing income.

The following are the studies used as approximations to need satisfaction:

Human development: http://hdr.undp.org/en/2016-report

Freedom: https://www.fraserinstitute.org/sites/default/files/human-freedom-index-2016.pdf

Happiness: http://worldhappiness.report/ed/2016/

Hunger: http://www.worldhunger.org/2015-world-hunger-and-poverty-facts-and-statistics/

Health care: http://www.commonwealthfund.org/publications/fund-reports/2017/jul/mirror-mirror-international-comparisons-2017

Another relevant index concerns countries' commitments to reduce inequality, from Sweden at the top to Nigeria on the bottom: https://oxfamilibrary.openrepository.com/oxfam/bitstream/10546/620316/31/rr-commitment-reduce-inequality-index-170717-en.pdf.

Piff et al. 2012 document that upper classes are more unethical than lower classes.

On the capabilities approach, see Sen 1999, Nussbaum 2000, and Robeyns 2016. Gough 2015 argues that needs are more fundamental than capabilities in shaping responses to climate change.

Radcliff 2013 connects politics with happiness.

Meyer 2015 reviews justice concerning future generations.

Piketty 2014 examines wealth and the need for redistribution.

Morell 2010 shows the relevance of empathy to democracy.

Mullainathan and Shafir 2013 report studies on how scarcity affects cognition. Cappelen et al. 2014 report different brain responses to inequalities depending on whether they correspond to different contributions to work effort.

Van Parijs and Vanderborght 2017 make the case for basic income. Segal 2016 proposes a basic income pilot program that Ontario implemented in 2017 but cancelled in 2019. Forget 2011 describes the Dauphin experiment.

PROJECT

Conduct three-analyses of concepts such as *equality*, *liberty*, and *democracy*. Develop better ways of measuring needs sufficiency in individuals and countries. Design international organizations that promote satisfaction of vital needs. Describe how power relations based on coercion, benefits, respect, and norms can affect justice.

8

Meaning

LIFE AND LANGUAGE

Major causes of disease and early death include tobacco use, poor diet, inactivity, and stress, but research has found that an equally important factor is lack of purpose in life. People who have senses of directions and purposes that keep them from wandering aimlessly through life have a much-reduced risk of heart attack, stroke, and Alzheimer's disease. Moreover, having a strong purpose in life is associated with better sleep, relaxation, and sex, along with less depression, hospitalization, and abuse of drugs and alcohol.

When people have purposes such as satisfying work and caring families, their lives are meaningful to them, which raises the classic philosophical question: What is the meaning of life? This question is both descriptive and normative, concerning what purposes people find worthwhile but also asking what purposes are ethically defensible. Nazis pursued goals such as exterminating inferior races, but their actions violated the vital needs of millions of people and therefore fail ethical standards of meaningfulness. Many people, such as Tolstoy in his later years, look for meaning and purpose in religious activities, but secular targets include the pursuit of love, work, and play. Through social cognitivism, natural philosophy should offer a plausible account of the meaning of life.

An oddity of professional philosophy in the last hundred years is that much more writing has been directed at the meaning of language than about the meaning of life. Philosophy is supposed to be the love of wisdom, and much more wisdom is needed to figure out why life is worth living than to work out details of language. Another oddity is that in English and some other languages (French, Spanish, Hebrew) the same word (e.g., "meaning") is used to cover both the meaning of words and the meaning of life. Many other languages use different words for what might seem to be different concepts. A third oddity is that the meaning of language and the meaning of life are never discussed together, as if they are completely separate issues.

This chapter proposes theories of meaning in language and in life that display surprising interconnections. In strong contrast to supernatural theories of both kinds of meaning, my accounts are rooted in natural mechanisms that are mental, biological, and social. Both kinds of meaning are examples of multilevel emergence, in which significance derives not from the heavens or from one simple thing or process but from complex interactions of molecules, neurons, mental representations, and social interactions.

After outlining fundamental questions concerning meaning and alternative answers to them, I consider the meanings of words and sentences in language from the perspective of the Semantic Pointer Architecture. As chapter 2 described, semantic pointers show how neural representations can be connected both with the world and with each other, providing a two-dimensional account of the meaning of brain processes that carries over into a rich account of language that accommodates both sense and reference. But linguistic meaning is not just in individual minds because it depends also on social interactions. The semantic pointer theory of communication provides the connection between the biological operations of meaning in individual brains and the social processes of meaning development. The result is a fully natural, three-dimensional account of the meaning of language, covering relations that are word to word, word to world, and word to group.

Analogously, I show how the meaning of life is not supernatural but rather rooted in the operations of emotional brains, where emotions are understood as semantic pointers tied to vital needs. In keeping with the discussions of morality and justice in chapters 6 and 7, the meaning of life also has an important social dimension resulting from the interactions we have with other people and the ways we care about them. So my account of the meaning of life is also multidimensional, tying together the world, individual brains, and interactions among people based on communicative interactions. Biological, mental, and social mechanisms generate human needs whose satisfaction or attempts at satisfaction make human lives meaningful without mystical aspirations.

ISSUES AND ALTERNATIVES

For both language and life, the fundamental question is: where does meaning come from? For language, this breaks down into questions about the meaning of particular portions, including words, sentences, and larger structures such as stories. How do words, sentences, and stories get their meaning? For the lives of human beings, there are also meanings operating at different scales. What makes your life meaningful at precise moments, over stretches of time such as a year, and for your whole existence?

It seems that the meanings of stories depend on the meanings of sentences, which depend on the meanings of words; but what is the nature of that dependency? Are there similar dependencies in meaningful lives so that the meaning of your whole existence breaks down to the meaning of particular moments? For both language and life, perhaps the meaning of the whole is greater than or at least different from the aggregate of the meanings of the parts, through processes that are mysterious until the underlying mechanisms are identified.

The major alternative ways of answering these questions rely on taking meaning to be either natural or supernatural. According to Plato, the meaning of language derives from supernatural sources in his heavenly forms. Similarly, many religious thinkers have assumed that the meaning of life is ultimately supernatural, based on the will and plans of their gods. Existentialists such as Camus reacted to doubts about the existence of God by wondering whether lives could be meaningful at all.

The task for natural philosophy is to work out how meaning can arise both locally and globally from physical processes. It seems mysterious how people can take arbitrary symbols like "cat" and use them to say things about the world that make sense to other people. Even more mysterious is how people adrift from religion can understand their lives as significant and valuable.

The lure of simplicity has inclined philosophers to look for just one source of meaning, for each of language and life. An example of a one-dimensional theory of meaning of language sees words as getting their meaning just from one kind of relation such as those with other words, somehow scaling up to cover the meaning of larger linguistic entities such as stories. Similarly, a one-dimensional theory of the meaning of life emphasizes a single factor such as love or work as what makes human lives valuable.

In contrast, this chapter develops accounts of the meaning of language and life that combine several dimensions, including important social aspects of both. Neither the meaning of language nor the meaning of life reduces to mentality alone, because both must also consider the world and other people. Such complexity makes meaning hard to pin down, but clarity comes from identifying the mechanisms that interweave multiple sources of meaning. The semantic pointer theories of thought and communication provide the relevant mechanisms in both cases.

LANGUAGE AND MENTAL REPRESENTATION

Should we use language to understand mind or mind to understand language? Analytic philosophy historically assumes that language is basic and that mind would fall into place if proper use of language was appreciated. Modern cognitive science, however, rightly judges that language is just one aspect of mind of great importance in human beings but not fundamental to all kinds of thinking. Countless species of animals manage to navigate the world, solve problems, and learn without using language, through brain mechanisms that are largely preserved in the minds of humans. So there is no reason to assume that language is fundamental to mental operations.

Nevertheless, language is enormously important in human life and contributes substantially to our ability to cooperate with each other in dealing with the world. Our species *homo sapiens* has been astonishingly successful, spreading from a small area of Africa around 100,000 years ago to cover the entire planet. This success depended in part on language, first as an effective contributor to collaborative problem solving and much later, only around 10,000 years ago, as collective memory through written records. Hence a theory of mind must include a theory of language, and philosophy of language remains an important but not primary part of philosophy.

Words and the World

Many philosophers have noticed that meaning has important dimensions that operate within language and also connect it with the world. The distinction is captured by a variety of terminological contrasts: sense/reference, connotation/denotation, and intension/extension. For example, the word "cat" has reference because it denotes cats in the world but also has sense because it is linked to other words such as "animal" and "pet." From a purely linguistic point of view, it is hard to see how these two aspects of meaning get mingled in a single word and how the meanings of words can contribute to the meanings of larger structures such as sentences and stories.

Eliasmith's Semantic Pointer Architecture provides solutions to both these problems. From the perspective of theoretical neuroscience, both word-to-word and word-to-world meaning arise from underlying brain processes. Semantic pointers are patterns of neural firing that are connected to the world through ongoing sensory and motor interactions. Learning enables neural firing to be shaped by sensory inputs from the eyes and other sense organs and also to be interactively affected by the use of the body to modify the world.

Reference is not a matter of some single baptismal event in which someone decides that the word "cat" refers to cats but rather an ongoing process of a person interacting with actual cats or visual and auditory images of them. In some cases, patterns of neural firing may be innate because they resulted from neural wirings that proved useful for survival in previous generations, for example in ensuring fast detection of food or predators.

But meaning is not as purely sensory-motor as overly empiricist theories of meaning have insisted but rather can be built up through relations of words to other words. Semantic pointers explain the construction of meaning as resulting from iterations of binding, as in the statement that cats eat mice, which binds the neural representations for *cat, eat,* and *mice.* Semantic pointers arise not merely by neurons becoming tuned to the world but also by the bindings of semantic pointers into new ones that become farther removed from the world through the neural mechanism of convolution. As chapter 2, *Brain–Mind*, and *Mind–Society* argue, language is both embodied and transbodied, operating through interactions of the brain with the world and with the capacity of the brain to build more remote mental representations such as *catnip* and *catnap.* Transbodied bindings allow the projection of meaning to concepts that totally lack reference to the world, such as *Cheshire cat.*

The mechanism of convolution explains how word-to-world and word-to-word meanings get combined, with the pattern of neural firing for *cat* affected both by a person's sensory-motor experiences of cats and by the associations between "cat" and other words. The various sensory and motor contributors to neural firing are combined by convolution with the verbal aspects, producing a neural pattern that encapsulates the two-dimensional meaning of the word.

Binding by convolution also explains how the meaning of sentences emerges from the meanings of words. As *Brain–Mind* details for both beliefs and images, brains can build up complex structures by recursive binding, as in sentences with high degrees of embedding such as "the black cat chased the brown mouse." The meaning of language is compositional because the underlying brain processes are compositional.

Words and Society

Many philosophers and linguists have noticed that the meaning of language does not just come from individual minds confronting the world. People rarely make up their own languages but rather learn how to speak the languages that are practiced by their families and friends. Both the word-to-word and word-to-world dimensions of meaning are substantially affected by interactions with other people. We learn important conceptual facts such as that cats are mammals by listening to what other people say, and we connect concepts with things in the world when people point out instances to us.

Therefore, meaning of language has a third, social dimension that depends on interactions with other people. For example, I have never seen an aardvark but have acquired the concept *aardvark* through other people's descriptions and depictions of it. Attention to social aspects of language raises an important problem for cognitive science, concerning how meanings can get from the head of one person into the head of another person, operating within large linguistic communities. Meanings locked into only one person would be useless for the kinds of cooperation and collaboration that make humans successful as a species. Purely mental mechanisms are not enough to account for social aspects of meaning, and it might seem that attention to brain mechanism only makes the problem harder, because there is no direct way for one person's brain to affect another's.

Fortunately, the semantic pointer theory of communication developed in *Mind–Society* (chapter 3) fills the gap between neural activity in one person's brain and neural activity in the brains of others. The neural firings of one person's brain affect the neural firings of others through multimodal communication. When people speak to you, you not only hear the sounds they produce but also see their facial expressions, gestures such as pointing, and body language. As *Mind–Society* (chapter 12) maintains, the key to transferring some approximation of neural activity from the brain of one person to another is to use available resources to produce approximately the same bindings of concept-to-concept and concept-to-world information compressed in semantic pointers. The relevant information can include emotional values as when a child learns that cats not only look and sound a certain way but can also be viewed as charming or alarming. Stories acquire emotional meaning not just as the aggregate of the emotions in individual sentences but through the interaction of the sentences with each other and the embodied attitudes of the storyteller. Such communicative interaction produces the word-to-group dimension of linguistic meaning.

The three natural dimensions of linguistic meaning show that it is not the product of supernatural entities such as Platonic forms or other abstractions but rather results from the interactions that people have with the world and with each other. The meanings of words derive from the meaning of concepts through neural patterns of firing that merge sensory and motor inputs with other neural patterns formed in part through interpersonal exchanges.

The apparently isolated neural processes in different people's brains cannot be transferred exactly, but they can be approximated by forms of communication that combine the utterance of words and speech acts with a wide range of nonverbal information including pointing. Then meaning is clearly not a thing or content but rather a complex process that requires a combination of mental and social mechanisms, all of them interacting with the molecular and neural mechanisms that make brains work.

Meanings are emergent because they belong to entities such as concepts, words, sentences, and stories without being just aggregates of the meanings from their components. For example, "cat" has meaning not found at all in "c," "a," and "t." More richly, the meaning of a sentence goes beyond the meanings of the words in it because it depends on the interactions of concepts; and a whole story is not simply the sum of the meanings of the sentences in it, because it depends on the interactions of the sentences such as their inferential relations. The two sentences "John loves Mary" and "Mary loves John" have different meanings because what matters is not just the aggregate of the meanings of "John," "loves," and "Mary" but rather their interactions. Similarly, the meaning of a novel such as *Uncle Tom's Cabin* is not just the sum of the individual sentences but arises from interactions that generate cognitive and emotional interactions to social situations such as slavery. I now show how the meaning of life, although very different from the meaning of language, is similarly a process of multilevel emergence.

THE MEANINGS OF LIFE

Nobody wonders whether language has any meaning, because it is obvious that people generally understand each other, with identifiable exceptions such as nonsense poems and postmodernist literary theory. In contrast, some philosophers and other writers have seriously worried about whether life has meaning. Doubts arise from the demise of religious views that placed meaning in the hands of God and from the general despair that occurs in people's lives because of suffering and the occurrence of mental disorders such as depression. Whether life has meaning is not just a philosophical issue, because medical research finds that people with a greater sense of purpose have longer lives, better sleep, and less risk of heart disease, stroke, and Alzheimer's disease.

Fortunately, there are good reasons for maintaining that people's lives can be highly meaningful, rendering doubts about the meaning of life as pointless as general doubts about the meaning of language. As with morality in chapter 6 and justice in chapter 7, the key to understanding the meaning of life is the concept of vital needs. Meaning comes from satisfying human needs and also from attempting to satisfy needs. Just as morality and justice are concerned with meeting the needs of other people, not just one's own, a major component of having a meaningful life is striving to contribute to satisfaction of the needs of other people.

Nihilism

Nihilism, the view that life is meaningless, is the counterpart of skepticism in epistemology, the claim that people possess no knowledge at all. Skepticism usually arises because of an illusory view of knowledge as requiring certainty through absolute foundations. Realization that such foundations are unavailable provokes the slide into skepticism. As chapter 3 showed, this slide is avoidable through more reasonable and empirically grounded ways of achieving knowledge by procedures that generate reliable coherence.

Nihilism often arises because of an illusory view of life based on religion. Spiritual leaders such as those in Christianity and Islam have strong interests in convincing people that ultimate meaning derives from the eternal glory that devotion to their doctrines will bring, not from daily actions. Then the key to having a meaningful life is to downplay the vicissitudes of everyday existence and attend to the long-term prospects of divine reward. Some religious organizations have allied with governments to combine allegiance to the church with acceptance of oppressive regimes. *Mind–Society* (chapter 8) analyzes the numerous emotional and social attractions of religious systems.

As soon as people appreciate that evidence is better than faith, the reassurance that comes from thinking of themselves as part of God's plan evaporates, and people can feel lost in the universe. Writers like Kierkegaard, Dostoyevsky, and twentieth-century existentialists suffered dread that without God life becomes bereft of meaning. However, just as the cure for skepticism is finding a better epistemology, so the cure for nihilism is finding a better perspective on life that is coherent with ethics and justice based on vital needs.

Does Everything Happen for a Reason?

When people have to cope with difficult situations in their lives, they sometimes reassure themselves by saying that everything happens for a reason. For some people, thinking this way makes it easier to deal with relationship problems, financial crises, disease, death, and even natural disasters such as earthquakes. It can be distressing to think that bad things happen merely through chance or accident. But they do.

The saying that everything happens for a reason is the modern, New Age version of the religious saying: "It's God's will." The two sayings have the same problem—the lack of evidence that they are true. Not only is there no good evidence that God exists, as chapter 4 argued, but we have no way of knowing what it is that he (or she) wanted to happen, other than that it actually did happen. Did God really will that hundreds of thousands of people die in the 2010 earthquake in Haiti, one of the world's poorest countries? What could be the reason for this disaster and the ongoing suffering of millions of people deprived of food, water, and shelter? Why do people find it reassuring that the earthquake happened for a reason such as the will of God, when such terrible events suggest a high degree of malevolence in the universe or its alleged creator? Fortunately, such events can alternatively (and with good evidence) be viewed as the result of accidents, and possibly even of chance.

The idea that chance is an objective property of the universe was advocated in the nineteenth century by the great American philosopher Charles Sanders Peirce, who called this doctrine *tychism*, from the Greek word for chance. Scientific support for the doctrine came in the twentieth century with the development of quantum theory, which is often interpreted as implying that some events such as radioactive decay are inherently unpredictable. Some cancers are caused by genetic and environmental factors, but random errors in DNA replication are also a major contributor.

Even if events that affect human lives do not happen by quantum chance, many of them happen by accident, in the sense that they are the improbable result of the intersection of independent causal chains. For example, the thousands of earthquake deaths in Haiti came about because of the results of several causal chains, primarily (a) the historical events that led to millions of people living near Port-au-Prince and (b) the seismic events occurring in the tangle of tectonic faults near the intersection of two crustal plates. These deaths were accidental in that the intersection of the unconnected causal chains was unpredictable. Neither history nor seismology are random, but their intersections often are so unforeseeable that we should call them accidental.

The doctrine that everything happens for a reason has intellectual variants. The German philosopher Hegel maintained that in historical development the real is rational and the rational is real. Similarly, before the 2008 meltdown in the financial system, it was a dogma of economic theory that individuals and markets are inherently rational (*Mind–Society*, chapter 7). Some naïve evolutionary biologists and psychologists assume that all common traits and behaviors must have evolved from an optimizing process of natural selection.

In history, economics, biology, and psychology, we should always be willing to consider evidence for the alternative hypothesis that some events occur because of a combination of chance, accidents, and human irrationality. For example, Keynes attributed financial crises in part to "animal spirits," by which he meant the emotional processes that can make people swing between irrational exuberance and pessimistic despair.

But if the real is not rational, how can people cope with life's disasters? Fortunately, even without religious or New Age illusions, people have rich psychological resources for coping with the difficulties of life. These include cognitive strategies for generating explanations and problem solutions and emotional strategies for managing the fear, anxiety, and anger that naturally accompany setbacks and threats. Psychological research has identified ways to build resilience in individuals and groups, such as developing problem-solving skills and strong social networks. Life can be highly meaningful even if some events are just accidents, as I now show.

Needs and Meaning

Merely satisfying wants and desires is no road to a meaningful life. There are many things that I enjoy, such as dark chocolate, red wine, and baseball games. But I would hardly consider my life to be meaningful if it consisted only of accumulation of pleasurable experiences. Chapter 6 made the crucial distinction between instrumental wants and vital needs, which are much more fundamental as prerequisites of human life. Vital needs include not just the biological ones for food, water, shelter, and health care but also the psychological needs for relatedness, competence, and autonomy. Life's meaning accrues not just from the satisfaction of these needs but even from their pursuit, which provides people with purposes.

Hunter-gatherers and homeless persons have to spend much of their days in pursuit of the food that is crucial for the continuation of their lives. For most people who are not so desperate, a greater proportion of life can be devoted to psychological goals. Relatedness is a vital need satisfied through a variety of personal affiliations, including romantic partners, families, friends, and work associates. *Mind–Society* (chapter 4) describes how minds become interdependent in that their actions influence each other's emotions. Sometimes these effects are negative, when other people make you miserable, but interactions from conversation to laughter to hugs to sex can provide the crucial contacts that satisfy people's needs for belonging.

The psychological need for competence can be satisfied by the pursuit of different kinds of achievements and accomplishments. For many people, work is the best way of satisfying this need, as long as the work is viewed as intrinsically valuable. Mind-numbing, tedious, and oppressive jobs obstruct the need for competence. There are three kinds of work: jobs, careers, and callings. Mere jobs help to satisfy biological needs for food and shelter, without contributing to the satisfaction of psychological needs. In addition to thwarting competence and achievement, jobs may diminish the vital need for autonomy, although even stultifying jobs can provide relatedness to people who share the worker's predicament.

In contrast, careers and callings furnish appreciated accomplishments. The difference between these two kinds of work is that callings are activities that one would do even if they did not provide a career's financial rewards which yield food and shelter. Callings deliver more motivation and engagement than careers, which are good for satisfying all biological needs and some psychological needs but lack the intense involvement of callings. Work amounts to a calling for people who would find it difficult to do anything else. Exemplars of callings are professions such as being a doctor, professor, or therapist, fascinating pursuits such as being an astronaut, and artistic activities such as being a painter, musician, or novelist. Callings may not be financially lucrative, but they often come with high degrees of autonomy.

The need for competence can be satisfied in ways other than work, for example by hobbies and recreational activities. Becoming proficient at woodworking, playing an instrument, producing a show, or shooting a basketball can satisfy the need for competence without being part of a career or calling. These activities fall more under the category of play rather than work, but play is not just a frivolous activity of children—it can help people of all ages enjoy and value their lives by accomplishing goals.

Autonomy does not require the absolute free will untrammeled by causal influences that would contradict known mechanisms of mind. Rather, the two main requirements for autonomy are freedom from external coercion and from internal mental illness. Coercion arises from social mechanisms of power coming from threats of physical harm but also from subtler forms of domination through social norms, as described in *Mind–Society* (chapter 6). The limitation on autonomy from mental illness results from breakdowns of molecular, neural, psychological, and social mechanisms laid out in *Mind–Society* (chapter 10).

Autonomy requires more than just the freedom to make rational choices that maximize preference satisfaction, because preferences may derive from defective goals. Rather, autonomy should enable pursuit of other vital needs for relatedness and competence, not wants that result from hedonistic desires or social manipulation.

Life Is a Constraint Satisfaction Problem

Satisfying psychological needs can require much balancing, for example when needs for relatedness or competence conflict with needs for autonomy. For example, a romantic relationship may help to satisfy the need for belonging but diminish autonomy because of requirements to accommodate the other person. There is no recipe or simple algorithm for how to solve the constraint satisfaction problem that requires trade-offs among relatedness, autonomy, and competence.

Individuals differ in how they achieve balance among needs at different stages of life. Children, for example, have a high degree of relatedness when they are being cared for by their families, a low degree of autonomy initially, and only a modest amount of competence that can be provided through play and school. Adolescents acquire greater autonomy and capacity for competence as they get older, but degree of relatedness can depend on hard-to-control circumstances concerning family functionality and romantic relationships. Older people who retire can secure a high degree of autonomy but may suffer from shortages of competence and relatedness.

Moreover, cultures differ in the extent to which they emphasize different psychological needs. Western society has an individualist bias to autonomy and competence over relatedness, whereas Asian societies tend to stress the importance of relatedness and social roles.

In my 2010 book, *The Brain and the Meaning of Life*, I used the slogan that the meaning of life is love, work, and play. Behind the slogan is the recognition that autonomy, relatedness, and competence are vital needs, and the observation that love, work, and play are major contributors to the satisfaction of these needs. As chapter 6 argued, these needs differ from arbitrary wants because they are tied into biological and social mechanisms. You cannot change these mechanisms, but finding activities that work with them to accomplish nonarbitrary goals can provide a meaningful life, without theological direction or the illusion that everything happens for a reason.

Meaningful goal pursuit does not always have to be aimed at the big three vital needs of autonomy, competence, and relatedness. The need for competence can be achieved through the subgoal of productive work, which can generate a whole series of sub-subgoals. For example, my subgoal today is to work on revising this chapter, which is a meaning-providing objective even though it is remote from the general need for competence and the even more general issue of life's meaning.

The Needs of Others

Most discussions of the meaning of life are individualistic, concerned with how a particular person can achieve significance. I claimed that a meaningful life arises from the pursuit of vital need satisfaction. But I have also maintained that natural philosophy needs to be inherently social, not by reducing the individual to the social but by connecting mental mechanisms with social ones. Accordingly, chapter 3 considered social dimensions of epistemology, chapter 4 defended the reality of groups, chapter 5 rejected reduction of the social to the mental, chapter 6 argued that people have an obligation to care about the needs of others, and chapter 7 took into account social needs in the consideration of justice. Similarly, the problem of the meaning of life is transformed by taking the social more seriously.

My interpretation of the meaning of life is already social to a large extent, because one of the vital needs that people can aim to satisfy is relatedness, which requires connections with other people. Emotional dependency on another is not an affliction to be fled, as ultra-individualism insists, but rather an ineliminable aspect of normal human emotional mechanisms. But relatedness can still be viewed egoistically, if you are only concerned with your own need for belonging without attention to the needs of other people.

I argued in chapter 6 that people are not only obligated to care about the vital needs of others; they naturally do so through three modes of empathy, using mirror neurons, analogies, and multimodal rule simulations. These capacities point toward another important way in which people's lives become meaningful, through caring about the needs of others and helping them to meet their needs. A major part of the meaning that accrues in most people's lives results from caring for others, not just as a feeling for them but also as concrete acts of helping them to satisfy their own vital needs.

For many people, meaning comes from raising children. I was initially dismayed when I read that couples without children are by and large happier than couples with children, because of all the emotional and financial strains of being a parent. Then I realized that the point of having children is not to make yourself happy but to make them happy, or more generally to help them develop as human beings who can satisfy their own needs. Then meaningfulness is different from happiness when it focuses on meeting the needs of others. Your life acquires value by enhancing the value of the lives of others by love, friendship, and compassion.

Purpose stems not just from pursuing your own goals but also by helping other people pursue theirs.

Meaningful lives come with positive emotions such as happiness, love, pride, gratitude, and compassion. But negative emotions can also contribute to meaning, for example the sadness that comes from bereavement for a loved one and the angry indignation that motivates efforts to overcome oppression.

A survey of almost 400 Americans provided evidence that clearly discriminates between happiness and meaning. It found that feeling happy is correlated with a life that is pleasant and free from difficult events such as bad health, but this kind of feeling good is not tied with a sense of purpose. Spending time with people you love is correlated with greater meaning but not with happiness. People who describe themselves as givers in their relationships had less happiness but more meaning compared to people who describe themselves as takers.

A giving relationship is one where the primary concern is with satisfying the needs of others. Besides parenting, such caretaking can occur in other kinds of relationships. Looking after a sick or disabled spouse is not a pleasant job, but it undeniably adds purpose to one's life. There are caring professions such as medicine and education where dedicated people devote themselves not just to exercising their own competence but also to caring deeply about the needs of other people, helping them to become full human beings.

Therefore, the meaning of life has a social dimension stronger than just having relationships with other people. Relationships are particularly meaningful when they do not just satisfy your own emotional and physical needs but also when they furnish the opportunity to help satisfy the needs of others. Being a giver rather than a taker may not make you happier, but it can help to provide you with a meaningful life. Caring for the vital needs of others as well as yourself distinguishes ethical purposes from the illegitimate goals of Nazis, psychopaths, and narcissists.

Meaning as Multilevel Emergence

The analogy proposed earlier between the meaning of language and the meaning of life is useful for determining what makes a whole life meaningful rather than just particular parts of it. I argued that the meaning of language is both compositional and multilevel. It is compositional because the meanings of words contribute to the meanings of sentences, which contribute to the meanings of stories. Emergence occurs because the meaning of larger structures is not simply an aggregate of the meanings of parts but rather depends on the interactions of the parts.

Similarly, the meaning of a whole life is not just the sum of the meaningfulness of particular episodes. An event such as a meeting between two people can indeed be meaningful through its contribution to their vital needs. For example, the first encounter between two people who eventually become romantic partners is full of meaning, because it led to some satisfaction of their vital need for relatedness. But such moments have even more significance when part of an overall relationship that consists of many moments over months, years, or decades.

Looking at life as a whole generates the difficult question: Overall, does my life add up to a high degree of satisfaction of vital needs for me and for the other people I care about? Answering this question requires not simply adding up the various needs that have been satisfied but also considering ways in which balancing various constraints adds up to a good solution to the constraint satisfaction problem of how to live your life. Sometimes people need to trade off love for work and work for love, balancing autonomy and competence against relatedness. The mark of a meaningful life is not just the aggregate of love satisfaction and work satisfaction but depends on how the balance has been carried out through interaction between love and work. Hence the meaning of life emerges from its components, just like the meaning of language.

The meaning of life also exhibits emergence of an even more complicated kind, because, like language, it depends on mechanisms operating at different levels. *Brain–Mind* (chapter 10) describes how language depends on mechanisms operating at molecular, neural, psychological, and social levels. Language is molecular because there are genes such as *FOXP2* that are implicated in people's ability to process syntax, and language is neural because the mechanisms for meaning and composition are best explained by the operations of semantic pointers. But language is also psychological and social, because people use it to interact with each other in ways that approximately transfer mental representations such as beliefs and attitudes.

Similarly, meaningful lives depend on molecular mechanisms such as the neurotransmitter/hormone oxytocin that mediates human interactions and on neural mechanisms such as the formation of multimodal images (pictures, sounds, touches) that people have of each other. Psychological mechanisms include inferences about how to live and how to understand other people. By virtue of the vital need for relatedness and the ways that the needs of others provide valuable pursuits, the meaning of life depends on social mechanisms of interpersonal communication. As for language, the emergence of rich meaning requires interactions of all of these levels of mechanisms, for example when a friendly communication generates trust by stimulating oxytocin.

The wrong way to understand complexity of this kind of system is to privilege one level, attempting to reduce the social to the psychological, or the psychological to the neural, or the neural to the molecular. The vital needs whose satisfaction contributes to the meaning of life operate at four interacting levels. The biological needs for food and water are largely physiological, but equally vital needs for shelter and health care are more complicated because they depend on people's interpretations of housing and disease. Meeting all of these needs requires the cooperation of human beings, making them highly social and dependent on the representations that people have of themselves and the world, tied to underlying neural and molecular processes. The meaning of life is enhanced by the meaning of language when we use narratives to understand our joys and tragedies.

Even more clearly, you cannot satisfy your psychological needs for relatedness, competence, and autonomy without the functioning of mechanisms that include neurotransmitters (molecular), semantic pointers (neural), belief acquisition (psychological), and conversation (social). Hence, both the meaning of life and the meaning of language display multilevel emergence depending on interacting mechanisms.

Obvious differences remain. The two kinds of meaning attach to different entities at different scales: words versus actions, sentences versus life stages, and stories versus whole lives. There can be life without language in the case of brain damage and meaningful language without meaningful life in the case of severe depression. Normally, however, language and life interact to support meaning in both realms.

THE MEANING OF DEATH

Some psychologists have asserted that people are heavily motivated by fear of their own mortality. This claim may well describe large numbers of people, not just Woody Allen, but is it normatively correct? Is it rational to fear death? How might this philosophical question be given an evidence-based answer?

It is still commonly believed that being rational is at odds with being emotional, but emotions such as fear can often be quite reasonable. For example, if a hurricane is predicted in the area in which you live, it is rational to fear the damage that can result, and evidence is accumulating to support fears about drastic declines in the environment resulting from climate change. On the other hand, fear about some potential event is irrational when there is no evidence that the event will actually threaten a person's well-being. Developing a strong fear of the earth being hit by a huge asteroid is currently irrational because there is no evidence that an asteroid strike is imminent. Is death like the hurricane or like the asteroid strike?

More than 2,000 years ago, the Greek philosopher Epicurus constructed an argument against fearing death that has since become even more plausible: "Death, therefore . . . is nothing to us, seeing that, when we are, death is not come, and, when death is come, we are not." Epicurus was one of the first atomists who believed that everything consists of material entities and that there are no souls that survive death. If your life ends at death, then you have nothing to fear, because there will be no YOU to experience pleasure or pain. It's all over when it's over.

Of course, there are other aspects of dying that are worth dreading, such as disease, disability, unfinished projects, and the distress of people who care about you. But from the philosophical perspective that there is no life after death, death itself is nothing to fear.

As chapter 2 reviewed, evidence is mounting that Epicurus was right that minds are material processes rather than supernatural souls. Cognitive neuroscience is rapidly developing experiments and theories that support the claim that brain mechanisms provide the best explanation of people's capacities for perception, reasoning, language, and even consciousness. If the mind requires the brain, then there is no mind to experience suffering of any kind when the brain stops functioning at death. Hence Epicurus was right that there is nothing to fear. If there were any good evidence that life does survive death, then we would have to reject Epicurus' conclusion, but phenomena such as near-death experiences and séances can easily be explained away.

The fear of death persists as a vestige of religious views that proclaim that life on earth is just a fragment of the existence of an eternal soul. Then religion becomes a solution to a problem that it itself created: You may be able to decrease your fear of death by believing that you have found the right religion that will ensure that your afterlife will be pleasant. Thus religion allows a person to careen from the fear-driven inference that death is threatening to the motivated inference that it will not be so bad in the afterlife. Of course, the happy resolution assumes that you have picked the right religion.

If I believed that life survives death, then I would be terrified at the prospect of an eternity of suffering, because I would have no way of knowing which religious beliefs to adopt. In addition to the different main religions such as Christianity, Islam, Hinduism, Buddhism, and Judaism, there are many variants, including dozens of different versions of Catholicism, Protestantism, and Islam. Guessing wrong could lead not only to problems in this life but to eternal punishment. Moreover, it is entirely possible that the "right" religion has yet to be invented.

This variety is one of the flaws in Pascal's famous wager that it is better to believe in God, because if religion turns out to be true, then you get eternal reward, instead of suffering eternal punishment. This wager assumes that you know what religion to bet on. In contrast, let me offer Thagard's wager: it is better not to believe in God, because then you do not have to suffer through a lifetime of worrying about death and finding the right religion! Happily, this wager fits perfectly with rapidly developing evidence that the mind is material. Hence both inference to the best explanation and inference to the best plan support the conclusion that death should not be feared.

SUMMARY AND DISCUSSION

Philosophical problems about the meaning of language and the meaning of life turn out to have interesting commonalities. Neither has plausible solutions that draw on supernatural entities such as abstract meanings, possible worlds, and divine plans. Rather, both can be approached by looking at mechanisms at four different levels: molecular, neural, mental, and social. Meaning is not a thing but a process that depends on interactions of parts occurring at multiple levels, resulting in multilevel emergence.

The Semantic Pointer Architecture illuminates the neural mechanisms that operate in languages and valuable lives. For words, the basic units of meaning in language, semantic pointers integrate two important dimensions of meaning, the relation of words to the world and the relation of words to other words. Words are meaningful because their mental representations as concepts are brain processes that combine sensory-motor interactions with the world and interactions with other concepts. Words are just marks on paper or sounds in the air, but their mental usefulness results from brain mechanisms that use convolution to unite world-derived neural firings with word-derived firings.

Convolution, the compression and binding mechanism that produces semantic pointers, shows how the meaning of sentences can emerge from the meaning of words. Sentences inherit the meaning of the words used to build them but have added significance because of the interactions produced by convolution. Similarly, whole stories can mean more than just the sum of the sentences in them, because of interconnections produced by inference and implication.

The meaning of language is not just two-dimensional, word-based and world-based, because it also has crucial social contributions for learning and applications. This third dimension results from the social mechanisms by which semantic pointers in one brain can modify the semantic pointers in another. The communicative interactions among people are not only verbal but also physical in the form of gestures, facial expressions, body language, and tone of voice. Three-dimensionally, the meaning of language is word-to-word, word-to-world, and word-to-group, where all relations result from semantic pointer mechanisms in brains.

The meaning of life is also three-dimensional, requiring people to interact with language, the world, and other people. You have to operate in the world in order to satisfy your biological needs for food, water, shelter, and health care. These vital needs are not merely desires but are conditions for having any desires at all. The psychological needs of relatedness, confidence, and autonomy are not so starkly prerequisite but are nevertheless required to stave off mental illness and early death and to have a shot at satisfying less central desires.

People's lives become meaningful through the pursuit of vital needs satisfaction, including the social needs tied to relatedness and competence. The nihilistic view that life is meaningless ignores how human nature furnishes people with goals worth pursuing. For natural philosophy, human nature consists of the mental and social mechanisms that explain how people function in the world and with each other.

Just as language is not an individual process with meaning confined to one brain, so the meaning of life is not simply a matter of satisfying your own individual vital needs. A major source of meaning in people's lives is helping to satisfy the needs of other people, both in personal relationships and also in work through beneficial professions. The meaning of your life is not just the sum of individual episodes but the life-long process of balancing your needs for competence, relatedness, and autonomy while ensuring that satisfaction of your biological needs keeps you alive.

A good life requires the application of values and other emotional mental states. To consider your life meaningful, you need to have emotional reactions to its occurrences that are connected to your fundamental values, which are also emotional brain processes. These emotional reactions combine internal perception of physiological changes with appraisals of the relevance of the situation to your goals, integrated in your brain as semantic pointers. The interactions with other people that satisfy the need for relatedness are also emotional, through verbal and nonverbal communication combined by convolution into semantic pointers.

Hence the theory of mind advocated in chapter 2 is a key part of understanding the meaning of life as well as the meaning of language. Like language, life acquires meaning through complex, multilevel processes described by semantic pointer theories of cognition, emotion, and communication. Three-dimensional theories of meaning in language and life cohere well with materialist theories of mind, knowledge, and reality, and also with needs-based theories of morality and justice. Figuring out how to live a life that is ethical, just, and meaningful requires attention to the same biological and psychological basis for the vital needs of yourself and others.

For some people, a major source of enjoyment and meaning is the production and appreciation of art. The next chapter shows how the same mechanisms used to explain the operation of language and the pursuit of meaningful lives also apply to the power of art through beauty and other emotional responses.

NOTES

Strecher 2016 presents research on the health benefits of having a purpose in life.

Speaks 2014 reviews philosophical theories of meaning in language. See also *Brain–Mind* (chapter 10 on language). Eliasmith 2013, 2005 develops neurosemantics, and Plebe and De La Cruz 2016 have a different approach. Frankland and Greene 2015 present evidence concerning neural encoding of sentence meaning.

In formal philosophy, "two-dimensional" has a different meaning concerned with possible world semantics, which I avoid for both metaphysical reasons (chapter 4) and linguistic ones: contrary to the Tarskian approach to meaning, semantics should causally relate syntax to the world, not just to more syntax. See *Brain–Mind* for a systematic critique of syntax-first approaches to language and thought.

Metz 2015 reviews philosophical theories of meaning in life. See also Thagard 2010b. King, Heintzelman, and Ward 2016 cover psychological accounts. On purpose in life, see Kashdan and McKnight 2013. Benatar 2006 argues that it is better never to have been born.

Hegel 1952 said that the real is rational, but *Mind–Society* provides much contrary evidence.

On the difference between meaning and happiness, see Baumeister, Vohs, Aaker, and Garbinski, 2013. On the origins of happiness, see http://voxeu.org/article/origins-happiness. Oishi and Diener 2014 discuss policy implications of studies of happiness and life satisfaction. Deaton and Stone 2014 discuss the complex relationship between happiness and having children. On happiness internationally, see https://ourworldindata.org/happiness-and-life-satisfaction/.

Cacioppo et al. 2015 and Cole et al. 2015 describe the multilevel effects of loneliness from lack of relatedness.

Tomasetti and Vogelstein 2015 provide evidence for randomness in cancer.

Terror management theory concerns how thinking about death affects people: Pyszczynski, Greenberg, and Solomon 1999.

The Epicurus quote is from http://www.epicurus.net/en/menoeceus.html.

PROJECT

Deepen philosophical theories of meaning based on inferential role, conceptual role, and reference using semantic pointers. Build a computational model that integrates the three dimensions of linguistic meaning. Describe the mental and social mechanisms by which psychotherapists can help people to have more meaningful lives.

9

Beauty and Beyond

AESTHETICS

When Bertrand Russell was asked why aesthetics was the one area of philosophy that he never wrote about, he said it was because he did not know anything about art. His friends responded that it had not deterred him from writing about other subjects.

I have already offered interconnected theories for epistemology, metaphysics, and ethics, including political philosophy. The major remaining branch of philosophy is aesthetics, the theory of art. Can social cognitivism be extended to all of the arts, including painting, sculpture, photography, music, dance, literature, theatre, and film? Then its scope would expand across all of the humanities, not just philosophy. This expansion would complete the ambitious project in this *Treatise* to integrate the cognitive sciences (*Brain–Mind*), the social sciences and professions (*Mind–Society*), and the humanities. I previously addressed the natural sciences in my 2012 book, *The Cognitive Science of Science*.

This chapter shows that natural philosophy encompasses aesthetics thanks to the powerful resources of Eliasmith's Semantic Pointer Architecture. Semantic pointers have unprecedented capacity to cover the cognitive diversity of art, because rich patterns of neural firing can apply to the full range of multimodal representations, including pictures, sounds, shapes, motions, and touches produced by sensory-motor inputs and recursive bindings.

Moreover, the semantic pointer theory of emotions enables natural philosophy to embrace all of the powerful emotional impacts of art, including beauty with its dimensions of joy, happiness, and awe. The emotional effects of art extend to many other emotions, including sadness, fear, and surprise, operating in audiences and in creative artists. Semantic pointers cover both the embodied, sensory-motor aspects of such emotions and the goal-oriented, appraisal aspects important to creators and appreciators. Art often functions as a kind of emotional communication between artists and their audiences, through approximate transmission of semantic pointers: appreciators end up with patterns of neural firing for perceptions and emotions similar to those of the artists.

Another theoretical idea valuable for understanding art is cognitive-emotional coherence, understood as parallel constraint satisfaction among elements that cover not only verbal representations such as sentences but also the full range of sensory and emotional images. Consider what makes a clothing ensemble appealing. The parts such as a jacket, shirt, and pants need to be individually attractive, as mentally represented by bindings of shape, color, and pattern with the resulting emotional reactions. But for the ensemble to work as a whole, there needs to be coherence among all of the pieces of clothing, satisfying constraints of fashion such as what colors go together and what patterns are complementary. Similarly, good food combines high-quality ingredients that complement each other in taste, texture, and appearance.

Coherence has long been recognized as important for art with respect to features such as unity and harmony. An ancient theory contends that beauty consists of unity in diversity, which minds can recognize by neural mechanisms of emotional coherence through parallel constraint satisfaction. Other important aesthetic emotions, including sadness, fear, horror, anger, surprise, shock, and interestingness, also exemplify emotional coherence, because the brain needs to put together a combination of sensory interpretation and emotional valuation that satisfies many constraints. These constraints range from perceptual ones concerning what colors, shapes, sounds, and flavors go together to goal-based ones tied to human interests such as love and play. When the unified semantic pointers outcompete other representations, the result is the conscious aesthetic experience of a work of art that integrates sensory and emotional responses.

A theory of aesthetics needs to apply across all of the arts, showing the universality of beauty and alternative emotional reactions. This chapter develops the basic theory with respect to painting, showing how artistic creativity produces new representations by combining old ones through the neural mechanism of binding by convolution. The same account of beauty, other aesthetic emotions, and creativity is then applied to music. Extensions to other areas of art are left for others to explore, although I make some brief remarks about empathy in fiction and film.

Applying ideas about cognition, emotion, and coherence to art might sound like a purely descriptive enterprise, just cognitive science rather than philosophy. Philosophy has an essential normative dimension, asking not only how things are but how they ought to be, and a theory of aesthetics should similarly be able to establish objective norms about good and bad art. I claimed that the primary goals of epistemology are truth, explanation, and human benefit and that the primary goal of ethics and justice is satisfying vital needs. If reasonable goals are identified for art, then different works can be evaluated with respect to how well they contribute, in line with the normative procedure sketched in chapter 1. As with moral judgments, the integral emotional aspect of aesthetic judgments does not prevent them from being objective, because emotions include cognitive appraisals that can be done well or poorly with respect to justifiable goals.

The creation and appreciation of art depend on the cognitive and emotional operations of individual minds, but art is also social because of the communications that take place among artists and their audiences. First, there is the basic communication that takes place from an artist to an appreciator. Second, artists rarely work completely alone but depend on a community of artists for inspiration and emotional support. Third, appreciators of art engage in social interactions with each other, pointing out salient aspects of works of art and sharing their emotional reactions in ways that influence each other. Appreciation of art is often a social enterprise, in gallery openings, musical concerts, and theaters. Hence a full theory of art needs to incorporate social communication along with cognitive and emotional mechanisms.

ISSUES AND ALTERNATIVES

The following are the central questions of aesthetics: What is art? What is beauty? What is the value and power of art? What is the nature of aesthetic judgment and value? What is taste? Are aesthetic evaluations subjectively relative to individuals and to cultures, or is there something objectively valuable about different forms of art?

The question "what is art?" sounds like a request for a definition, but the semantic pointer theory of concepts implies that a better answer comes by a three-analysis. Table 9.1 provides a start, outlining the exemplars, typical features, and explanations that operate across the arts. It lists classes of exemplars rather than specific ones, but it is easy to insert instances that almost everyone agrees are good examples of art. Some of my personal favorites include da Vinci's *Mona Lisa*, Beethoven's *Ninth Symphony*, Rodin's *The Thinker*,

TABLE 9.1

Three-Analysis of the Concept *Art*	
Exemplars	Paintings, sculptures, songs, dances, novels, plays, poems, films
Typical features	Created product, artist, audience, sensory qualities, emotional impact, interesting parts, coherent wholes
Explanations	Explains: why people appreciate some kinds of products, and why some people occupy their lives in creative pursuits
	Explained by: human creativity and emotion based on mental and social mechanisms

King's College Chapel at Cambridge University, Tolstoy's *War and Peace*, Eugene O'Neill's *Long Day's Journey into Night*, Yeats' poem "Aedh Wishes for the Cloths of Heaven," and the movie *Casablanca*. Such examples pin down the concept of art to some extent, even in the absence of a definition that provides necessary and sufficient conditions.

Exemplars lack generality, which can be provided by identifying features that are typical of art even though there may be exceptions. Works of art are products of human imagination, sometimes in the form of physical objects such as paintings but sometimes also in more intangible entities such as musical compositions. There is no need to elevate this feature into a necessary condition, because art such as a lovely piece of driftwood might be simply found. Typically, however, art has both an artist (or, occasionally, a group of artists) responsible for production and another group of people who are intended to be the appreciative spectators or audience. Works of art are creative in that they are novel, valuable, and surprising, as discussed in *Brain–Mind* (chapter 11).

All works of art have sensory effects on human beings through what we can see, hear, or touch. If cooking is considered an art form, then smell and taste can be included. Artistic products also have an emotional impact, ranging from the mild cognitive emotion of interestingness to the overwhelming reaction of awesome beauty. Typically, artworks consist of attention-grabbing parts that add up to a coherent whole with emergent sensory and emotional properties.

The concept of art helps to explain many important aspects of human life, including why some people devote their lives to the production of works of art and also why art contributes so much to the lives of people who appreciate it. The creation and appreciation aspects of art cry out for deeper explanations based on mental and social mechanisms that generate imaginative inspiration and responsive approval. This three-analysis of art is not meant to capture the assumptions of ordinary people but rather to point to understanding of beauty that can be developed through a naturalistic account.

Table 9.1 displays my bias toward cognitive/social explanations of art, but alternative approaches need to be considered. One answer to the question "what is art for?" would be the same as the nihilist one for the meaning of life. Maybe art has no purpose or function and so is not for anything at all. Perhaps it is just a silly pastime of misguided idlers who waste their lives producing art or indolently enjoying it. As with the meaning of life, this desperate view can be put aside in favor of attempts to provide a more promising answer.

One account of the purpose of art fits with religious views such as medieval Christianity and fundamentalist Islam. Then the fundamental purpose of art is to extol the greatness of God, and many great works such as Michelangelo's paintings in the Sistine Chapel and Bach's oratorios have had religious themes and motivations. However, in recent centuries religion has played much less of a role of in the inspiration, production, and appreciation of art, so a purely religious theory of art is no longer sustainable. Rather, religion can be understood as one of the many social and psychological forces that has inspired different kinds of art over the centuries.

Some cultural commentators suppose that the function of art is purely social, through power relations and political purposes such as engaging the masses. My multilevel mechanism approach is fully open to the recognition that art has important social aspects but connects them descriptively and normatively with mental processes in individuals. As I hope is clear in my discussions of knowledge, morality, justice, and meaning, explanations of human activities demand the interaction of mental and social mechanisms.

Another alternative to my social-emotional approach would be a purely cognitive one based on cold, rational judgments. Perhaps an aesthetic judgment is just another proposition, to be evaluated for truth and falsity in accord with its fit with the world and a priori intuition. This approach is implausible because aesthetic reactions are highly emotional as evident in people's use of concepts like *beautiful, astonishing, awesome, grotesque*, and so on. Equally implausible, however, is a narrowly antirational stance that finds nothing in aesthetic appreciation besides raw emotion with no cognitive component. Both purely cognitive and purely emotional accounts make art appreciation and production too individualistic, neglecting the social mechanisms that produce communication.

To show that the social cognitivist approach to natural philosophy surpasses these alternatives as a general theory of art, I need to demonstrate its descriptive and normative force in many domains. I begin with a discussion of beauty in painting.

BEAUTY IN PAINTING

Painting is only one kind of art, and beauty is only one of the emotional reactions that painting generates. Nevertheless, when people think of art they most commonly think of beautiful paintings by artists like Rembrandt, da Vinci, van Gogh, or Monet. The ancient Greeks disputed whether beauty is primarily a matter of harmony and proportion or rather of splendor in the form of color and light.

Semantic pointers show how to combine both these aspects by integrating multimodal representations, coherence, and emotions. The human brain appreciates beauty as a combination of harmony and splendor, of both unity and diversity. I now show how this works through an analysis of van Gogh's famous painting, *The Starry Night*. A similar account could easily be given for many other paintings widely regarded as beautiful, such as Botticelli's *Venus*, da Vinci's *Mona Lisa*, or Monet's *Water Lilies*.

The Starry Night

If you are not familiar with this painting, do a Web search to find a color image. Start by considering the many unusual objects in it. The largest object is the green cypress tree on the left. Although not beautiful in itself, its swirling dynamics make it interesting enough to attract attention. The large moon is a brilliant yellow, strong in both color and light, surrounded by a glowing ring. The eleven stars are smaller and less bright but notable even though they display no clear pattern. The swirling blue sky is an object in itself, distinctive in its energetic swirls and shades of cobalt blue. The hills in the background have an even darker shade of cobalt blue. All of these objects mentioned were visible from van Gogh's asylum room in Provence where he stayed for a year after cutting off his ear. However, the town, including the church steeple portrayed in the bottom-right part of the picture, was not visible from his window but was added from his previous experience of Dutch villages.

The beauty of *The Starry Night* is not just the attractiveness of the individual parts, even though the tree, moon, stars, sky, hills, and church are each lovely. What makes this painting astonishingly beautiful is the way in which they all come together, with the mixture of moon, sky, and stars swirling above the more stable land. How does the mind put it all together? Each object by itself has only a moderate emotional impact, but the overall image generates in many spectators such descriptions as marvelous, gorgeous, and sublime.

The neural mechanisms required for the appreciation of the beauty of such paintings are complex, operating in numerous interconnected brain areas, but they can be understood using semantic pointers and coherence. Individual objects must be perceived using bindings of separate features, including shape, color, and light intensity. For example, perception of the moon as a bright yellow crescent requires a pattern of neural firing that results from bindings of the crescent shape, the bright yellow color, the lighter yellow background, and the darker blue background of the sky.

Seeing the whole sky requires grasping relations between the moon and the stars, using the capacity of semantic pointers to separate the ingredients of spatial and other relations. For example, the relation that the moon is in the sky is not just the aggregate of the firing of neural populations for *moon*, *sky*, and *in*. Using techniques described in *Brain–Mind*, different semantic pointers are generated for *the moon is in the sky* and *the sky is in the moon*, keeping track of what is in what. The same mechanisms enable the brain to see the sky in relation to the tree, hills, and the village, so that perceiving the painting as a whole requires recursive bindings.

However, there are countless paintings and other collections of objects that are produced by recursive binding but that fail to evoke beauty. Looking out my study window, I see trees and buildings that together are attractive but hardly beautiful. What makes *The Starry Night* beautiful as a whole? Isolated semantic pointers can account for local splendor and diversity produced by shape, color, and light but do not explain what makes the whole painting harmonious, unified, and beautiful.

Because of people's experiences with landscapes, there are numerous constraints that generate expectations. *The Starry Night* satisfies many of these constraints, because it places the stars in the sky above the land. But the view from my window also satisfies such basic constraints without approaching beauty. The key differences are first that the ingredients in my window view are not so individually striking, and second that the juxtaposition is not so surprising as seeing all at once the moon, stars, tree, and village. In van Gogh's painting, everything fits together in a way that is both familiar and new.

The Starry Night has large variety in the particular ingredients but great unity in the way that the colors and shapes come together. There is color unity from the use of similar shades in the sky and the hills. The yellow in the stars is very similar to the yellow in the moon, and there is some overlap between the green of the tree in the green in the background to the village. The dynamic swirls in the sky are shapes that accentuate and intensify the swirls in the tree. Therefore, spatial and color constraints are all well satisfied, making the picture highly coherent.

Emotional Coherence

Chapter 8 described how language is compositional and meaningful thanks to the bindings of semantic pointers, with concepts bound into sentences and combined into coherent stories. Similarly, paintings are compositional thanks to the bindings of features into objects and relations and their combination into coherent scenes that satisfy multiple perceptual constraints.

Mere cognitive coherence, however, is not enough to generate the emotional response of beauty. According to the semantic pointer theory of emotions, feelings result from brain mechanisms that recursively bind neural representations of internal bodily states, cognitive appraisals of goal relevance, and multimodal representations of the situation. For beauty, the representation of the painting is produced by visual binding, but what produces the emotional reaction?

For beautiful paintings, there are two primary sources for the joy or awe that people experience. First, there are the local emotional reactions to the ingredients in *The Starry Night*, such as the attractive blue swirl in the middle of the sky and the pleasant little church. These ingredients produce an aggregate pleasure that simply adds up the total of the pleasures of the parts. Second, and just as important, there is the overall coherence that generates positive emotions. The human mind takes inherent pleasure when diverse elements come together coherently in a beautiful object, whereas in ugly scenes the elements produce negative emotions because they fail to fit together. A work of art is sublime if its beauty is so great that it inspires other emotions such as awe and a sense of grandeur.

Thus beauty results from attractive parts combined in a coherent whole, generating an overall reaction of emotional coherence. Like emotions in general, this reaction combines cognitive representation and appraisal with physiological changes that contribute to the conscious experience of the feeling of beauty. *The Starry Night* accomplishes in its viewers a superb combination of lovely ingredients, cognitive coherence through constraint satisfaction, and overall emotional coherence generating positive feelings. Aesthetic reactions and judgments therefore employ the same neural mechanisms as moral judgments described in chapter 6.

Is beauty a property of a painting or of a person's perception of it? The answer is both, because the painting has physical properties such as color, shape, and organization, which have a predictable causal effect on people who share at least some cultural background with the painter. Judgments of beauty have changed over the millennia during which humans have produced paintings, going back around 40,000 years to cave paintings in Europe and Indonesia. During this time, human anatomy and physiology does not appear to have changed substantially, so culture undoubtedly does influence people's judgments of beauty. Nevertheless, the basic neural mechanisms of shape and color perception across cultures are sufficiently similar that we can seek objectivity in the interactions between depictions of objects and the brain processes of aesthetic evaluation.

These commonalities suggest that there can be at least some objectivity in judgments of beauty, enabling aesthetics to be normative as well as descriptive. Some things really are ugly, such as the star-nosed mole and the blobfish, which regularly show up on lists of ugliest animals. Rather than unity of attractive parts, these animals have unsightly features with poor overall fit. In contrast, paintings are intrinsically beautiful if they generate for most people the positive feelings that go with emotional coherence based on attractive ingredients. I am not suggesting a purely bottom-up appreciation of beauty from parts to wholes, because emotional reactions to wholes can help to focus attention on parts that change sensory inputs. Like perception in general, aesthetic appreciation is both top-down and bottom-up, through parallel processing that works through the simultaneous operation of billions of neurons (*Brain–Mind*, chapter 3).

Undoubtedly, cultural context affects aesthetic judgments, as when early impressionist paintings by Monet and Renoir were rejected by Paris art shows. Chapters 6 to 8 argued that objectivity concerning morals, justice, and the meaning of life comes from universal human needs, and the end of this chapter discusses cases where the production and appreciation of art contributes to the autonomy, competence, and relatedness of artists and audiences.

Multilevel Emergence and Social Causes of Beauty

My discussion so far has been theoretical because there are currently no experimental ways of identifying the operation of semantic pointers and coherence in the brain. But brain scan studies suggest that experiences of beauty correlate with activity in the orbitofrontal cortex, an important emotion area that is also involved in decision making and moral judgments. Like emotions and other cognitive processes, however, there is no single module for beauty and aesthetic judgments in the brain, which likely engages many other areas important for emotion and cognition, such as the nucleus accumbens (for pleasure) and the visual cortex (for perception). The burgeoning field of neuroaesthetics needs more than brain scanning search for neural correlates and should also investigate the mechanisms by which neurons in multiple brain areas generate aesthetic experience.

At the molecular level, there is evidence that the opioid system connected with pleasure is involved in aesthetic judgment, because men given morphine judge attractive women to be more beautiful, although this might confound aesthetics with sexual desire. I expect that viewers of beautiful paintings also engage dopamine circuitry relevant to pleasure, as has been shown for music.

My discussions of multilevel emergence in earlier chapters highlighted not only the molecular, neural, and psychological levels but also the social one, so we can look for group mechanisms connected to the appreciation of beauty. For objects and colors, there is an important social dimension because people are often attracted to things they find familiar, through what psychologists call the mere exposure effect. What people are exposed to and become familiar with results from the interactions they have with other people, including families, friends, and influential commentators. For example, appreciation of the objects in *The Starry Night* depends on recognition of crescent moons and church steeples that results in part from social influences.

More directly, when a painter is working for a patron, as Leonardo da Vinci did for the Medici, there can be emotional communication resulting from the artist's intention to produce something that the patron will appreciate and enjoy. Even Vincent van Gogh, who was isolated and sold few paintings in his life, cared about the reactions of his brother to the paintings that he produced. Artists often take pains concerning the framing and presentation of their work, for example when Mark Rothko was highly particular about the lighting of his paintings. So painting is a social process involving the communication of emotional judgments of the artist to the people who view it. Painters cannot expect viewers to appreciate their work with exactly the same perceptions and emotions that went into their creation, but they can hope to generate some approximation of these.

In addition, the perception of beauty can be a social process when artists show others their works in progress and get responses concerning the overall beauty of the painting as well as of the particular ingredients. For example, van Gogh established a studio in Arles and was joined by Gauguin. In the early 1900s in Montmartre, a group of artists that included Picasso and Matisse socialized and worked together. Appreciation of art can also be a social process when audiences in a gallery can help each other to pick out different aspects of it and thereby increase the emotional experiences of the individuals. *Mind–Society* (chapters 6 and 8) describes similar kinds of emotional communication operating in politics and religion.

Because of the involvement of molecular, neural, psychological, and social mechanisms in the generation and appreciation of beauty, it qualifies as an example of multilevel emergence. Other emotional reactions to painting, even the ugliness that operates in paintings such as van Gogh's *The Potato Eaters*, are similarly multilevel.

Beauty operates in many other domains besides painting. The concept *beautiful* can apply to physical objects such as faces and cars; to musical construction such as songs, symphonies, and dances; and also to verbal constructions such as poems. The beauty of music can also be explained by the mechanisms of emotional coherence resulting from the satisfaction of constraints among attractive elements. Chapter 10 describes how mathematics can similarly generate judgments of beauty as emotional coherence.

Although beauty is the most sought-after and discussed aspect of art, it is far from the only source of emotional engagement. The many counterexamples to the claim that art is only concerned with beauty point to other emotions that are generated by art. Beauty is not always a happy experience, as shown by the existence of sublimely sad paintings and music.

OTHER EMOTIONS IN PAINTING

Beauty is undoubtedly important in painting and the other arts, but the intense engagement that motivates artists and audiences can be accomplished by many other emotions. Like music and literature, painting can inspire joy, sadness, grief, fear, horror, disgust, surprise, shock, mirth, anger, disturbing ugliness, desire, and lust. These emotional reactions are not always exclusive, because there can be mixtures of emotions such as wonder and delight in *The Starry Night* and admiration and worry in van Gogh's self-portraits. The semantic pointer theory of emotions easily accommodates mixed emotions because neural patterns combine diverse cognitive appraisals and physiological perceptions that generate feelings best described using more than one word or concept.

All of these emotions arise in the same way as beauty in *The Starry Night* but with very different results. Visual perception binds shape and color and lighting into objects, which are then bound into larger scenes. Because of the emotional associations of the objects and their relations, the brain reaches an interpretation that is cognitively and emotionally coherent. The specific interpretation depends on physiological changes invoked by the objects and scenes but also by the appraisal of the situation with respect to the goals and previous experiences of the perceiver.

The emotional result can be far from beauty. To take an extreme example, consider Willem de Kooning's *Woman III*, which sold for more than $100 million in 2006. (Images of this painting and all the others mentioned in this chapter can easily be found by Web searches.) The central image identifies as a person with a face, but both are hideous rather than beautiful. The eyes are mismatched and distorted, while the nose is contorted and enlarged. The shape of the mouth conveys a demonic grin rather than a smile. The rest of the body is also grotesquely distorted. The colors are unpleasant, especially the putrid yellow, so overall this painting is disturbingly ugly.

Then why is it considered art? It is more than just interesting, a mild emotion that is primarily cognitive based on a moderate degree of goal relevance. The sheer ugliness also suggests anger toward women that may have originated with de Kooning's difficult relationship with his mother. Ugliness can also blend into disgust, the feeling of repulsion of things that are contaminated or unclean, with a strong visceral response that generates intense physiological change. Despite its hideousness, and in fact because of it, de Kooning's work is widely recognized as an intensely engaging piece of art. Similarly, Picasso's *Guernica* is not pretty but brilliantly displays horror and anger at the carnage of the Spanish Civil War.

Ugliness can also blend into horror, as in Francisco Goya's painting *Saturn Devouring his Son*. The individual parts of the painting are disturbing, including Saturn's demonic eyes and the son's bloody shoulders. The horror comes from the relation of these two objects, with the son's head inside Saturn's mouth. This painting is unpleasant and horrible but also fascinating because of the intensity of the aggression that it displays. The dark colors and the intense grip that Saturn has on the body of the devoured son all contribute to the emotional coherence of the piece that enables it to combine multiple negative emotions.

Similarly horrible, Vincent van Gogh's *Skull of a Skeleton with Burning Cigarette* was an arresting image of death and decay long before the association between smoking and cancer was known. Caravaggio's *Medusa* and the right damnation panel of Hieronymus Bosch's *The Garden of Earthly Delights* also contain gruesome images that express and generate such negative emotions as horror and disgust. The bold, bleak, and sometimes grotesque paintings of twentieth-century artists Francis Bacon and Lucian Freud sell for millions of dollars, even though they are more repulsive than beautiful, stimulating unnamed emotional experiences of puzzlement and shock.

Less intense than horror, fear is another negative emotion that can draw people's attention. Edvard Munch's *The Scream* conveys and inspires intense fear. Salvador Dali's *The Face of War* expresses the fearsomeness of war and generates the same emotion in people who view it. Caravaggio's *David with the Head of Goliath* depicts decapitation that naturally makes people afraid but also delivers a sense of triumph and pride in David's face.

Sadness and grief are also negative emotions that are often depicted in art. Mantegna's *Lamentation of Christ* is one of many paintings showing religiously important deaths. The tragic death of the French revolutionary Marat was painted by both Munch and Jacques-Louis David. Van Gogh's *At Eternity's Gate* evokes both beauty and sorrow through the arresting image of a man cradling his head.

Besides the negative emotions of ugliness, horror, fear, anger, disgust, sadness, and grief, there are positive emotions inspired by paintings that can be differentiated from beauty. Pieter Bruegel the Elder's *The Wedding Dance* conveys the excitement and happiness of a public celebration. The physical movements in Henri Matisse's dance paintings also express and evoke happiness. Many landscape paintings such as John Constable's *The Cornfield* capture the serenity as well as the beauty of outdoor scenes. Paul Barthel's *Happy Children* inspires the viewer to share the children's delight. Diego Rivera's Detroit Industry Murals encapsulate the commercial excitement and workers' pride in the Ford Motor Company factories.

Many erotic paintings convey desire and lust, which are usually positive but can flow into emotional negativity when desire is unsatisfied and lust is violent. *In Bed, the Kiss* by Henri de Toulouse-Lautrec is an intimate portrayal of two women embracing. The middle panel of Bosch's *Garden of Earthly Delights* shows people engaged in pleasurable activities. Michelangelo's *Leda and the Swan* portrays a woman and a bird but is thought to suggest the painter's lust for his male assistant.

Most paintings are variations on styles and topics that are culturally accepted. But occasionally artists appear who have dramatically different approaches that generate surprise, an emotion that combines physiological arousal with judgments of incoherence. People naturally expect new art to be coherent with their expectations based on previous experience, but highly original artists violate these expectations and generate surprise. When the surprises are unpleasant, new kinds of art can be viciously criticized, for example when the French Impressionists first presented their work.

In extreme cases, paintings are not only surprising but shocking, surprise that is so intensely negative that it can border on disgust or scorn. Picasso's early Cubist paintings were shocking in this way, as were the later even more abstract paintings by artists such as Mark Rothko. Paintings can generate surprise in various ways, through unusual components or arrangements that are incoherent with each other and with prior expectations. M. C. Escher's work is striking because of its elegant display of geometrically paradoxical scenes. Scorn can also be the emotional reaction of critics to art they consider to be kitsch, such as black velvet paintings of Elvis Presley.

Humor is an emotional reaction that arises from the presentation of material that is surprising through incoherence with previous expectations but also generates a new coherent interpretation. Some paintings naturally inspire laughter, as in the surrealist painting *L.H.O.O.Q.* of the Mona Lisa with a mustache by Marcel

Duchamp. Equally amusing are paintings by Salvador Dali and René Magritte. Trompe l'oeil paintings such as *Escaping Criticism* by Pere Borrell del Caso generate mirth as well as appreciation of beauty.

There are hundreds of emotion words in English, and you can find examples of paintings that exemplify many emotions by doing a Web search with just "painting" plus the emotion word. I have only given a few examples of paintings that illustrate the role of some particularly emotional important emotions for painting. For each of these, it would not be hard to provide the kind of detailed analysis that I did for *The Starry Night:* identify the components of the image that bind shape and color, the overall coherent scene that results from the relations among the components, and the cognitive appraisal that along with physiological changes results in an emotional response.

According to Tolstoy, art begins when people call up emotional experiences with the purpose of communicating them to others. Such artists engage in reverse empathy, where one person sets out to make another person feel similar emotions. Instead of regular empathy where you try to understand people by putting yourself in their shoes, reverse empathy is the attempt to make someone else empathize with you. Painters who pour their emotions into a work of art by producing something that is beautiful, horrible, or tragic are accomplishing the transfer of their own emotions to other people. Hence paintings are a kind of social communication where approximations to semantic pointers are transferred not by words or body language but by visual images.

The abstract expressionist Mark Rothko proclaimed:

I'm interested only in expressing basic human emotions—tragedy, ecstasy, doom, and so on—and the fact that lots of people break down and cry when confronted with my pictures shows that I *communicate* those basic human emotions.... The people who weep before my pictures are having the same religious experience I had when I painted them.

Not all painters consciously intend to influence the emotions of other people, but without such influence no one would pay any attention to their work.

Mind–Society provides examples of nested emotions, which are emotions about emotions. Nested aesthetic emotions include longing for beauty, comfort in splendor, fear of ugliness, and revulsion for horror. Thanks to recursive binding, the semantic pointer theory of emotions explains nested emotions, as well as mixed emotions such as awe as a combination of wonder and fear.

I have shown that emotional influence goes well beyond beauty to include many other negative and positive emotions. Similarly varied emotional communications take place within communities of artists when they inspire, support, criticize, and ridicule each other. Contrary to the solitary stereotype, most artists rely heavily on friends and associates for benefits that range from the personal to the aesthetic. Hence the emotional impact of art comes from multilevel emergence, through mechanisms ranging from the dopamine response for loveliness to the emotional contagion of artists appreciating each other's work.

CREATIVITY IN PAINTING

All of the paintings mentioned so far are creative, possessing the typical features of creativity of being new, valuable, and surprising. Novelty comes from different directions, including new subject matter, new combinations of images, and new techniques. *Brain–Mind* (chapter 11) describes neural mechanisms for creativity that lead to new objects and new methods of producing objects. *Mind–Society* (chapter 13) examines social mechanisms that contribute to creativity through the interactions of individuals. These neural and social mechanisms also explain creativity in painting.

My discussion of beauty showed how perception of a painting generates a coherent interpretation of multiple objects, where mental representations consist of neural firing patterns in the form of semantic pointers. But how does the artist produce the picture for the perceiver to observe? Artistic imagination requires the production of images that have not existed before, and it might seem mysterious how creative products can result from biological processes in brains.

The ancient Greeks thought that creativity results from inspiration by the Muses, nine goddesses who inspire people to produce beautiful new paintings, songs, poems, and other pieces of art. Natural philosophy of art aims for a scientific explanation of creativity that describes the mechanisms by which creative products arise. What generates semantic pointers that are new, valuable, and surprising?

In painting *The Starry Night*, van Gogh was far from the first to depict a sky, but the novel beauty of the painting results from different objects combined in different ways. His brain and hands put together the vibrant moon by combining the crescent shape, the bright yellow, and the surrounding light yellow. He was doing much more than copying what he saw out of the window of his room, instead generating a new combination of images. Painting is not simply the bottom-up procedure of translating a perception into a drawing but also requires the top-down procedure of retrieving images from memory.

Brain–Mind (chapter 3) describes how visual and other kinds of imagery work by regenerating patterns of neural firing that result from stored synaptic connections learned from previous experiences. New images can be generated by producing new neural patterns by transforming old ones through such operations as intensification, juxtaposition, combination, and rotation. Imagining a bright yellow crescent moon requires activating the images of a crescent moon and bright yellow, and then combining them into a unified object in the artist's mind. The basic neural mechanism for this is the same as occurs repeatedly in perception: binding by convolution of firing patterns for yellow, crescent, and moon.

Producing a larger and more complicated image with the moon, stars, tree, and village is just more binding, applied recursively by additional operations of convolution. Because the brain is a parallel processor with billions of neurons firing at once, the imagining of a complex scene does not have to be done step by step but can occur in a short period of time, less than a second. Artists do not have to do the imagining just in their own heads, because they can also work with external aids including sketches and the painting itself. Many artists work with preliminary sketches that suggest the overall structure of the image to be produced but can also alter their plan as they perceive how the painting is coming along. So perception and imagination interact, with images resulting from what is stored in the head of the artist, preliminary sketches, and engagement with the material that is rising on the canvas.

While painting, artists are constantly evaluating, making judgments about whether the work is looking good and approaching completion. Such valuations are inherently emotional, with the painter feeling happy when the desired results are being achieved but sometimes frustrated or even angry or depressed when the painting is not working out. Emotions then drive revisions that instigate changes in shape, color, and organization. So painting is a combination of cognitive operations of image generation by binding and emotional operations of evaluation and motivation. Twentieth-century nonrepresentational art, such as Pollock's drip paintings, abandons depiction but still aims at emotions such as beauty, surprise, joy, awe, intrigue, and wonder.

Just as artists use emotional evaluations to judge the progress of a painting, so viewers and prospective buyers can evaluate the painting based on their goals. If the viewer wants the painting to be beautiful, then it will only generate positive emotions such as happiness if the images in the painting come together in a way that is emotionally coherent in itself or with envisioned décor. However, some viewers will be engaged by other kinds of emotional reactions, such as shock, horror, disturbing ugliness, pride via social status, and greedy anticipation of financial gain.

The primary evaluation by the viewer is whether a painting is emotionally engaging or not, at least generating some emotions, positive or negative. Paintings that are derivative as slight variations on the work of other artists, clichéd, or kitsch will generate boredom, aesthetic death by yawning. To be fully appreciated, a creative work of art has to be moderately surprising in the cognitive-emotional sense of not being completely coherent with previous expectations. Like music composers and other artists, successful painters must produce works that are familiar enough to be comprehensible but new enough to be interesting, walking a fine line between coherence and incoherence.

Creativity in painting is not just producing new objects but also developing new techniques that can be responsible for producing whole series of paintings, such as the masterworks that van Gogh produced in the last two years of his life. He learned some of his techniques from the Impressionists but developed his own approach such as using thick applications of paint. *Brain–Mind* (chapter 11) discusses procedural creativity, the development of new methods that can be mentally represented by multimodal rules built out of semantic pointers. Such rules are not just verbal expressions such as "if you apply paint, then use thick lines." Rather, multimodal rules are *if–then* structures where the *if* and *then* parts are semantic pointers that include sensory features and motor operations. Applying paint is a kinesthetic operation, requiring the muscles of the arms and fingers; the representation of a thick line needs to be partly visual, based on perception and recollection of what thick lines look like. Hence, in line with the discussion of knowledge-how in chapter 3, procedural creativity in painting requires generation of neurally represented rules such as <paint> → <use-thick-lines>, where the brackets indicate semantic pointers rather than words.

In the history of art, procedural creativity has introduced new methods such as perspective drawing in the Renaissance, Impressionist use of different kinds of brushstrokes, and the radical use of color and shape by abstract expressionists. Procedural creativity helps to produce works that are new and surprising, although they may raise doubts about the value of the resulting pieces that violate established constraints and expectations.

Leonardo da Vinci's procedural creativity was evident in the original methods that he described in his voluminous notebooks, such as (a) basing visualizations on a knowledge of human anatomy, (b) using geometrical perspective and other mathematical techniques, (c) depicting light and shade in ways based on scientific principles, and (d) displaying mental events in his subjects by means of their physical gestures. Georgia O'Keeffe developed a dramatic new way of producing abstractions of familiar objects that are infused with explosive emotions, erotic energy, and female sensibility.

For natural philosophy, creativity in painting is not divine intervention or an inscrutable black box but rather a set of neural mechanisms for generating new representations, procedures, and experiences by means of perception, memory, binding, emotional evaluation, and multimodal rules. Social mechanisms can also contribute to creativity, when a painter gets suggestions from people about how to make a work in progress more novel and emotional engaging. Even if artists are not directly influenced during the production of a particular painting, they remain indebted to interactions with teachers and friends that provided images and techniques available for new creations. Because the molecular processes such as dopamine circuitry are also important to emotional evaluation of artworks, creativity in painting is yet another case of multilevel emergence. Molecular, neural, psychological, and social mechanisms interact to produce works that are expressive and engaging.

A theory of artistic creativity and other answers to aesthetic questions should apply to all forms of art, not just painting. I now show that the same neural mechanisms that I have used to explain beauty, emotional engagement, and creativity in painting also apply to music.

BEAUTY IN MUSIC

Painting suggests that the purpose, value, and power of art result from emotional engagement that arises from integrating elements into a coherent whole. One form of emotional engagement is beauty, where emotional coherence among attractive elements generates a strongly pleasant conscious reaction. Other forms of emotional engagements invoke positive emotions such as desire and mirth but also negative emotions such as grief and horror. This account of aesthetics potentially applies to all other kinds of art, but I just sketch how it applies to music.

The concept *beauty* is typically applied to visual objects such as paintings, sculptures, photographs, faces, and cars. But songs and other musical pieces can also be beautiful, as well as sad, funny, and scary. How does the brain perceive and create music, using neural mechanisms for binding, emotion, and coherence?

Music is practiced in all human cultures but is rare and much simpler in other species such as songbirds. Psychologists are still debating whether the ubiquity of music in human societies is the result of innate dispositions that wire the brain especially for music or just a side effect of other mechanisms such as language and movement. Research finds enormous cultural variation in music production and appreciation, with different peoples enjoying different kinds of instruments, rhythms, and melodies. But the ubiquity of perception, language, and facial expressions implies that the basic mechanisms of binding and emotion operate in all human brains.

For specific Western examples, consider Beethoven's symphonies and the most popular songs by the Beatles such as "Yesterday," "Hey Jude," and "Here Comes the Sun." The beauty of these pieces does not arise simply from the individual notes but rather depends on the way in which they get combined into chords, melodies, and entire compositions. Music perception is simultaneously cognitive and emotional, as people react to the pleasantness or unpleasantness of the tones and their combinations.

Just as neural processes bind shapes and colors into objects, they can bind notes with different pitches into chords, which are two or more notes heard simultaneously. According to David Huron, pitch is a high-level representation that results from several processing stages. Different frequencies of sound excite different places along the basilar membrane in the ear's cochlea. Each place causes the firing of different sensory neurons in the auditory cortex corresponding to relative location, with firing rates proportionate to the input frequencies. The neural representation of pitch results from combining the location and the firing rates.

Melodies are sequences of notes that join pitches and rhythms over a short time period. Just as a sentence can be formed by binding nouns, verbs, and other parts of speech into a recognizable whole, so a melody can be formed from notes related to each other by their place in scales. Binding these notes together over a short time produces recognizable melodies such as the opening notes of the Beatles "Yesterday" or Beethoven's *Fifth Symphony*: da-da-da-DUM. Melodies can then be combined into whole songs or other compositions, using memory to keep track of organized sequences.

Depending on the listeners' cultural background and personal experiences, particular notes can have emotional values attached to them, just as do shapes and colors that go into paintings. For example, in Western music, the note C is pleasant for most people, so the key of C major (where the scale begins at the note C) is one of the most common signatures in Western music. Popular records combine notes into appealing ensembles, for example with the C major chord consisting of C, E, and G, where E and G are consonant (sound well together) with C. Melodies in the key of C produce appealing sequences such as C C G G A A G, better known as "Twinkle, Twinkle Little Star." Different musical traditions develop patterns of notes, chords, and melodies that guide people's expectations and help them to recognize patterns that generate pleasure.

A note or a chord possesses little beauty on its own, but when combined into a melody an organized group of notes can have a larger emotional impact that evokes adjectives like "beautiful," "sublime," and "lovely." Beethoven's *Fifth Symphony* and

"Yesterday" are beautiful in their short melodies but also the way in which these melodies are combined and altered as part of the larger whole. Each piece offers enough similarity in its melodies to be appreciable as a whole but with enough variation to keep it interesting. The variations can introduce diversity in rhythm, tempo, and key, but there has to be enough overall connection that the piece forms a unity.

My discussion of unity and diversity in painting described how beauty results from multimodal representations, coherence, and emotion. The sounds in a musical piece are not represented by words (ignoring song lyrics for the moment) but by neural firings in auditory cortex. Like visual patterns of neural firing, combinations of sound representations use binding by convolution. As with visual objects, binding also generates emotional reactions from a combination of physiological responses and cognitive appraisals based on expectations of what will be heard next.

What makes a melody beautiful? If beauty is emotional coherence, and coherence is constraint satisfaction, we can ask what constraints are satisfied in an enjoyable melody. From the innate biology of audition and from cultural experience, people approach music with expectations of what sounds fit together with each other. Melodies can gain coherence by means of familiar rhythms, arrangements of ascending and descending notes, compatible timbres (sound quality of notes), and varied repetitions that blend them into larger pieces such as symphonies.

In addition to the binding of sounds that takes place in instrumental pieces, songs require another kind of binding with words. The melody for "Yesterday" came to Paul McCartney while asleep, and he and John Lennon struggled to find appropriate lyrics for it. McCartney's initial version used the feeble language "Scrambled eggs, baby I love your legs," rather than the ultimate, poignant line "Yesterday, all my troubles seemed so far away." The emotional power of the song comes from the combination of the sad verbal message with the melancholic melody. Many other examples such as the gorgeous second movement of Beethoven's *Seventh Symphony* show that beauty can derive from sadness as well as from happiness, as long as a piece achieves emotional coherence among attractive elements.

OTHER EMOTIONS IN MUSIC

The semantic pointer theory of emotions describes the mechanisms by which the brain integrates perception of physiological changes with cognitive appraisals of the goal relevance of a situation. But it does not assume that physiology and appraisal are always equally important. For more intellectual emotions such as excited ambition, appraisal may be a large contributor, whereas there may be a larger impact of physiological changes for more visceral reactive emotions such as fear provoked by snakes.

Similarly, different kinds of art may involve different balances of physiological perception and cognitive change. For verbal forms of art such as dramatic plays, cognitive appraisal may operate more strongly than physiological change, whereas for less verbal forms of art such as painting and instrumental music that are driven strongly by visual and auditory inputs, there may be a greater role played by physiological change. Music inspires many physical changes through such actions as dancing and marching.

The last section described how music can be beautiful, but it can also generate related emotions such as ecstasy, rapture, exhilaration, elation, and even a sense of spiritual transcendence. I now provide examples of pieces of music, both classical and popular, that evoke the full range of emotions produced by paintings.

Happy Music

There are many happy pieces of classical music, such as Beethoven's luscious *Sixth (Pastoral) Symphony*, which has upbeat and exhilarating melodies. The fourth movement of Beethoven's *Ninth Symphony* melds glorious music with Schiller's exuberant poem *Ode to Joy*. Johann Sebastian Bach's *Badinerie* is similarly fast-paced and uplifting, as are parts of Vivaldi's *Four Seasons* and Handel's *Water Music*. Happy Western music is usually written in major keys, although this association may be culturally acquired because such music also tends to be faster and have higher pitches, both of which are associated with energetic movements.

Philip Johnson-Laird and Keith Oatley identify these musical associations of emotions:

- Happiness: medium tempo, loud, wide range of pitches in melody, major scale, consonant.
- Sadness: slow tempo, soft, low pitch, small range of pitches, minor scale, mildly dissonant.
- Anxiety: rapid tempo, moderate volume, low pitch, minor scale, dissonant.
- Anger: rapid tempo, loud, high pitch, minor scale, dissonant.

Popular music usually combines melodies with words to produce a coherent effect by combining a happy tune with happy words. The following are some examples of songs that combine upbeat tempos with words that encourage positive appraisals: Bruno Mars' "Marry You," Perry Como's "Accentuate the Positive," Lisa Minelli's "Cabaret," Taj Mahal's "Cakewalk into Town," King Harvest's "Dancing in the Moonlight," The Beach Boys' "Fun Fun Fun," The Beatles' "Here Comes the Sun," and Dire Straits' "The Walk of Life."

Occasionally, popular songs combine upbeat tunes with emotionally negative lyrics. Examples include Bruce Springsteen's "Ain't Good Enough for You" and Ann Murray's version of "Snowbird," which have lively, catchy tunes yet disgruntled lyrics. The band Arcade Fire combines lively rhythms with cultural satire that expresses disdain for social and technological changes.

Mirth and humor are uncommon musical emotions, but they do occur in both classical and popular music. Haydn's *Surprise Symphony* (number 94) makes people laugh by a sudden loud chord at the end of a quiet opening new movement. Beethoven's frantic *Rage Over a Lost Penny* is more amusing than angry. Classical music can also be funny when played with unusual instruments or when adapted for parodies, as in the opera sequences in Bugs Bunny cartoons and P. D. Q. Bach's *Pervertimento for Bagpipes, Bicycles, and Balloons*.

Parodies can also provide humor in popular music, for example in Weird Al Yankovic's takeoffs on various modern forms. For example, his hilarious song "White and Nerdy" is a musical analog of the gangster rap song "Ridin," carrying over the basic tune but replacing tough Black guys with a geeky, young White man. Other songs gain humor just from the verbal subject matter, such as "A Boy Named Sue" by Johnny Cash.

Desire, lust, and other emotions associated with sex and romance are easier to convey with words and pictures that with music, but some purely instrumental pieces are connected with positive emotions that range from affection to love. Some composers such as Beethoven have produced compositions with love interests in mind, as with his enchanting *Für Elise*.

More readily identifiable are thousands of popular songs about love, affection, and desire. Some of the more explicit include Marvin Gaye's "Sexual Healing" and "Let's Get it On," and lusty rap songs like R. Kelly's "Ignition (Remix)." Sometimes songs also convey sexual frustration, as in the Rolling Stones "Satisfaction."

The songs mentioned all have an element of surprise in them, but sometimes music can even be shocking. The initial Paris audience jeered at Igor Stravinsky's ballet *Rite of Spring* with its unusual tones, rhythms, and frenetic choreography. Atonal music such as the compositions of Arnold Schoenberg are shocking in their dismissal of Western conventions. Once lyrics are included, it is easy to be shocking by saying things that people do not want to hear, as in the political aggression of some punk rock, evident even in the name of the band Dead Kennedys.

Sad Music

As painting examples showed, art can be beautiful without being happy, and there are many classical works that powerfully convey sadness. Examples include Mozart's *Requiem Mass*, Beethoven's *String Quartet No. 14*, and Mahler's *Symphony No. 9*. Sadness is even more remarkably portrayed in popular music.

Some years ago, I was surprised that my favorite music streaming service has a playlist called Sad Songs. I wondered: Why would anyone want to listen to a bunch of sad songs? I assumed that sad songs make you sad, and everyone would rather be happy than sad. But when I listened to the playlist, I discovered that it includes many of my favorite songs, and I really enjoyed listening to it. The cultural traditions of American blues and Portuguese Fado move people with mournful songs. What is it that makes sad songs so appealing?

When you experience sad songs, paintings, or literature, you reproduce some of the same physiological reaction that occurs when you yourself are in a bad situation, but the cognitive appraisal is very different. After all, the sad song, tragic drama, or horror movie is not about you and therefore is not a threat to your own goals. In fact, the realization that the current state of your life is not sad, tragic, or horrific can give you a feeling of relief and perhaps even happiness that you are not suffering in the same way that the singer or actor is. So the emotional engagement that the sad song provides comes without personal threat.

The top 10 of *Rolling Stone Magazine's* list of 500 best songs display an amazing array of emotions, much more diverse than happy and sad. Many songs exhibit multiple emotions, and 2 of the 10 combine positive and negative ones. Aretha Franklin's magnificent "Respect" is positive in expressing pride and confidence in demanding respect for herself and also negative in indicating resentment that she has not always been treated that way. "Hey Jude" by the Beatles is sad in recognizing that the person to whom it is addressed is suffering (reportedly John Lennon's son who was upset about his parents' divorce). But it is also hopeful and positive in providing the message that you can take a sad song and make it better.

Other songs on the list are more resoundingly negative. Number 1 on the list is my own personal favorite, Bob Dylan's "Like a Rolling Stone," which intensely displays resentment, anger, and gloating at the misfortune of the person addressed. The song "Satisfaction" by the Rolling Stones portrays social resentment as well as sexual frustration. Marvin Gaye's "What's Going On" indicates sadness and distress about political and environmental degradation. Nirvana's "Smells Like Teen Spirit" conveys angst and disillusionment.

On the positive side, John Lennon's "Imagine" inspires hope for a better world, and the Beach Boys "Good Vibrations" evokes happiness in a good relationship.

Chuck Berry's "Johnny B Goode" and Ray Charles's "What'd I Say" portray happiness about good music and romance.

All of these songs combine original music, appropriate lyrics, and superb performances to evoke intense emotions. It does not matter whether a song is happy or sad, only whether it has an emotional impact on the listeners. People are happy to like sad songs, just not boring ones. Live performances are often more compelling than recordings because the visual embodiment of the music by the musicians provides even more emotional engagement than the melodies alone. Moreover, musical and theatrical performances can establish an interaction between entertainers and the entertained that enhances the emotions of both.

Disturbing Music

Other negative emotions conveyed by music include disgust, anger, disturbing ugliness, and fear, although these are not as common as sadness. I do not know of any classical pieces that serve to generate disgust, but some incompetent performances can be viewed by discriminating listeners as repugnant if they strongly violate aesthetic norms. Popular music can more easily be disgusting in its lyrics, and there is even a variety of hip-hop music called horrorcore. The most repulsive lyrics I have encountered are in King Gordy's "Mr. 187," which pointlessly and hideously combines pedophilia and necrophilia.

Anger is harder to detect in purely instrumental music, although some people find it in the powerful opening of Beethoven's *Fifth Symphony* and in Shostakovich's *Symphony No. 5*. But anger is common and palpable in popular music concerning both the personal and the political. Personal anger is often directed against unfaithful lovers, as in Carrie Underwood's song "Before He Cheats" and in CeeLo Green's aggressive "Fuck You." Political anger was expressed in folk songs of the 1960s such as Bob Dylan's "With God on Our Side." Songs have also been important parts of political movements based on anger, defiance, and hope, such as the communist "Internationale," the Nazi "Horst-Vessel-Lied," and the Chilean "El pueblo unido jamás será vencido." Later musical genres sometimes combine both personal and political anger, as in the anarchist strains of punk rock, the nihilist tirades of heavy metal, and the insubordination of hip-hop.

Music rarely aims to be disturbingly ugly, although I have doubts about atonal music, punk rock, heavy metal, and hip-hop. Mathematician Scott Rickard claims to have produced the ugliest music ever in a piece that is perfectly random, so that absolutely no pattern can be discerned in the notes. I find it less ugly than the genres just mentioned, mostly just boring, which is a very different emotional reaction. Ugliness is jarring incoherence, not just lack of coherence.

Finally, consider highly negative music that produces fear and horror. The scariest movie music I have heard is the theme from the movie *Jaws*, although it would have been less terrifying without the accompanying image of the voracious shark. Many other movies have soundtracks that accentuate fear using shrieking noises, unpleasant cords, and sudden high notes, as in the strident tones of the musical accompaniment to the shower scene in *Psycho*.

Independent of movies, there are pieces of classical music that evoke fear or at least anxiety, such as Mussorgky's *Night on Bald Mountain* and Wagner's *Ride of the Valkyries*. In popular music, there are songs such as Michael Jackson's "Thriller" that use both rhythms and lyrics to revive primitive fears about the unknown. "The End" by the Doors and heavy metal songs also try to suggest the imminence of doom and destruction. Both musically and lyrically, Nirvana's "Smells like Teen Spirit" recapitulates the grungy anxiety of teenage life. In both music and painting, art matters to people because it engages emotions.

CREATIVITY IN MUSIC

As in painting, musical creativity requires the combination of mental representations construed as semantic pointers, but in music the representations are auditory images rather than mental pictures or words. For example, Beethoven's powers of auditory representation were so great that he continued to compose after he had become too deaf to hear the piano. Producing music requires combinations that are synchronic (combining notes into a chord, or voices into harmony) and diachronic (producing a melody as a series of notes or chords). Combination of representations is also required to imagine how different instruments will play a piece together. Combination of auditory representation works the same way as combination of visual representations in painting: binding by convolution of patterns of neural firing to produce new semantic pointers.

In addition to combinations of auditory representations, songs require integration of auditory and verbal representations, connecting lyrics to the music, as in the last movement of Beethoven's *Ninth Symphony* and in all of the Beatles' hits. Some of the best of the Beatles' songs combined McCartney's tunes with Lennon's wordplay. Visual representations may also have operated in the background, as in the Beatles' song "Lucy in the Sky with Diamonds" and Beethoven's evocative *Pastoral Symphony (No. 6)*.

New combinations do not arise randomly but rather are produced in the context of attempts to generate new compositions. External associations in which verbal or visual representations inspired musical compositions operated in Beethoven's mind with influences such as women, religion, history (e.g., Napoleon), and literature. The Beatles' songs were inspired by a wide variety of associations, including relationships, other personal experiences, and drug trips. Occasionally, composers think analogically, as when Beethoven based some of his passages on folk melodies and reused portions of his previous work. Much of the Beatles' early work was admittedly analogical, as they tried to copy and modify parts of works by their favorite singers/songwriters such as Chuck Berry, Roy Orbison, Smokey Robinson, and later Bob Dylan. One of the Beatles' songs, "Because," resulted from an inversion of Beethoven's "Moonlight Sonata"!

Musical creativity is primarily oriented toward producing new pieces, but some musicians have also produced new methods that can be understood as multimodal rules. Examples of such procedural creativity include Bach's new approaches to harmony and counterpoint, the Italian invention of opera in the sixteenth century, the Beatles' development of new electronic recording techniques, and the hip-hop manipulation of dual turntables.

Overall, the creative processes of Beethoven, the Beatles, and other musicians included inspiration by external associations and analogies to generate new neural combinations into semantic pointers.

EMPATHY IN LITERATURE AND FILM

Other areas of art are wide open to the explanations of beauty, other emotions, and creativity that I have provided for painting and music. The study of literature can also draw on the discussions of language in chapter 8 and in *Brain–Mind* (chapter 10). But I only address here one important aspect of the aesthetics of literature and film: the contribution of empathy to people's enjoyment and appreciation of books and movies.

In *The Passionate Muse*, Keith Oatley provides an insightful account of the role of emotions in literary fiction. He summarizes:

> Fiction is based on narratives in which characters act on their intentions and encounter vicissitudes. Readers enjoy entering into the lives of characters, following their projects, and coming to empathize with them as their plans progress or meet obstacles. Readers enjoy, too, meeting characters with whom they sympathize, and being reminded of emotional episodes in their own lives.

Similarly, Roger Ebert describes empathy in film:

> The movies are like a machine that generates empathy. It lets you understand a little bit more about different hopes, aspirations, dreams and fears. It helps us to identify with the people who are sharing this journey with us.

My favorite novels, plays, and films all include at least one character with whom I can empathize to some extent.

But how does such empathy work in the brain? Chapter 6 describes three modes of empathy based on verbal analogies, mirror neurons, and multimodal rule simulation. All of these can contribute to engagement with fictional works, but I think the third is by far the most important. You may gain some emotional identification with characters by verbal mappings between their situations and yours such as recognizing a similar predicament, or by having a gut reaction that mirrors their fall off a cliff. But ongoing involvement with a character works best with multimodal simulation.

Recall that multimodal rules are mental representations such as <insulted> → <hurt>, where the words in brackets indicate semantic pointers using nonverbal representations. For example, <insult> can include the tone of voice, facial expression, and obnoxious gesture that goes with an abusive remark, and <hurt> is the felt emotional response. Empathizing with an insulted character in a novel or movie involves running in your own mind the rule so that you can appreciate more directly how being insulted is hurtful.

As a fictional plot unfolds, series of multimodal rules can then lead to succession of emotions in the audience that dynamically correspond to those of the characters. For example, the rule <insulted> → <hurt> might chain with the rule <hurt>→ <seek revenge> to help understand why a character is acting vengefully. Writers of novels, plays, and film scripts need to be adept at verbal and visual descriptions that enable audiences to run such mental simulations that enable them to understand and care about the characters. Camus said that "The first thing for a writer to learn is the art of transposing what he feels to what he wants to make others feel." For example, my favorite movie *Casablanca* brilliantly engages the watcher with chains of emotions that include love, regret, sadness, fear, anger, and pride.

SUMMARY AND DISCUSSION

My naturalistic theory of aesthetics condenses into the following principles:

1. The main mental and social functions of art are the expression and transmission of emotions, in relationships among creative artists and their appreciators.
2. Artistic emotions are semantic pointers in brains that integrate sensory representations with combinations of physiological changes and cognitive appraisals.
3. The central emotional response to art is beauty, resulting from pleasurable emotional coherence through unity in diversity of sensory representations.
4. Art generates other important emotional responses, including interest, shock, sadness, fear, anger, and disgust.
5. Art is good or bad depending on the intensity and quality of the emotions that it generates in audiences and artists.

Because the term "expressivism" is used in aesthetics to mark theories that emphasize emotions, this theory might be called "neural expressivism." Unlike other versions of expressivism, the theory is neural because of its incorporation of the semantic pointer theory of emotions.

The fifth principle is probably the most contentious. As noted in the discussion of emotional values in chapter 6, emotions can be rational if the cognitive appraisal of a situation takes into account vital human needs rather than capricious wants. Art serves the needs of artists when producing it increases their autonomy, competence, and relatedness. Autonomy comes from acting independently of existing conventions to produce something that is new, valuable, and surprising. Producing such works satisfies the competence need of the artists for achievement, and communications with other artists, spectators, and audiences satisfies the need for relatedness. Compared to satisfaction of these psychological needs, the satisfaction of biological needs by the usually meagre financial returns of art is only a small contributor to emotional reactions.

Similarly, art can offer valuable contributions to the needs-related emotions of its appreciators. Collectors, spectators, and audiences benefit from emotional engagement with art, both beautiful and otherwise, in ways that enhance their autonomy, competence, and relatedness. Selecting art that fits with personal preferences rather than conventional norms contributes to autonomy and self-identity. Propagating successful art boosts competence by satisfying achievement goals such as demonstrating good taste and social influence. Enjoying art in the company of artists and other appreciators increases relatedness. With such contribution towards vital needs, the emotional engagement of art is not capricious but rather can generate objective judgments about its objective value.

Art occurs at the social intersection of mind and world, when creators and appreciators use their brains to generate and perceive works that stimulate emotions. Neural representations consist of patterns of firing that encode sensory inputs that include the visual ones in painting and the auditory ones in music. Binding by convolution generates complex combinations of sensory-motor-verbal-emotional representations that enable the creation, perception, and evaluation of artworks.

I showed how this way of thinking about art illuminates painting and music, but it would not be hard to extend it to many other artistic domains. Sculpture is like painting in having a large visual component but adds elements of touch and implied motion that can be accommodated by more complicated kinds of binding into semantic pointers. Photography is primarily visual and therefore subject to the same kinds of explanation that I gave for painting. Dance combines several modalities, including hearing music, seeing physical action, and kinesthetic appreciation of the motions of dancers.

Literature is produced by words in different forms—poems, novels, and plays—but it is also amenable to multimodal explanations based on semantic pointers. Good writing often works through powerful visual images produced by descriptions, similes, and metaphors, as in Romeo's proclamation that Juliet is the sun. Literature also can stimulate other kinds of images, such as the sound of someone's voice, the taste of a fine meal, or the emotional tone of an argument.

The semantic pointer theories of emotion and communication explain the feelings of artists and appreciators, fueling the social interactions that foster the production and appreciation of art. Table 9.2 reviews the emotions discussed in relation to painting and music and extends them to literature. Sometimes this impact is accomplished by empathy, when writers engage readers by helping them to feel something like the imagined emotions of fictional characters.

Good movies are even more intensely multimodal, because they combine visual experiences, kinesthetic reactions tied to the motions of bodies, musical accompaniments, and powerful emotional responses, including awe, joy, sadness, and horror. Understanding how movies produce these reactions requires a unified theory of perception, cognitive interpretation, and generation of emotions, all provided by semantic pointers.

As a scientific hypothesis, these theories of cognition and communication might be wrong. They would be wrong if dualism is true and human minds operate by special spiritual stuff rather than through neural mechanisms. They would be wrong if behaviorism or radical embodied cognitive science were correct that explanations of mind do not need to invoke mental representations. They would be wrong if the best theory of mental representations was based on a language of

TABLE 9.2

Emotions in Painting, Music, and Literature

Emotional reaction	Painting	Music	Literature
Beauty	van Gogh, *The Starry Night*	Beethoven, *Fifth Symphony*	Yeats, "Aedh wishes for the cloths of heaven"
Joy and happiness	Breughel, *The Wedding Dance*	Beethoven, *Ninth Symphony*	Wordsworth, "I wandered lonely as a cloud"
Mirth and humor	Magritte, *The Treachery of Images*	Haydn, *Surprise Symphony No. 94*	Russo, *The Straight Man*
Desire and lust	Hayez, *The Kiss*	Marvin Gaye, "Sexual Healing"	Nabokov, *Lolita*
Surprise and shock	Picasso, *Les Demoiselles d'Avignon*	Stravinsky, *The Rite of Spring*	Roth, *Portnoy's Complaint*
Sadness and grief	Mantegna, *Lamentation of Christ*	Beethoven, *Seventh Symphony*, second movement	Shakespeare, *King Lear*
Disgust	de Modena, *Inferno*	King Gordy, "Mr. 187"	Burroughs, *Naked Lunch*
Anger	de Kooning, *Woman III*	Bob Dylan, "With God on Our Side"	Larkin, "This Be the Verse"
Disturbance and ugliness	Francis Bacon, *Three Studies for a Portrait of Henrietta Moraes*	Twisted Sister, "O Come All Ye Faithful"	Ellis, *American Psycho*
Fear and horror	Goya, *Saturn Devouring His Son*	Wagner, *Ride of the Valkyries*	Shelley, *Frankenstein*

thought in which the fundamental operations of mind are verbal rather than multimodal. They would be wrong if future cognitive scientists construct new theories of neural representation and processing that fit better with biological and psychological evidence than does the Semantic Pointer Architecture.

Currently, however, the arts provide wonderful examples of how human brains generate and use powerful mental representations that mingle words, pictures, sounds, and emotions. Semantic pointers show how this mingling can be accomplished by billions of neurons. Neuroaesthetics can incorporate scanning data that identify brain areas active when people are feeling beauty and other aesthetic emotions with complementary theories that provide mechanistic explanations of the causes of aesthetic feelings. Art is beautiful and has other emotional impacts because of how our brains work and how we communicate with each other.

Like all philosophy, aesthetics is normative as well as descriptive and explanatory. Because the major point of art is emotional engagement, and because emotions have an evaluative component through the contribution of cognitive appraisal, we can probe whether people's emotional reactions are psychologically and sociologically appropriate. Somebody who thinks that a swastika is beautiful can be advised of the horrible impact of the visual shape that occurred when it was co-opted by the Nazis. Similarly, superficially appealing tunes such as the "Horst-Wessel-Lied" can be cognitively reinterpreted once their social roles are understood. People's aesthetic experience and education can be broadened to lead them to revise their judgments, so that garish paintings and trivial songs no longer seem so appealing.

Natural philosophy in the form of social cognitivism provides a descriptive and normative theory of art, consonant with the ethical theory developed in chapters 6 to 8. Aesthetics is not the same as ethics because it concerns sensory works rather than actions in general and because it deals with special emotions such as beauty. But the two fields share emotional responses that can be evaluated based on coherence with human needs. This continuing emphasis on coherence as constraint satisfaction also connects with the coherentist epistemology in chapter 3, and my neural-naturalist approach to the arts complements the brain-based materialism of chapters 2 and 4.

This account of aesthetics also links with the discussions in the next and final chapter in surprising ways. Is the autonomy of artists and audiences that I claim is enhanced by art compatible with challenges to the existence of free will? Does my characterization of beauty in painting and music transfer to the widely recognized mathematical concept of beauty? Finally, is art a capability only of human beings, or can it also be accomplished by computers and by other animals?

NOTES

The Bertrand Russell story is from Schilpp 1980, p. 88.

Levinson 2003 and Gracyk 2012 are good overviews of aesthetics. Bullot and Reber 2013 discuss the psychology of art. On neuroaesthetics, see Chatterjee and Vartanian 2014; Huston, Nadal, Mora, Agnati, and Conde, 2015; Kirsch, Urgesi, and Cross, 2016; and Starr 2013. Currie, Kieran, Meskin, and Robson 2014 discuss how aesthetics can draw on science. Noë 2015 is skeptical.

On the definition of art, see Adajian 2012 and Lopes 2014.

Sartwell 2016 reviews the philosophy of beauty. Eco 2004 provides many historical examples. Aschenbrenner 1985 and Beardsley 1958 discuss coherence in art.

Ishizu and Zeki 2011 find activation in the orbitofrontal cortex with the experience of beauty. Thagard 2000 discusses beauty and surprise in terms of emotional coherence.

On creativity and the brain, see Zaidel 2014; Vartanian, Bristol, and Kaufman 2013; Thagard and Stewart 2011; Thagard 2014a (from which this chapter borrows a few paragraphs), 2014b, and *Brain–Mind* (chapter 11).

Discussions of emotions and art include Tolstoy 1995, Collingwood 1997, Robinson 2005, and Perlovsky 2014.

Chelnokova et al. 2014 discuss molecular mechanism for beauty involving internal opioids. Salimpoor et al. 2015 describe how musical pleasure involves complex interactions between dopamine systems and cortical areas.

The Rothko quote is from Rodman 1961, pp. 93-04. Schama 2006 (p. 437) says many other painters assume the participating presence of the public. Studies of the social contexts of art include Roe 2015 and Smee 2016. *Mind–Society* discusses the role of music in politics and religion. Zajonc 1968 introduced the mere exposure effect.

Kania 2012 reviews the philosophy of music.

On the psychology of music, see Huron 2006 and Hallam, Cross, and Thaut 2011. Schiavio, Menin, and Matyja 2014 consider the role of embodied simulation in music appreciation. Johnson 2007 develops embodied aesthetics.

On music and emotion, see: Huron 2015; Huron, Anderson, and Shanahan 2014; Johnson-Laird and Oatley 2016 (quote from p. 87); Juslin and Sloboda 2011; Juslin 2013; and Sievers, Polansky, Casey, and Wheatley 2013, who describe how music and movement support emotion. Swafford 2013 discusses what makes melodies beautiful. Sachs, Damasio, and Habibi 2015 analyze when sadness evoked by music is pleasurable. Menninghaus et al. 2017 explain why negative emotions are powerful for attention, involvement, and memorability.

On the brain and music, see Koelsch 2012, Levitin 2006, and Brattico and Pearce 2013.

Villarreal et al. 2013 study neural correlates of musical creativity.

Scott Rickard's random composition is described at https://www.classicfm.com/music-news/ugliest-music/.

Brown and Jordania 2011 review cultural universals in music such as pitch and scales. McDermott et al. 2016 identify cultural variations in reactions to consonance and dissonance.

For cognitive and neural approaches to literature, see Jacobs 2015; Oatley 2012, 2016 (quote from p. 15); Starr 2013; and Zunshine 2015. The Camus quote is from Carroll 2013, p. 155. Thagard 2011 takes a cognitive-emotional approach to allegory.

French and Wettstein 2010 collect philosophical examinations of film and the emotions. Zacks and Magliano 2013 discuss film and cognitive neuroscience.

The Ebert quote about film and empathy is from http://ew.com/article/2014/07/09/life-itself-movie/.

PROJECT

Do a three-analysis of *beauty*. Extend neural expressivism to many other aesthetic realms such as sculpture, photography, dance, film, drama, and poetry. Consider alternative theories of literary interpretation and explain why there is no affective fallacy. Debunk the so-called art instinct on the grounds that art appreciation emerges from social influences that build on interactions of innate mechanisms for perception and emotion.

10

Future Philosophy

LOOKING BACKWARDS AND FORWARDS

How does it all hang together? Systems of philosophy have been out of fashion since the nineteenth century. Rather than follow Hegel in constructing integrated systems of epistemology, metaphysics, ethics, and aesthetics, philosophers have preferred to concentrate on narrower topics. Major philosophical approaches, including analytic philosophy, phenomenology, and religious philosophy, have preferred the popular academic strategy of saying more and more about less and less.

Undaunted, this book has delved into all major areas of philosophy, except for formal logic, which I consider a minor branch of epistemology. I started with philosophy of mind, not because it provides axioms from which the rest of philosophy can be derived but because it draws on new scientific advances that have implications for all the other areas of philosophy. Understanding how minds work contributes directly to establishing plausible theories of knowledge, reality, morality, justice, meaning, and beauty.

The resulting system of philosophy, social cognitivism, contains no a priori truths but instead gets its plausibility from both local and global coherence. Local coherence comes because each topic draws on scientific evidence drawn from psychology, neuroscience, and other relevant fields. Global coherence comes because the philosophical doctrines defended in this book fit together in a harmonious whole. Figure 10.1 provides a value map of philosophy as I view it, indicating the interconnections of the defended philosophical doctrines.

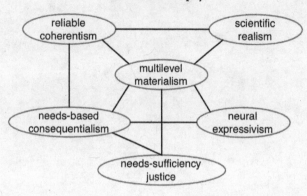

FIGURE 10.1 Natural philosophy as a coherent system. Ovals indicate positive evaluations, and solid lines indicate mutual support.

Figure 10.1 shows how philosophy hangs together, not just metaphorically but by coherence relations tied to scientific evidence. The concluding sections of each of chapters 3 to 9 indicated connections among the philosophical positions. Multilevel materialism in the philosophy of mind fits well bidirectionally with reliable coherentism in epistemology. Understanding the brain as operating with neural mechanisms of parallel constraint satisfaction supports and is supported by the view that knowledge is based on reliable coherence. Both of these views fit with scientific realism as the most plausible approach to metaphysics.

There are internally coherent alternatives to my system of social cognitivism, such as religious philosophies that espouse faith and supernaturalism, but these are incompatible with centuries of accumulated evidence. Another alternative would be complete skepticism, relativism, and nihilism, but these conflict with a wealth of plausible theories and values. Figure 10.2 expands Figure 10.1 to show incoherence with some less plausible alternatives.

Value-oriented areas of philosophy should aim to produce interconnected theories, of morality, justice, the meaning of life, and art. In my approach, these all work with the naturalistic view of values as emotional processes that can aspire to objectivity because cognitive appraisal is subject to rational scrutiny with respect to vital human needs, even if physiological reaction is not. All of these value areas cohere with each other and with the view of mind as brains operating with integrated cognitions and emotions. The interdependency of cognition and emotion in the brain also shows how epistemology can incorporate value as an important dimension.

The resulting approach to philosophy is inherently social rather than individualistic because it recognizes that minds, knowledge, and some aspects of reality are all affected by social mechanisms as well as mental ones. Philosophers

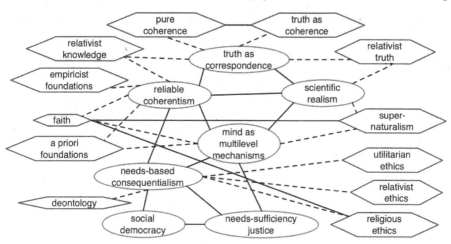

FIGURE 16.2 Natural philosophy and alternatives. Ovals indicate positive evaluations, and hexagons indicate negative ones. Solid lines show mutual support, and dotted lines show incompatibility.

have long feared that allowing an incursion of society, minds, and especially emotions into philosophical deliberations would be the death of rationality. But objective norms in epistemology, ethics, political philosophy, and aesthetics cannot be extracted from pure reason. Instead, the project of establishing and defending norms requires reflections on how minds and societies work and sometimes fail to work in the pursuit of vital needs. Laying out mechanisms and appreciating the circumstances under which they break down is the key to establishing defensible and usable norms in all of these areas of philosophy. This book extends *Brain–Mind's* treatment of cognitive science and *Mind–Society's* treatment of social sciences and professions to illuminate philosophy and the arts.

I have tried to present plausible, interconnected answers to the major questions of philosophy but have neglected some important issues. The purpose of this concluding chapter is to address three important philosophical questions that remain unresolved despite relevant advances in cognitive science. I cannot now solve these problems but suggest how solutions might be found through future advances in the cognitive and other sciences.

I begin with the ancient problem of free will, which arose in connection with morality in chapter 6. I now connect it with the theories of mind, reality, and morality already developed and point to how outstanding issues might be resolved by future progress in the sciences. Then I look at the difficult case of mathematics, which deserves a theory of knowledge and reality that is coherent with the ones presented in chapters 2 to 5. Not enough is known yet about how brains do math to help resolve key disputes in the philosophy of mathematics, but I indicate some needed advances.

Finally, I sketch how to overcome the limitations of this *Treatise* arising from its exclusive focus on human beings. Acute philosophical issues arise in relation to nonhuman animals whose mental capacity is increasingly becoming appreciated, for example dolphins, chimpanzees, crows, and even octopuses. What do they know and experience, and what are our resulting moral obligations to them? Even more up in the air are questions about the mental abilities and moral roles of machines that are becoming increasingly sophisticated through developments in artificial intelligence and robotics. I do not try to answer these questions about animals and computers but speculate about how answers might be found.

FREE WILL

The question of whether people have the ability to freely choose among different courses of action has been important in philosophy since the ancient Greeks. The early mechanists such as Epicurus realized that viewing mind as working by interacting atoms generated a threat to free will, so they considered that atoms might swerve unpredictably to make free will possible. Later, Christian theologians wrestled with how free will could be compatible with the omnipotence of God who created everything and knows everything that will happen.

The issue has become even more acute in recent decades because of advances in psychology and neuroscience. As the mechanisms of mind are increasing well understood, there seems to be less and less room for free will. Some eminent scientists have come forward as skeptics about free will, maintaining that it is an illusion to be abandoned along with exploded ideas about soul and immortality. In response, some philosophers who are aware of recent scientific advances have stood up as free will defenders.

It is important to be clear about what is being denied and defended. Alfred Mele helpfully distinguishes three different conceptions of free will: supernatural, ambitious, and modest. Supernatural free will belongs to nonmaterial souls that are capable of immortality and can operate with considerable independence from the physical and biological mechanisms. Ambitious free will requires that agents have open to them alternative decisions that are compatible with past history and the laws of nature. Modest free will requires only that people have the ability to make rational, informed decisions and act on them when they are not subjected to undue force. Mele's own view is that humans do have modest free will, possibly have the "deep openness" required for ambitious free well, but probably lack supernatural free will.

My view is that supernatural free will is incompatible with current evidence and theories about minds and brains, making it just as dispensable as souls and immortality. Ambitious free will is probably unsustainable given increasing knowledge about how the brain works, because people's choices are not as unconstrained as deep openness requires. I suspect that what Mele calls modest free will survives but is so diminished that it hardly amounts to being free. I explore the issues using philosophical theories previously developed in the book concerning mind, knowledge, reality, and meaning and suggest how future neuroscience might deal with modest free will.

Mind

Chapter 2 argued that the best explanation of evidence about the mind is that all mental processes result from multilevel mechanisms interacting via neurons, molecules, mental representations, and social communications. Hence there is no need to postulate nonmaterial souls that are the traditional bearers of free will. In Christian theology, the doctrine of free will played an important role in explaining how there could be so much evil in a world created by an all-knowing, all-powerful, and all-caring God. Assuming that people have free will accounts for evil produced by people, although it still has difficulty with natural disasters such as diseases, earthquakes, and tsunamis. Chapter 6 provided biological explanations for evil human actions, and medicine and geography can adequately explain natural disasters. So supernatural free will is gratuitous.

Ambitious free well is more interesting, because it does seem to most of us that we have available deep openness to alternative courses of action. In 2016, I made an important decision that seemed to come with full freedom. I decided to retire from teaching in order to concentrate on writing this *Treatise*. I was under no external pressure, because mandatory retirement no longer exists in the University of Waterloo, and I was not under any internal pressure from any recognizable mental illness. It therefore seemed that it was open to me either to retire or to continue teaching, whichever I wanted. Similarly, people make important and unimportant decisions every day, in ways that seem unimpeded by natural causation.

But knowledge of the mental and social mechanisms sketched in this book, and described in more detail in *Brain–Mind* and *Mind–Society*, challenges this confidence about the deep openness of decisions. True, I would not have retired if I had chosen otherwise, but the neural mechanisms of choice are becoming increasingly well-known. My mental representations of the actions *retire* and *don't-retire* were not just words in my head but rather patterns of neural firing resulting through binding into the action *I will retire in 2016*. Decision making is a process of parallel constraint satisfaction, in which the brain accepts or rejects different encoded mental representations based on their interactions with others in a process that is both cognitive and emotional. In the end, I decided to retire because the neural representation of the option of retiring and becoming a full-time writer satisfied more cognitive and emotional constraints than the alternative choice to stay a professor.

My decision only seemed compatible with everything that has already happened and with the laws of nature because of my not knowing the full mechanisms of molecules and neurons that actually produced the decision. Practically, people will never know what is going on in all of the billions of neurons in their brains whose interconnections have been shaped by years of learning and unknown genetic structure, along with chemical influences of important molecules such as dopamine, serotonin, cortisol, and oxytocin. Nevertheless, psychological and neurological evidence continues to accumulate that the brain works mechanistically, making it more likely that deep openness is as much an illusion as supernatural free will. No one experiment refutes free will, but many experiments point to the conclusion that people have much less freedom than they think they do.

Nevertheless, we might be able to hang onto modest free will, which requires only being able to occasionally make rational, informed decisions. My decision to retire certainly seemed like it was rational and well-informed, because I took more than a year to make it, read broadly about the costs and benefits of retirement, and even adopted the unusual practice of taking a weekly vote on whether I wanted to retire. The weekly vote is an excellent way to avoid the impetuousness of sudden decisions and the perils of predicting future emotions. My votes considered the advantages of retirement such as the freedom to write and travel when I wanted and avoidance of the slings and arrows of outrageous bureaucrats. They also considered the costs such as reduced income and less contact with interesting students and colleagues. My decision also had a social element, because I discussed it in advance with friends and family members who were equipped to tell me if I was being an idiot. It therefore seemed that in this particular case modest free will reigned, even if it is swamped in other cases such as whether to finish a whole bar of dark chocolate.

Modest free will is compatible with experimental results in neuroscience that suggest that the brain has already made decisions well before they become available to consciousness, so that people's feeling that they are consciously making simple decisions is biologically unsustainable. Cognitive and social psychologists have also identified many limitations on the freedom with which people think they are acting, when in fact we are subject to cognitive biases and social distortions.

Mele rightly points out that these experiments only show that people have much less free will than they think, not that there are no important cases where modest free will operates. In my retirement case, my deliberation stretched over years and was free of external coercion and internal disruption, so in this case I was rational and well-informed, and free at least in some weak sense. My decision certainly displayed the vital need of autonomy, which only requires that actions are self-chosen in ways consistent with one's own interests and values, not the supernatural freedom that could only be possessed by a nonmaterial soul.

Nevertheless, although my retirement decision was made in a rational and well-informed way, how much freedom did I have to make the decision in this way? My approach to the problem of retirement drew on decades of experience studying the strengths and weaknesses of people making decisions, including teaching students about it in critical thinking classes. This experience reflects all the different kinds of learning that go into establishing connections between neurons that determine how they contribute to the process of parallel constraint satisfaction. My temperament and inclinations to make decisions in this way arose just as all personality dimensions do, through a combination of genetics, epigenetics, early childhood learning, and decades of subsequent experience, where learning is just modifying synaptic connections.

Hence, I chose freely to retire, but I did not choose freely how I chose. I could not help making the decision in a rational, well-informed way, as opposed to an impulsive way driven by fleeting emotions more affected by molecular and physiological changes than by rational evaluations. At other times, like everyone else, I sometimes succumb for situational reasons to impulse over rationality. Hence my "modest" free will here seems too modest to deserve the name "free."

In the discussion of action in *Brain–Mind* (chapter 9) and the examination of legal responsibility in *Mind–Society* (chapter 11), I advocate abandoning the term "free will" in favor of the looser notion of "freeish will." All that is required for freeish will is that people be independent of external social coercion such as someone holding a gun to their heads and from deviant internal neural causation such as diseases caused by breakdowns in mechanisms. Freeish will suffices for the nonmetaphysical autonomy that is crucial for moral responsibility, just societies, and meaningful lives.

But the issue of modest free will versus freeish will is far from settled, because there is enormous room for deeper understanding of decision, intention, action, and consciousness that can only be gained by improved theories with richer mechanisms. *Brain–Mind* (chapter 9) provides a theory of intentions as semantic pointers that interact with emotions, which are also semantic pointers, to generate actions. This theory fits perfectly with the explanation of consciousness as semantic pointer competition, providing a promising suggestion about how conscious intentions actually work.

These accounts of intention, action, and consciousness need to be thoroughly integrated with each other and developed further in line with experimental findings from neuroscience and psychology. Because autonomous actions are supposed to be *self*-chosen, the integration should include the theory of the self as multilevel mechanisms provided in *Brain–Mind* (chapter 12).

The result would be a much richer account of what goes on in the brain when cognitions, intentions, and emotions interact to cause decisions, consciously or unconsciously. It would fill a major gap in current psychology that is obsessed with "dual process theories" that are not actually theories because they do not specify mechanisms. The dual processes are supposed to operate in two "systems" that are (a) fast, automatic, and unconscious or (b) slow, deliberate, and conscious. All of these accounts are sketchy about how the systems work and worse than sketchy about what determines which system people are adopting at any given time.

In contrast, the account of intentions as competitive neural firings provides a plausible suggestion about how these systems operate and hence about when and how people think using conscious deliberations that are candidates for modest free will. *Brain–Mind* (chapter 9) gives a preliminary explanation of how neural mechanisms produce thinking fast and slow, with transitions between automatic and deliberate modes, but much theoretical and experimental research is needed.

As with my retirement decision, even if people choose more freely when they are using the slow, deliberate, and conscious system, it is much less plausible that people can choose to be in this system. Future neural explanations of automatic and conscious decision making should clarify considerably what is actually happening when people act badly and well. I therefore expect that future research will shed considerable light on the nature of will that is modestly free or freeish. Already enough is known to kiss goodbye to free will that is supernatural and ambitious with respect to deep openness.

Knowledge

The problem of free will is also interesting for epistemology. How can we justify claims that people have free will of various sorts? For some free will defenders with little interest in scientific evidence, it is just intuitively obvious that people have free will, but chapter 3 argued that intuitive obviousness is more a matter of psychological and social history than epistemic merit. Mele presents free will skeptics as trying to give deductive arguments from experimental results to the nonexistence of free will, but arguments from scientific evidence are inductive and abductive rather than deductive.

In keeping with coherentist epistemology, the best procedure is to determine whether the hypothesis that people have free will is part of the best explanation of human action. Because of advances in neuroscience in its growing partnership with psychology, the emerging explanations come from neural mechanisms, which are increasingly applying to a wide range of evidence about human decision and action. The current state of cognitive science suggests that there are better available explanations of actions than those based on supernatural or ambitious free will, so we can eliminate them from further consideration unless some new evidence comes to light that revives them. For example, if solid and convincing evidence suddenly appeared that there are dead people in heaven making choices using their incorporeal souls, then neural theories of decision making would evaporate.

It remains to be seen how much free will is compatible with future, improved theories of the neural causes of decision and action. But the methodology for evaluating claims about free will survive: consider which hypotheses about free will are part of the overall best explanation as determined by explanatory coherence that takes into account all available evidence and alternative hypotheses.

Reality

Metaphysics is also relevant to the problem of free will because of the issue of determinism. If everything has a cause, then human actions have causes, which seems to be incompatible with humans having free will. The first question is whether determinism is true, and chapter 4 allowed for the possibility that causation is merely probabilistic, perhaps because of quantum indeterminacy. However, the element of chance introduced into human decision making is not encouraging for any kind of free will, supernatural, ambitious, or modest. We want our actions to be produced by our rational decisions, not by chance.

A more interesting question comes from consideration that the causation of human action is a case of multilevel emergence. Perhaps emergence from one level to another could introduce a hint of free will, as there are many cases where wholes have properties not found in their parts because they result from the interactions of the parts, not mere aggregation. Could free will be a property of a whole person even though the parts in the form of molecules, neurons, and brain areas are governed by causal mechanisms? Perhaps this emergence could come from top-down causation because social interactions are important for the causal processes by which people make their decisions, when people are heavily influenced by the attitudes and behaviors of others.

But multilevel emergence provides little encouragement for the hypothesis of even modest free will. The ability of people to make occasional decisions that are rational and well-informed results jointly from social influences, interacting mental representations, neural firings, and chemical reactions. Is there any reason for supposing that these add up to free will through multilevel emergence? Attributions of emergence need to be based on evidence whose explanation requires recognition of special properties of wholes. Current brain mechanisms do not require the addition of free will as an emergent property of minds, although perhaps future investigation will justify the addition.

Morality and Meaning

The importance of the problem of free will for philosophy comes more from ethics than metaphysics. Christianity needs free will to explain the existence of evil and to allow room for judgments of praise and blame. Free will is required for assessing people as committing sins and deserving eternal punishment. Even without this religious background, legal systems need to make decisions about responsibility in order to determine fair and effective means of dealing with criminals. The most aggressive free will skeptics also deny moral responsibility on the grounds that we cannot blame people for choices that they did not freely make. Hence trials, punishment, imprisonment and the whole legal system become morally unjustifiable, with ensuing chaos. The explanation of evil in chapter 6 suggests that there will always be psychopaths.

To avoid legal anarchy, we need a good psychological theory of how people make decisions and actions and a good ethical theory of what makes actions right and wrong. My account of decision making by emotional coherence and of actions by causation through intentions and emotions is a start on the psychological theory. Chapter 6 provided a candidate for the ethical theory, needs-based consequentialism. What becomes of moral responsibility when we combine the neural determinants of action with the view that actions are right or wrong based on how well their effects satisfy vital human needs?

The happy answer is that holding people morally responsible for their actions and duly punishing them can be fully justified. First, having a culture in which people are held responsible for their actions is a useful way to ensure that most people behave in ways that are consistent with human needs. Societies that have high degrees of trust and social accountability tend to have less crime and other threats to human need satisfaction. Second, when people transgress and horribly violate the vital needs of other people through actions like murder and rape, society has good consequentialist reasons for removing them as threats to others for appropriate lengths of time.

Justifications for punishment then include both (a) preventing additional harm by the perpetrators and (b) deterring them and others from committing further crimes. The traditional justification for punishment based on retribution for freely committed sins goes out the window, but there still remain ample grounds for punishment in severe cases. The psychopaths and pedophiles discussed in chapter 6 may have problematic brains that make them less than free, but their threats to the need satisfactions of other people outweigh the cost to the criminal of incarcerating them.

Similarly, neural explanations of human behavior that challenge free will pose little menace for the meaning of life when it is construed as pertaining to vital needs rather than supernatural connections. The elimination of gods from the scientific picture of the world is only a threat to the meaning of life if meaning is assumed to emanate from religion. My account of the meaning of life in chapter 8 based on vital needs satisfaction is fully compatible with all recent developments in psychology, neuroscience, quantum physics, and other branches of science. Even without free will, you can still have love, work, and play to help satisfy your needs for relatedness, competence, and autonomy.

Some philosophers and psychiatrists maintain that advances in brain science have *no* implications for philosophical issues about morality and the meaning of life. But I want philosophy to improve in step with scientific progress rather than by consulting intuitions that are merely a reflection of current prejudices and lack of knowledge. Surveying ordinary people about what they think about free will and determinism merely pools ignorance. Instead, we can draw on increasing grasp of minds, brains, and societies to achieve better plans for how people can thrive.

This progress requires improved theories that demand substantial conceptual change. The reconceptualization of humans as biological animals that began with Darwin has to continue with increasing understanding of the biological determinants of decisions and actions. Some concepts have to be eliminated, for example soul, immortality, and supernatural free will. New concepts have to be introduced, for example in the understanding of decisions as the result of parallel constraint satisfaction carried out by interacting neurons in the prefrontal cortex. There may have to be additional differentiations of kinds of actions to help figure out which ones are the result of freeish deliberation. The brain revolution can illuminate the problem of free will when philosophy realizes that it needs to ally with science.

At the same time, investigations of free will should benefit from improvements in all areas of philosophy. Better philosophy of mind, epistemology, metaphysics, and ethics will all contribute to better answers to questions about free will. For example, an improved understanding of causality beyond the proposal in chapter 5 should mesh with improved neurocomputational models of conscious intention and action to deepen understanding of the extent to which actions are free.

MATHEMATICAL KNOWLEDGE AND REALITY

Since Plato, many philosophers have looked to mathematics to provide a guide for philosophy, which could share in the certainty and necessity that mathematics seems to provide. This influence has been unfortunate, because it provides a distorted image of how philosophy can accomplish its aims by gaining understanding of knowledge, reality, morality, meaning, and beauty. The chapters of this book exhibit the advantages of taking science as more central to philosophy than mathematics. Moreover, the view of mathematics as possessing a priori certainty and necessary truths is open to challenge.

Nevertheless, no system of philosophy would be complete without an account of mathematics that explains its practice and applicability to the world. Mathematical knowledge has great practical importance shown in applications to technology and science, but it also seems to possess a pristine elegance that makes it less open to challenge than other fields. There is currently no good account of how mathematics can have these seemingly contrary properties of being both practically useful and amazingly solid.

Social cognitivism should be able to furnish a natural philosophy of mathematics, but sadly there are serious gaps in current knowledge about how mathematics works in minds, society, and the world. I hope these gaps will be filled by future research in science and philosophy and try to point in some useful directions. I propose conjectures about mathematical knowledge and reality that need to be defended by showing that they are superior to currently available alternatives:

1. Mathematical knowledge develops in human minds through the construction of semantic pointers for concepts and rules.
2. This knowledge results from a combination of embodied sensory-motor learning (e.g., integers) and from transbodied bindings (e.g., irrational and imaginary numbers).
3. Mathematical knowledge is not simply cognitive, for emotions also contribute to mathematical discoveries and judgments such as appreciation of beauty.

4. Like other kinds of knowing, mathematics is not a purely individual enterprise but also depends on communities that provide support, inspiration, and communication.
5. Mathematical knowledge applies to the world in the same way that scientific knowledge does, as approximate correspondence to reality.

Philosophers and mathematicians have long feared that the objectivity of mathematics would be threatened by the incursion of psychological factors, but the opposite is true, just as in epistemology and ethics. Understanding how mathematics operates in human minds, including emotional and social factors in addition to cognitive ones, enhances rather than undercuts the epistemic qualities of mathematics. However, attention to the history of mathematics with a psychological perspective does challenge some exaggerated views of its certainty and necessity.

Issues and Alternatives

Mathematics raises challenging questions to be answered by competing philosophical theories. Why does mathematics seem more certain than other kinds of knowledge? Why do many philosophers think that mathematical truths are necessary, true not only of this world but of all possible worlds? Why is proof a special way of achieving knowledge? How is mathematics, which seems to transcend the world in the form of necessary truths, nevertheless applicable in practical spheres such as technology? How do finite human minds grasp mathematical ideas about infinity, for example concerning the apparently infinite number of integers and the apparently even larger infinity of real numbers? How do mathematicians generate mathematical knowledge? What are numbers anyway? A good theory of mathematics should be able to answer these questions, or at least to show why the question is inappropriate based on false presuppositions, as I will argue for the questions about necessity and certainty.

Competing philosophies of mathematics include Platonism, logicism, formalism, fictionalism, social constructivism, empiricism, structuralism, and Aristotelian realism. Platonism is the view that numbers and other mathematical objects such as sets exist in abstract form independent of human minds and the physical world. Many mathematicians find Platonism appealing because it explains their sense that they are grasping eternal realities that do not depend on the vagaries of mind and world. Platonism would explain the apparent certainty and necessity of mathematics but has trouble with why the abstractions of mathematics are often so useful in dealing with the world of agriculture and computers. Platonism also seems to require dualism, because only a nonmaterial soul could somehow grasp the nonmaterial reality of completely abstract entities. Brains are just not up to the job.

To counter the metaphysical extravagance of Platonism, many alternative philosophies of mathematics have been developed. One popular view in the twentieth century was logicism, which claims that all of mathematics can be reduced to formal logic. The hope was that logic, with basic rules such as modus ponens and obvious mathematical principles such as $X = X$, could be used to generate all truths of arithmetic. This hope was dashed by Gödel's proof that any formal system that is both consistent and adequate for arithmetic is incomplete: there are statements that cannot be proved or disproved in that system.

Formalism is a philosophy of mathematics that treats mathematics as a set of marks on paper, a syntactic exercise that uses proofs to derive formulas from other formulas. This approach has problems with both the certainty of mathematics and its applicability to the real world. Similarly, the claim of fictionalism that math is just make believe has trouble explaining its usefulness and convincingness. Another skeptical view is social constructivism, which emphasizes the cultural aspects of mathematics.

Empiricism claims that all knowledge derives from the senses (chapter 3). From this perspective, mathematical truths such as $4 + 5 = 9$ and the Pythagorean theorem are just generalizations about observed reality. As a matter of empirical fact, four objects added to five objects make nine objects. This view explains the practical applicability of mathematics to the world but has trouble dealing with the apparent certainty and necessity of mathematics, especially as it concerns abstract entities such as infinite sets.

Structuralism is a philosophy of mathematics that denies the reality of objects such as numbers but maintains that math describes abstract structures that are the basis for mathematical truths. For example, $4 + 5 = 9$ is a truth about the structure of arithmetic, not about particular numbers. Like Platonism, structuralism sees mathematics as concerned with abstract entities and therefore has the same difficulty of figuring out how mathematics applies to the physical world. Like structuralism in the philosophy of science dismissed in chapter 4, structuralism in the philosophy of mathematics has the problem of figuring out how there can be structures and relations without objects that are related. In set theory, a relation is a collection of ordered objects, which makes no sense if there are no objects to order.

The current philosophy of mathematics that fits best with what is known about minds and science is James Franklin's Aristotelian realism, inspired by Aristotle's claim that concepts are about the world rather than abstract forms postulated by Platonic realism. According to Aristotelian realism, mathematics is the science of quantity and structure, just as physics, chemistry, and biology are sciences about the world. To explain how mathematical knowledge develops, Aristotelian realism needs to merge with a neuropsychological account of mind such as the Semantic Pointer Architecture.

Math in the Mind

The first step toward how mathematics works in the mind, the world, and society is a theory of the mental representations and processes required for the development and learning of mathematics. Platonism obviously fails as the psychology of mathematics because the abstract nature of mathematical objects and truths would divorce them from explanation based on mental mechanisms such as neural processing: there is no apparent way how neurons could interact with numbers as abstract ideas. Plato thought that all mathematical concepts and in fact all concepts are innate, with experience merely reminding people about abstract ideas that their souls knew before birth.

This view is problematic for many reasons, from the lack of evidence that souls exist before birth to the difficulty of understanding how experience could reactivate such knowledge. Perhaps some basic concepts about objects and quantity are innate but not numbers like 10^{100} and relatively new areas of mathematics such as calculus and abstract algebra. More plausibly, mathematical knowledge consists of learned concepts and beliefs that include axioms and theorems, comprehensible as semantic pointers.

Mathematicians are mature humans with sophisticated linguistic capabilities, so it is tempting to think of mathematical knowledge as a matter of language. But some primitive numerical abilities are found in nonhuman animals and nonvocal human infants. For example, fish, bees, pigeons, and rats can all detect differences in small numbers of objects, as can infants only six months old. Stanislas Dehaene argues that animals and humans have the basic ability to perceive and manipulate numbers. Other theorists prefer to explain nonlinguistic numerical abilities as resulting from a more basic sense of magnitude, with size more basic than counting. Either way, we need to recognize that people's ability to do mathematics is not just symbol manipulation but depends on more primitive sensory and motor abilities.

Such multimodal abilities are explained by viewing mental representation as semantic pointers, which accommodate verbal information bound into larger packages, where these packages can also include sensorimotor inputs. Groups of neurons can detect visual patterns such as small groups of numbers and bind these into larger representations such as *three cats* and *many dogs*. As chapter 2 argued for concepts in general, the sensory and motor component of semantic pointers allows them to account for embodied cognition in which thoughts are strongly affected by sensory and motor representations. The Semantic Pointer Architecture has been used to detect visual numerals and to do simple reasoning about numbers represented by convolutions of numeral shapes and relational information about number order.

Moving beyond sensory-motor inputs, can the semantic pointers explain important mathematical concepts such as *number, set, infinity*, and *shape*? Google's dictionary defines a number as "an arithmetical value, expressed by a word, symbol, or figure, representing a particular quantity and used in counting and making calculations and for showing order in a series or for identification." This definition would be more impressive if arithmetic were not circularly defined by the dictionary as the branch of mathematics dealing with numbers.

Alternatively, Table 10.1 presents a three-analysis of the concept of *number* in line with the theory that concepts are patterns of neural activation capable of integrating exemplars, typical features, and explanations. It might seem that all numbers are of equal importance, but most people are much more familiar with low numbers such as 2 and 3 as well as powers of 10: 149,356 is still a number, but it is not a standard example for most people. Mathematicians become familiar with many numbers that are mathematically interesting, such as π (3.14159 . . .), **e** (2.71828 . . .), and googol (10^{100}).

What counts as a number has expanded over the history of mathematics. Some cultures have only a limited range of number concepts such as *one, two,* and *many*, whereas the ancient Greeks developed ideas about the natural numbers that can grow very large. But they lacked the concept of zero as a number, which was a

TABLE 10.1

Three-Analysis of the Concept *Number*

Exemplars	2, 3, 10, 100, π
Typical features	Expressed by words such as "two," represents a quantity in a collection of units, used in counting, used in calculations, ordered
Explanations	Explains: patterns of things in the world, spatial and temporal structures, ability to count and calculate
	Explained by: mental mechanisms of representation and manipulation

development in ancient India. Early mathematicians were suspicious of negative numbers such as –2, and later extensions to imaginary numbers such as the square root of –1 and infinite sizes of infinite numbers have provoked much consternation. The ancient Greeks did not want to believe that there are irrational numbers that cannot be expressed as the ratio of integers, but proof of the irrationality of the square root of 2 showed that the concept of number has to expand beyond familiar exemplars.

The typical features such as counting apply to the most familiar numbers but become more problematic when extended to more abstract quantities such as π and **e**. Quantity does not serve well as a rigorously defining feature of number, since quantity is often defined in terms of number. Concepts of counting and quantity have important sensory-motor aspects, because people often count by pointing to one thing after another, using visual, tactile, and motor representations. Then the typical features of numbers are not purely verbal but connected to sensory-motor representations that can be bound into concepts of number used for more abstract calculations such as using π to figure out the area of a circle. The ordering aspect of numbers is a typical feature is embodied when numbers are viewed as arranged on a line.

The concept *number* makes explanatory contributions, for example in tracking changes in the world such as how two rabbits can quickly generate large numbers of rabbits. Numbers also explain practices such as being able to divide things among people, including fractional divisions such as pieces of orange. The explanatory role of numbers gets even more impressive when it is connected with mathematical principles in physics and other fields that deal with quantities. My attempt to explain numbers in terms of their mental uses is highly contentious, with many issues to be worked out.

Nevertheless, even in the absence of a tight definition, we can understand people as having a concept of number that combines exemplars, typical features, and explanations. This combination is explained as a neural process by the semantic pointer theory of concepts. I am not claiming that numbers are semantic pointers, just that concepts of numbers are semantic pointers. The relation between things and concepts of things was already discussed in chapter 3 and is further elucidated in the section on math in the world.

The discussion of the sensory and motor aspects of quantity, counting, and order might give the impression that all mathematical concepts are embodied and learned by experience. But we have seen that the semantic pointer theory of concepts also explains how concepts can be transbodied, going beyond sensory-motor representations by recursive binding. Contrary to empiricist philosophy of mathematics, concepts are not restricted to generalizations from observations but can also be constructed by combination using the process of convolution of patterns of neural firing.

Like concepts about theoretical entities discussed in chapter 4, mathematical abstractions can be formed by combining existing concepts. For example, the concept of prime number does not have to be learned inductively from examples, although it may have some exemplars such as 3, 5, and 7. Rather, *prime number* comes from combining concepts of number, divisibility, and negation. Conceptual combination carried out by the neural process of convolution quickly takes us beyond the limitations of embodiment.

Axioms and Theorems

Concepts say nothing about the world without being combined into beliefs. The number 4 alone makes no claims, but the belief that 4 + 5 = 9 asserts that if you have four things and five things then you have nine things. Mathematical truths do not have to be just generalizations about the observable world, because conceptual combination by binding into semantic pointers allows construction of beliefs that usefully transcend the senses. Arithmetic does serve to characterize quantities in the world but can also be based on different kinds of coherence. When mathematical principles are part of scientific theories, they gain explanatory coherence from how well they jointly explain evidence. But math is often different from science in consisting of axioms and theorems rather than hypotheses and explanations.

Like arithmetic, geometry contains general statements, but they usually concern shapes rather than quantities. The Pythagorean theorem can be expressed in various ways, for example in the words that in a right triangle the square of the hypotenuse is equal to the sum of the square of the other two sides. Alternatively, the theorem can be represented jointly by the diagram in Figure 10.3, which shows the lengths of the respective sides, and the equation: $a^2 + b^2 = c^2$. This theorem requires several concepts including *triangle, right angle,* and *hypotenuse* but combines them into a new claim supported both by observations in the world and by proof from axioms. The words, equation, and image are complementary, and all can be formed into mental representations as semantic pointers using the techniques spelled out in *Brain–Mind*.

Because of the historical influence of Euclidean geometry and later views such as logicism, mathematical knowledge is often construed as a system in which theorems are derived from a small set of axioms and definitions. Ideally, the axioms should be self-evident and the methods of proof for establishing

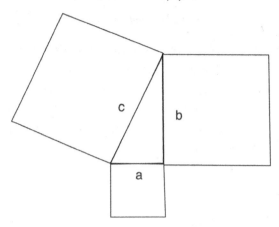

FIGURE 10.3 Visual representation of a right triangle.

theorems should be deductively precise, resulting in a whole body of knowledge that is unassailable. The section on coherence will suggest that this unassailability is exaggerated.

Nevertheless, philosophy of mathematics has to take seriously the role of axiomatic systems in mathematical practice. How can we understand axioms, theorems, and deductive inference within the Semantic Pointer Architecture? Axioms are easily understood as multimodal rules with an *if–then* structure, where the *if* and *then* parts can be verbal or sensory-motor. For example, Euclid's first postulate, that a straight line can be drawn between any two points, becomes the verbal rule: if you have two points, then you can draw a line between them.

But mental representation of this postulate is not simply a matter of binding linguistic entities into larger linguistic entities, because the semantic pointers that get bound together into the larger postulate can include sensory-motor aspects. For example, *draw between* is a motor activity as well as one that is visually perceived. Both the points and the line connecting them are also naturally encoded visually. So geometrical axioms and the theorems that depend on them are multimodal, as naturally captured by multimodal rules consisting of semantic pointers.

Outside of geometry, axioms and theorems are more likely to be just verbal, for example the first two axioms of Peano arithmetic: 0 is a natural number, and for any natural number x, $x = x$. The crucial link between axioms and theorems is proof, a series of logically correct steps from the axioms and definitions to the theorem. Logical steps include those performed by modus ponens: P and *if P then Q*, therefore Q. Eliasmith has shown that the Semantic Pointer Architecture can carry out modus ponens, but simulations of other aspects of mathematical proof are daunting challenges for future work.

Metaphors and Analogies

Contrary to the impression given by axiomatic systems, mathematics has much substantial cognitive flexibility that can be seen in the role of metaphors, analogies, and heuristics. George Lakoff and Rafael Núñez describe how arithmetic grows out of a metaphor of object collection, based on a mapping from the domain of physical objects to the domain of numbers. The actions of collecting objects of the same size and putting collections together map onto addition, and taking a smaller collection from a larger collection maps onto subtraction. Even mathematical ideas about infinity have metaphorical connections with processes that have beginning and extended states. Lakoff and Núñez describe mathematical knowledge as developing from a combination of embodied metaphors and conceptual blending. *Brain–Mind* (chapter 10) shows how metaphor and blending are naturally explained in terms of semantic pointers, but much theorizing and simulating is required to support the hypothesis that mathematical metaphors are the result of neural mechanisms.

Complex metaphors rely on analogies that map between two domains, and many historical examples show the value of analogical thinking for developing mathematical knowledge. For example, Descartes developed a new approach to geometry by analogy to algebra. *Brain–Mind* (chapter 6) proposes a theory of analogy based on semantic pointers, but the applicability of this account to mathematics needs to developed. Similarly, the heuristics that George Polya argued are important in mathematical thinking, such as decomposing a problem, should be translatable into multimodal rules with semantic pointers, but much research is required to carry out the translation.

Coherence and Necessity

Why does mathematics seem so much certain than ordinary empirical knowledge? It is easy to imagine crows that are not black but hard to see how $3 + 5$ could be other than 8. Chapter 3's arguments against a priori truths should extend to mathematics but leave the problem of explaining why mathematical truths seem so solid on grounds other than pure intuition.

My answer is that mathematical certainty is an extreme case of the high confidence that derives from multiple kinds of coherence. For example, we have three strong and interlocking reasons for thinking that the Pythagorean theorem will continue to hold. First, it matches innumerable observed cases where the size of hypotenuses has been calculated and turned out to conform to the theorem. That the theorem is actually true explains the evidence that it seems to be true.

Second, the Pythagorean theorem is deductively coherent with Euclid's axioms because of the proof that derives it from the axioms. There is no need to suppose that the axioms are self-evident, because they also achieve coherence with observations and the many other plausible theorems that can be inferred from them.

Third, the Pythagorean theorem shares in the explanatory coherence of the scientific and technological fields that employ it. For example, geography uses it to measure distances, and architecture uses it to construct buildings. For all of these kinds of coherence, the Pythagorean theorem is so centrally embedded in our web of belief that it has become hard to imagine that it is false.

Chapter 3 argued that imagination is not a good guide to truth because it is constrained by prejudices and ignorance. In contrast, the three kinds of coherence that support mathematics are each reliable in establishing sustainable truths, and pooling them creates even more reliability. Like geometry, arithmetic also benefits from this triple coherence, because it coheres with observations, axiomatic proofs, and usefulness in science and technology.

Certainty is never absolute, in keeping with the emphasis on fallibility in chapter 3. There have been a few historical cases where established mathematical views have been overturned, for example Euclid's fifth postulate that implies that at most one line can be drawn parallel to a given line through a point not on the line. Einstein's theory of relativity acquired substantial evidential support while employing non-Euclidean geometry that rejects this postulate. The apparent certainty of mathematics comes from its overall coherence, not from the metaphysical guarantee of Platonism, the purely deductive coherence of logicism, or the triviality of formalism.

My coherentist justification of mathematical knowledge cannot accommodate another apparent feature of it, necessary truth. Math is supposed to be true of all possible worlds, not just this one, as shown by our inability to imagine its falsehood. But chapter 3 argued that such failures are often signs of ignorance rather than profundity, so the best strategy is what chapter 5 called eliminative explanation. Necessary truths in general can be explained away as an illusion resulting from insufficient understanding of the interactions of mind and world. The illusion can be replaced by appreciation of the objectivity and less-than-absolute certainty that comes from multiple kinds of coherence.

Emotions

Mathematics is supposed to be the epitome of cold reason, but emotions are important for mathematical cognition in several ways. First, according to G. H.

Hardy, a mathematician is a maker of patterns, whose importance are judged in part by their beauty.

> The mathematician's patterns, like the painter's or the poet's, must be *beautiful*; the ideas like the colours or the words, must fit together in a harmonious way. Beauty is the first test: there is no permanent place in the world for ugly mathematics.

The experience of ugliness is an unpleasant emotion to be avoided as much as the positive experience of beauty is to be pursued. Chapter 9 showed how beauty is an emotional response to the coherent unification of attractive diversity, and Hardy is right that mathematical beauty is the same. Elegance results from how basic axioms and efficient proofs unify diverse theorems.

Second, in order to pursue the thousands of hours it takes to become an accomplished mathematician, people have to find it at least interesting, an epistemic emotion accompanied by moderate physiological arousal and modest goal relevance. Third, in order to put in the much greater degrees of effort required to make substantial breakthroughs, mathematicians need to find their work exciting, a much stronger epistemic emotion. Fourth, emotions also operate in the social context of mathematics, for example when mathematicians evaluate each other's work as being surprising rather than dull, or as an exciting breakthrough rather than just a mundane extension.

Mathematics in the World

Some philosophers have argued that we should believe in numbers and other mathematical structures not for Platonist reasons but simply because they are indispensable for doing science. But what does indispensability amount to? Mathematical areas such as arithmetic and geometry are useful for generating explanations and producing calculations that are part of technology and science. Chapter 4 argued that technology contributes to inferring that science is about the world, and a similar argument works for math.

If the Semantic Pointer Architecture shows how mathematics can operate in minds and brains, why is math so amazingly useful in the world? Platonism faces the problem of figuring out why heavenly ideas are so useful in worldly applications ranging from building bridges to quantum computing. One solution to this problem would be to meld Platonism with idealism. If there is no world external to human minds, and if mathematical ideas are just abstractions in human minds, then the mind-to-world connection is unproblematic because everything is mind.

But I argued in chapter 4 that not everything is mind, which is a relatively recent addition to the universe according to available evidence. So the math-to-world connection remains problematic.

One straightforward solution to this problem would be the empiricist one that mathematical statements are just generalizations about the world, so it is automatic that they apply to the world. However, like the exaggeration that all mathematics is embodied, empiricism is incompatible with recognition that mathematics operates with many layers of abstraction such as sets, groups, and geometric manifolds that go far beyond ideas that fit with sensory and motor experience.

Formalism and logicism also have serious problems with applicability. If mathematics is just a system of marks, or consequences of logical truths, its applicability to the world remains unexplained. Fictionalism says that mathematical objects are fictions just like unicorns but is flummoxed about why arithmetic, geometry, and other branches of mathematics are so much more useful for dealing with the world than theories of unicorns.

Another misguided way of understanding the relation between mathematics and the world is to say that the world is fundamentally mathematical. This view originated with the followers of Pythagoras, who were so impressed by the applicability of mathematics to the world that they concluded that the world is nothing but mathematics, with numbers as the fundamental reality. The Pythagorean fallacy is to leap from the historical observation that mathematics has been enormously valuable in describing and explaining the world to the conclusion that reality is just mathematics.

Max Tegmark uses the following argument in his book *Our Mathematical Universe*.

1. There exists an external reality completely independent of humans.
2. Physics aims toward a complete description of this external reality in the form of a Theory of Everything.
3. The Theory of Everything would avoid words and concepts in favor of mathematical symbols.
4. Therefore, our external physical reality is a mathematical structure.

I agree with step 1, for the reasons given in chapter 4, but the rest of the steps are shaky.

Even if physics manages to come up with a theory that unites general relativity and quantum theory with a new account of quantum gravity, the result would not be a Theory of Everything. For reasons given in chapter 5's discussion of reduction and emergence, it is implausible that quantum gravity would tell us much about biochemistry, evolution, neuroscience, psychology, and sociology. Moreover, we currently have no idea what the theory of quantum gravity will look like, or if it will ever be achieved. The current candidates of string theory and loop quantum gravity use lots of math, but they also abound with concepts that help to make the symbolic equations meaningful. If quantum gravity was just math with uninterpreted symbols, people might be able to use it to make predictions, but they would still find it unsatisfying in its failure to provide explanations. The legitimate aims of science include understanding, not just prediction. That is why there are now more than a dozen competing interpretations of quantum theory from scientists and philosophers not content with its impressive predictive accomplishments.

The philosophy of mathematics needs a metaphysical theory that complements the epistemology based on semantic pointers and coherence. The strongest current candidate is Aristotelian realism, which is realist in that it takes mathematical claims to be true or false about the world. According to Franklin, there were mathematical properties and relations in the world long before people came along to develop mathematical knowledge. There were quantities of stars and other objects, and many other properties in relations such as symmetry and ratio. Mathematics is the science of quantities and structures in the world.

The main problem with this view is that much of mathematics does not seem to be about the world. For example, transfinite set theory developed from Cantor's creative proofs that there is an infinite number of infinite sets of different sizes. Many areas of mathematics developed in isolation from physical considerations of the world, for example group theory and the theory of complex numbers, which only later turned out to be useful in physical applications such as Schrödinger's equation in quantum theory.

I think that the applicability of mathematics to the world is best explained by two factors. First, much of mathematics is motivated by the desire to understand the world, from descriptive generalizations to explanatory ventures in physics. Hence many mathematical concepts such as those in basic arithmetic and geometry are at least partly embodied and tied to empirical generalizations with what chapter 8 called word-to-world meaning.

Second, conceptual combination allows mathematical concepts to transcend the world, with word-to-word meanings akin to those found in fictional novels, plays, and films. But just as such works sometimes turn out to reveal a lot about the world, for example in the psychological insights of Leo Tolstoy and Jane Austen, some of the transcendent concepts that mathematicians develop turn out to be useful for physicists and other scientists. Hence latterly valuable concepts like complex numbers and non-Euclidean geometry begin to acquire word-to-world semantics as well, just as Aristotelian realism requires.

My metaphysics of mathematics is thus a combination of realism and fictionalism, varying with different domains. For mathematical domains derived from sensory-motor experience or connected with scientific theories, realism rules. But for domains that merely reflect the creative power of conceptual combination, the products are best viewed as fictions, although they may eventually turn out to be connected with the quantities and structure in the physical world when they are incorporated into scientific theories.

So what are numbers? The psychological suggestion that numbers are concepts understood as patterns of neural firing is not plausible, because there seem to be far more numbers, perhaps infinitely more, than there are brain-bound concepts limited by measly billions of neurons. For now, we can only say how the concept of number functions, as neural patterns with the typical features laid out in the three-analysis in Table 10.1. I am not happy with any of the standard views that numbers are abstract entities or fictions, but a better account remains to be developed.

Mathematics in Society

Postmodernists deny both mind and world and see reality and mathematics as socially constructed. I have argued that math is tied to mind *and* world, but there are undoubtedly ways in which mathematics is importantly social. Historically, the problems that mathematicians work on have often been shaped by social circumstances with respect to enterprises such as agriculture, finance, business, technology, war, and science in general. Recognizing that mathematicians are affected by their social contexts is very different from saying that these contexts determine the content of their mathematics—the concepts, axioms, proofs, and theorems that they develop.

There have been social influences on mathematical controversies, for example in the battle between the continental approach to the calculus as developed by Leibniz versus British adherence to Newton's notation. But such controversies are resolved by mathematical criteria rather than social power.

Nevertheless, the practice of mathematics has identifiable social mechanisms. The proofs proposed by one mathematician have to be checked by other mathematicians who sometimes discover errors to be corrected. It takes much collective work to verify and improve complex proofs such as Andrew Wiles' proof of Fermat's Last Theorem. Like scientists, mathematicians sometimes collaborate to produce work that they would not be able to accomplish alone. An extreme example is Paul Erdős, who had hundreds of collaborators. Mathematicians, like the scientists in chapter 4 and the artists in chapter 9, often belong to emotional communities on which they depend for support and inspiration. Hence mathematics depends on social mechanisms of verbal and nonverbal communication.

I have tried to show how Franklin's Aristotelian realism can be enriched by Eliasmith's Semantic Pointer Architecture to provide a new approach to the philosophy of mathematics. To make this venture convincing, my speculations about mathematical concepts and theorems as semantic pointers have to be tested by the standard cognitive-science method of building computational models that simulate the thinking of mathematicians and ordinary people doing math. I hope that future philosophers and theoretical neuroscientists will pursue these tasks. The result would be a naturally superior alternative to the rationalism and empiricism that have dominated the philosophy of mathematics.

NONHUMANS: ANIMALS AND MACHINES

This book and its companions, *Brain–Mind* and *Mind–Society*, have been solely concerned with humans. But there are more than a million species of nonhuman animals on this planet, some displaying considerable intelligence in their abilities to communicate and learn to solve problems. (From here on, I use "animals" as short for "nonhuman animals" and "nonhumans" to cover machines as well.) Moreover, machines are becoming increasingly sophisticated, through rapid developments in artificial intelligence in applications such as voice recognition, machine learning, and robotic cars.

How do my philosophical conclusions concerning knowledge, reality, morality, justice, meaning, and beauty apply to animals and machines? If beasts and bots turn out to have the same mental abilities, social capacities, and needs as human beings, then should they be subject to the same standards of morality and justice?

It will take a whole book to review the current state and prospects of machine intelligence, in comparison with findings about animal intelligence. Here I only look at the philosophical significance of what is known about the capacities of nonhumans. Future investigations of more powerful machines and smarter animals will likely require revisions of these tentative conclusions.

Animal Minds

What does it take to have a mind? Panpsychists think that all physical objects have rudimentary consciousness, but chapter 2 argued against attributing mind or consciousness to objects that lack behavioral indications and mechanistic capabilities for sensation, perception, problem solving, learning, and communication. Some plants have simple systems of sensation and communication, for example when trees detect light, modify their growth systems to increase light, and send signals to other trees. But the cells that enable plants to detect and transmits signals lack the ability of neurons to encode and interpret environments in complex ways. So it would be a huge stretch to attribute minds to plants, let alone to rocks.

On the other hand, animals ranging from bees to elephants have large systems of neurons, from a few hundred in simple worms to billions in advanced mammals. Bees can learn the location of food sources and perform dances that tell other bees where to find them. Octopuses can solve complicated problems such as removing lids from jars and learning to navigate mazes. Crows can use tools such as twigs to solve problems of food retrieval that require multiple steps. Prairie dogs communicate with each other by signals that indicate the size and appearance of interlopers. Elephants display complex emotions including grief. Dolphins use clicks to communicate with each other. Therefore, by the same kind of inference to the best explanation that justifies attribution of minds to people, we can conclude that animals have minds too.

But how do these minds compare to those of humans? The answer must go beyond behavior to consider its underlying causes, which requires assessment of the mental and social mechanisms across species. To begin, we can ask whether animals have semantic pointers and recursive binding. Animals undoubtedly use neurons similar to ours to encode features of their environments, so they have basic mental representations consisting of neural firing patterns that can be generated by stored synaptic connections. Moreover, visual bindings have been detected in animals as primitive as newborn chicks. Some birds and mammals are able to detect relations among objects such as same and different. At least 10 species of animals have sufficient cognitive capacity to be able to identify themselves in mirrors. All these abilities suggest that animals have the mechanisms of neural representation and binding to produce semantic pointers.

Nevertheless, current evidence finds strong differences between humans and animals in their ability to deal with complex relations. As described in *Brain–Mind* (chapter 6), humans not only understand relations but also relations among relations, for example when they notice causal analogies between two different systems, such as the atom and the solar system. This noticing requires the brain to perform repeated bindings of bindings to produce semantic pointers built up out of other semantic pointers. Other kinds of high-level thinking that depend on recursive binding include: constructing and understanding complex sentences with embedded clauses, such as this one; abductive inference to nonobservable entities, such as gods and atoms; and long-term planning with alternative contingencies, such as taking a train if airports are closed.

Perhaps this larger binding capacity is just the result of having larger brains with more neurons, because binding by convolution takes a lot of neurons. However, elephants and whales have more neurons than humans but modest intelligence compared to us. Brainpower may also depend on the connectivity of neurons, kinds of neurons, and range of neurotransmitters.

So it remains unknown what makes human brains better at the recursive bindings needed for language and complex problem solving. On the basis of behavior and neuroanatomy, it may be reasonable to conclude that mammals have simple emotions found in humans such as happiness and fear. But a much bigger leap in the ability to represent oneself in society is required for social emotions such as guilt, shame, and fear of embarrassment.

Animals clearly have biological needs such as food and water, but they seem to lack the representational capacity for the human psychological needs for autonomy and competence. Relatedness is easier to identify, as many animals operate in social groups. But much more has to be known about animal brains in comparison with humans to determine whether they have needs that support the concerns about morality, justice, meaning, and beauty discussed for humans in chapters 6 to 9.

Machine Minds

Reflection about machines is both easier and harder. It is easier because we know how machines work since we build them, but harder because artificial intelligence and robotics are rapidly moving fields. The evolution of animal intelligence is extremely slow, whereas breakthroughs in artificial intelligence and robotics are frequent occurrences. The impressive developments within the past decade include Apple's Siri, IBM's Watson, Google's AlphaGo, and various companies' self-driving vehicles. None of these use semantic pointers and recursive bindings, but robots being developed by Chris Eliasmith's research group already do.

Computers currently display a modicum of creativity, for example in Chef Watson's production of original food recipes, AlphaGo's ability to generate moves in the game Go that surprise and impress experts, and programs that generate better art and music than most people. But machines and nonhuman animals still fall far short of the ability of some humans to produce solutions that are new, surprising, and hugely valuable. The discussions of art and music in chapter 9, along with the examinations of individual creativity in *Brain–Mind* (chapter 11) and collective creativity in *Mind–Society* (chapter 13), suggest that fully creative computers will require at least the following capacities:

1. Multimodal representations for vision, sound, and so on.
2. Conceptual combination into new ideas and hypotheses.
3. Analogy use to adapt and expand previous solutions.
4. Procedural creativity that generates new methods as multimodal rules.
5. Continuous evaluation of creative products, accomplished in humans by emotional reactions.

No current machines have the bindings of cognitive appraisal and physiological perception that constitute emotions, but machines continue to advance in both cognition and physical complexity. Machines currently do not have any vital needs crucial for life, because even the most sophisticated robots are not alive. Hence needs-based consequentialism and the attendant theories of justice and meaning can safely exclude machines from moral consideration. It remains to be determined how this might change if advances in hardware and software provide machines with such typical features of biological life as the ability to reproduce.

There are thus many open questions concerning the current capacities and future possibilities for machine and animal minds. I hope that the accounts of human minds and societies in this *Treatise* provide a useful specification of human capabilities that serves as an inspiration for further investigations.

SUMMARY AND DISCUSSION

The discussions in this concluding chapter are tentative, because not enough is currently known to provide solid solutions to problems about free will, mathematics, and nonhuman thinkers. But I hope the chapter serves to suggest avenues for future deliberation, advancing philosophy along with science on topics of great significance.

These topics provide further reason to think that philosophy still matters as science develops. Philosophy is not the clever solution of logical and linguistic puzzles, nor the futile attempt to extract profundity from introspection or word salad. Rather, it attempts and sometimes succeeds at answering general and normative questions about knowledge, reality, and morality by connecting with the most relevant scientific findings about mind, society, and the world.

In his book *Mind and Cosmos*, Thomas Nagel argues against attempts to naturalize mind, meaning, and value, which he thinks are as fundamental as matter and space-time in accounting for what there is. He advocates a nontheistic but also nonnatural teleology that supposes that the natural world has a propensity to give rise to beings smart enough to be ethical. The purpose of the universe assigned by this teleology is as evidence-free as theology and only gains plausibility from the great difficulty of using science to help explain the importance in human lives of mind, meaning, and value.

My *Treatise on Mind and Society*, encompassing *Brain–Mind* and *Mind–Society* as well as the present book, is a systematic attempt to overcome this difficulty. I have drawn on a recent theoretical breakthrough in neuroscience, Chris Eliasmith's Semantic Pointer Architecture, to develop systematic answers to a broad range of philosophical questions. We now have a theory about how brain mechanisms support the full range of mental operations, including emotion, creativity, consciousness, and communication. Hence we can develop new and plausible accounts of major issues in all areas of philosophy, from epistemology to aesthetics.

Philosophy, cognitive sciences, and social sciences will continue to evolve together, with philosophy making important contributions by virtue of its greater generality and normativity. Instead of retreating into an introverted role focused on its own history and methods, philosophy can be extraverted by attending to real-world problems, including humanistic concerns about the arts. In universities, philosophy departments can serve as intellectual hubs connecting science, technology, and the humanities.

Philosophy can benefit from developing new tools as well as new ideas. This book has shown the fertility of new methods for pursuing philosophical questions, including three-analysis for understanding concepts, value maps for displaying conceptual systems, and coherence construed as parallel constraint satisfaction. Such methods contribute to a philosophical procedure that collaborates with science while avoiding fruitless techniques of thought experiments, introspection, and pure reason.

I have explained mind as resulting from multilevel mechanisms that involve molecules, neurons, mental representations, and social interactions. The same mechanisms make sense of meaning, both the meaning of language that operates in human minds and the meaning of life more generally. Values operating across knowledge, morality, society, and art emerge not from some vague teleology or the will of God but from human nature rooted in the cognitions, emotions, and social relations that result from brain mechanisms.

The result is not as consoling and reassuring as the insupportable belief that everything happens for a reason. Sadly, it does seem that causality, chance, and accidents are part of the fabric of human existence. Nevertheless, people can achieve valuable and meaningful lives by attending to the vital needs of others and themselves, while striving with science and philosophy to understand the world and our place in it.

CONCLUSION: 12 RULES FOR PHILOSOPHICAL LIFE

As this book goes to press, Jordan Peterson's 12 *Rules for Life* is on bestseller lists, despite the commonplace nature of his rules, which boil down to simple injunctions such as take care of yourself and tell the truth. Perhaps the appeal comes from their embedding in a philosophical framework that combines obscure metaphysics, Christian mythology, existentialist angst, and narrow-minded individualism. As an antidote, let me finish with an alternative set of rules that summarize the major philosophical lessons of *Natural Philosophy*:

1. Base your beliefs on evidence, not faith, and consider alternative hypotheses (chapters 1, 3).
2. Understand your mind as a brain working with multiple biological and social mechanisms (chapter 2).
3. Be emotional but reasonable (chapters 2, 6).
4. Respect reality (chapter 4).
5. Explain using mechanisms, not just stories (chapter 5).
6. To be moral, act to meet the vital needs of yourself and others, including biological and psychological needs (chapter 6).
7. Equality does not have to be perfect but should cover everyone's vital needs (chapter 7).
8. Recognize that freedom is not absolute but requires respect for the freedom and needs of others (chapters 6, 7).
9. For a meaningful life, make it rich with love, work, and play (chapter 8).
10. Fear dying, but not death (chapter 8).
11. Enhance your life with beauty and other aesthetic emotions (chapter 9).
12. Act autonomously, but recognize that free will is limited (chapter 10).

Following these rules empowers an integrated approach to the central problems of philosophy concerning mind, knowledge, reality, morality, and beauty. Further integration comes from the many connections I have made between areas of philosophy and the cognitive and social sciences, as expounded in *Brain–Mind* and *Mind–Society*. Natural philosophy both benefits from and contributes to the general understanding of brain, mind, and society, including humanistic concerns with values and the arts. The sciences and the humanities hang together as a coherent whole, thanks to a unifying theory of how brains make minds.

NOTES

Free will skeptics include Gazzaniga 2009, Harris 2012, Libet 1985, and Wegner 2003. Free will defenders include Dennett 2003 and Mele 2014, 2017. I discuss free will in connection with action in *Brain–Mind* (chapter 9) and in connection with law in *Mind–Society* (chapter 11).

On the psychology and neuroscience of free will, see Glannon 2015; Inzlicht, Legault, and Teper 2014; Mele 2015; Nichols 2011; Turri 2017; and Shariff et al. 2014. Philosophers sometimes say that that you have free will if you could have done otherwise, which depends on a counterfactual *if*. Chapter 5 argues that such counterfactuals are neither true nor false, and their plausibility depends on underlying mechanisms, in this case about mental actions. Dual process theories are summarized by Kahneman 2011.

On the self, see *Brain–Mind* (chapter 12, based on Thagard 2014d) and Thagard and Wood 2015.

Horsten 2012 and Irvine 2009 are overviews of philosophy of mathematics. Hacking 2014 considers Platonism and alternatives. Kitcher 1983 takes an empiricist approach. Paseau 2013 reviews naturalism in the philosophy of mathematics. Brown 2009 critiques the view that mathematics is socially constructed. Franklin 2014 develops Aristotelian realism.

On the psychology and neuroscience of mathematics, see Dehaene 2011; Ferrigno, Jara-Ettinger, Piantadosi, and Cantlon 2017; Goldstone et al. 2017; Hyde and Spelke 2011; Lakoff and Núñez 2000; Landy, Allen, and Zednik 2014; Leibovich, Katzin, Harel, and Henik 2017; and Opfer and Ziegler 2012.

Hardy 1967 (p. 14) remarks on beautiful patterns in math. Zeki, Romaya, Benincassa, and Atiyah 2014 examine neural correlates of the experience of beauty in mathematics. Thagard 2000 (chapter 3) discusses mathematical coherence as constraint satisfaction. Polya 1957 examines mathematical heuristics. Bartha 2010 discusses mathematical analogies.

Colyvan 2015 reviews philosophical disputes about the indispensability of mathematics.

Tegmark 2014 claims that the universe is mathematical.

De Waal 2017 reviews animal cognition. Penn, Povinelli, and Holyoak 2008 argue for limitations in the cognitive abilities of animals. The number of neurons in different animals is listed at https://en.wikipedia.org/wiki/List_

of_animals_by_number_of_neurons. For a debate about whether animals have emotions, see https://www.psychologytoday.com/blog/hot-thought/201711/do-animals-have-emotions-debate.

On the current state of artificial intelligence and comparisons with animals, see my old course notes: http://cogsci.uwaterloo.ca/courses/cogsci300.html. I am currently writing a book called *Bots and Beasts: What Makes Machines, Animals, and People Smart?*

Nagel 2012 defends teleology against naturalism.

Peterson 2018 gives his rules for life. I discuss the flaws in Peterson's philosophy of life in two blog posts: https://www.psychologytoday.com/ca/blog/hot-thought/201802/jordan-peterson-s-flimsy-philosophy-life

and

https://www.psychologytoday.com/ca/blog/hot-thought/201803/jordan-petersons-murky-maps-meaning.

PROJECT

Develop a comprehensive theory and computational model of thought and action that clarifies the operation of will. Explain numbers by integrating psychological, neural, and metaphysical considerations. Systematically compare semantic pointers in machines, humans, and other animals. Answer the questions at the end of chapter 9 about art in connection with free will, mathematics, machines, and nonhuman animals.

REFERENCES

Abrams, M. (2012). Mechanistic probability. *Synthese, 187*(2), 343–375.
Adajian, T. (2012). The definition of art. In E. Zalta (Ed.), *Stanford encyclopedia of philosophy*. https://plato.stanford.edu/entries/art-definition
Armstrong, D. M. (2004). *Truth and truthmakers*. Cambridge, UK: Cambridge University Press.
Arneson, R. (2013). Egalitarianism. In E. Zalta (Ed.), *Stanford encyclopedia of philosophy*. https://plato.stanford.edu/entries/egalitarianism/
Aschenbrenner, L. (1985). *The concept of coherence in art*. Berlin: Springer.
Atkinson, A. B. (2015). *Inequality: What can be done?* Cambridge, MA: Harvard University Press.
Baillargeon, R., Kotovsky, L., & Needham, A. (1995). The acquisition of physical knowledge in infancy. In D. Sperber, D. Premack, & A. J. Premack (Eds.), *Causal cognition: A multidisciplinary debate* (pp. 79–116). Oxford: Clarendon Press.
Banich, M. T., & Compton, R. J. (2018). *Cognitive neuroscience* (4th ed.). Cambridge, UK: Cambridge University Press.
Baron-Cohen, S. (2011). *The science of evil: On empathy and the origins of cruelty*. New York: Basic Books.
Bartha, P. (2010). *By parallel reasoning: The construction and evaluation of analogical arguments*. Oxford: Oxford University Press.
Bashour, B., & Muller, H. D. (Eds.). (2014). *Contemporary philosophical naturalism and its implications*. New York: Routledge.
Baumeister, R. F., Vohs, K. D., Aaker, J. L., & Garbinsky, E. N. (2013). Some key differences between a happy life and a meaningful life. *The Journal of Positive Psychology, 8*(6), 505–516.
Beardsley, M. C. (1958). *Aesthetics: Problems in the philosophy of criticism*. New York: Harcourt, Brace & World.
Bechtel, W. (2008). *Mental mechanisms: Philosophical perspectives on cognitive neuroscience*. New York: Routledge.

Bechtel, W. (2017). Explicating top-down causation using networks and dynamics. *Philosophy of Science*, 84, 253–274.

Benatar, D. (2006). *Better never to have been born*. Oxford: Clarendon Press.

Berger, P. L., & Luckmann, T. (1966). *The social construction of reality: A treatise in the sociology of knowledge*. Garden City, NY: Anchor.

Bernhard, R. M., Chaponis, J., Siburian, R., Gallagher, P., Ransohoff, K., Wikler, D., . . . Greene, J. D. (2016). Variation in the oxytocin receptor gene (OXTR) is associated with differences in moral judgment. *Social Cognitive and Affective Neuroscience*, 11(12), 1872–1881.

Bernhardt, B. C., & Singer, T. (2012). The neural basis of empathy. *Annual Review of Neuroscience*, 35, 1–23.

Bethlehem, R. A., Allison, C., van Andel, E. M., Coles, A. I., Neil, K., & Baron-Cohen, S. (2017). Does empathy predict altruism in the wild? *Social Neuroscience*, 12(6), 743–750.

Blair, J., Mitchell, D. R., & Blair, K. (2005). *The psychopath: Emotion and the brain*. Malden, MA: Blackwell.

Bloom, P. (2013). *Just babies: The origins of good and evil*. New York: Crown.

Bloom, P. (2016). *Against empathy: The case for rational compassion*. New York: Ecco.

Blouw, P., Solodkin, E., Thagard, P., & Eliasmith, C. (2016). Concepts as semantic pointers: A framework and computational model. *Cognitive Science*, 40, 1128–1162.

Boroditsky, L. (2011). How language shapes thought. *Scientific American*, February, 63–65.

Bostyn, D. H., Sevenhant, S., & Roets, A. (2018). Of mice, men, and trolleys: Hypothetical judgment versus real-life behavior in trolley-style moral dilemmas. *Psychological Science*, 29(7), 1084–1093.

Boyd, R., & Richerson, P. J. (2005). *Not by genes alone: How culture transformed human evolution*. Chicago: University of Chicago Press.

Brattico, E., & Pearce, M. (2013). The neuroaesthetics of music. *Psychology of Aesthetics, Creativity, and the Arts*, 7(1), 48.

Broadbent, A. (2013). *Philosophy of epidemiology*. Basingstoke, UK: Palgrave Macmillan.

Brown, R. C. (2009). *Are science and mathematics socially constructed? A mathematician encounters postmodern interpretations of science*. Singapore: World Scientific.

Brown, S., & Jordania, J. (2013). Universals in the world's musics. *Psychology of Music*, 41(2), 229–248.

Bullot, N. J., & Reber, R. (2013). The artful mind meets art history: Toward a psycho-historical framework for the science of art appreciation. *Behavioral and Brain Sciences*, 36(2), 123–137.

Bunge, M. (2003). *Emergence and convergence: Qualitative novelty and the unity of knowledge*. Toronto: University of Toronto Press.

Cacioppo, J. T., Cacioppo, S., Capitanio, J. P., & Cole, S. W. (2015). The neuroendocrinology of social isolation. *Annual Review of Psychology*, 66, 733–767.

Cacioppo, J. T., Cacioppo, S., Cole, S. W., Capitanio, J. P., Goossens, L., & Boomsma, D. I. (2015). Loneliness across phylogeny and a call for comparative studies and animal models. *Perspectives on Psychological Science*, 10(2), 202–212.

Campbell, D. (1974). Evolutionary epistemology. In P. Schilpp (Ed.), *The philosophy of Karl Popper* (pp. 413–463). La Salle, IL: Open Court.

Caporael, L. R., Griesemer, J. R., & Wimsatt, W. C. (2014). *Developing scaffolds in evolution, culture, and cognition*. Cambridge, MA: MIT Press.

Cappelen, A. W., Eichele, T., Hugdahl, K., Specht, K., Sørensen, E. Ø., & Tungodden, B. (2014). Equity theory and fair inequality: A neuroeconomic study. *Proceedings of the National Academy of Sciences of the United States of America*, 111(43), 15368–15372.

Cappelen, H., Gendler, T. S., & Hawthorne, J. (2016). *The Oxford handbook of philosophical methodology*. Oxford: Oxford University Press.

Carroll, S. B. (2013). *Brave genius: A scientist, a philosopher, and their daring adventures from the French Resistance to the Nobel Prize*. New York: Crown.

Cerullo, M. A. (2015). The problem with Phi: A critique of integrated information theory. *PLoS Computational Biology*, 11(9), e1004286. doi:10.1371/journal.pcbi.1004286

Chalmers, D. (2010). *The character of consciousness*. Oxford: Oxford University Press.

Chatterjee, A., & Vartanian, O. (2014). Neuroaesthetics. *Trends in Cognitive Sciences*, 18(7), 370–375.

Chelnokova, O., Laeng, B., Eikemo, M., Riegels, J., Løseth, G., Maurud, H., ... Leknes, S. (2014). Rewards of beauty: The opioid system mediates social motivation in humans. *Molecular Psychiatry*, 19(7), 746.

Church, A. T., Katigbak, M. S., Locke, K. D., Zhang, H., Shen, J., ... Ching, C. M. (2013). Need satisfaction and well-being: Testing self-determination theory in eight cultures. *Journal of Cross-Cultural Psychology*, 44(4), 507–534.

Churchland, P. M. (2007). *Neurophilosophy at work*. Cambridge, UK: Cambridge University Press.

Churchland, P. M. (2013). *Matter and consciousness* (3rd ed.). Cambridge, MA: MIT Press/Bradford Books.

Churchland, P. S. (1986). *Neurophilosophy*. Cambridge, MA: MIT Press.

Churchland, P. S. (2002). *Brain-wise: Studies in neurophilosophy*. Cambridge, MA: MIT Press.

Churchland, P. S. (2011). *Braintrust: What neuroscience tells us about morality*. Princeton, NJ: Princeton University Press.

Clark, A. (2008). *Supersizing the mind: Embodiment, action, and cognitive extension*. Oxford: Oxford University Press.

Colaço, D., Buckwalter, W., Stich, S., & Machery, E. (2014). Epistemic intuitions in fake-barn thought experiments. *Episteme*, 11(2), 199–212.

Cole, S. W., Capitanio, J. P., Chun, K., Arevalo, J. M., Ma, J., & Cacioppo, J. T. (2015). Myeloid differentiation architecture of leukocyte transcriptome dynamics in perceived social isolation. *Proceedings of the National Academy of Sciences of the United States of America*, 112(49), 15142–15147.

Collingwood, R. G. (1997). *Outlines of a philosophy of art*. Bristol, UK: Thoemmes Press.

Colyvan, M. (2015). Indispensability arguments in the philosophy of mathematics. In E. Zalta (Ed.), *Stanford encyclopedia of philosophy*. https://plato.stanford.edu/entries/mathphil-indis/

Craver, C. F. (2007). *Explaining the brain*. Oxford: Oxford University Press.

Craver, C. F. (2015). Levels. In T. Metzinger & J. M. Windt (Eds.), *Open MIND*. Frankfurt: MIND Group. doi:10.15502/9783958570498

Craver, C. F., & Darden, L. (2013). *In search of mechanisms: Discoveries across the life sciences*. Chicago: University of Chicago Press.

Crockett, M. J., & Rini, R. A. (2015). Neuromodulators and the (in)stability of moral cognition. In J. Decety & T. Wheatley (Eds.), *The moral brain: A multidisciplinary perspective* (pp. 221–235). Cambridge, MA: MIT Press.

Currie, G., Kieran, M., Meskin, A., & Robson, J. (Eds.). (2014). *Aesthetics and the sciences of mind*. Oxford: Oxford University Press.

Dadds, M. R., Moul, C., Cauchi, A., Dobson-Stone, C., Hawes, D. J., Brennan, J., & Ebstein, R. E. (2014). Methylation of the oxytocin receptor gene and oxytocin blood levels in the development of psychopathy. *Development and Psychopathology*, 26(1), 33–40.

Daly, C. (Ed.). (2015). *The Palgrave handbook of philosophical methods*. Basingstoke, UK: Palgrave Macmillan.

Dammann, O., Poston, T., & Thagard, P. (in press). How do medical researchers make causal inferences? In K. McCain & K. Kampourakis (Eds.), *What is scientific knowledge? An introduction to contemporary epistemology of science*. London: Routledge.

Daniels, N. (2016). Reflective equilibrium. In E. Zalta (Ed.), *Stanford encyclopedia of philosophy*. https://plato.stanford.edu/entries/reflective-equilibrium/

Danks, D. (2014). *Unifying the mind: Cognitive representations and graphical models*. Cambridge, MA: MIT Press.

Darby, R. R., Horn, A., Cushman, F., & Fox, M. D. (2018). Lesion network localization of criminal behavior. *Proceedings of the National Academy of Sciences of the United States of America*, 115(3), 601–606.

Dawkins, R. (1976). *The selfish gene*. New York: Oxford University Press.

de Waal, F. B. M. (2017). *Are we smart enough to know how smart animals are?* New York: Norton.

Deaton, A., & Stone, A. A. (2014). Evaluative and hedonic wellbeing among those with and without children at home. *Proceedings of the National Academy of Sciences of the United States of America*, 111(4), 1328–1333.

Debowska, A., Boduszek, D., Hyland, P., & Goodson, S. (2014). Biological correlates of psychopathy: A brief review. *Mental Health Review Journal*, 19(2), 110–123.

Decety, J. (Ed.). (2014). *Empathy: From bench to bedside*. Cambridge, MA: MIT Press.

Decety, J., & Wheatley, T. (2015). *The moral brain: A multidisciplinary perspective*. Cambrdige, MA: MIT Press.

Dehaene, S. (2011). *The number sense: How the mind creates mathematics* (2nd ed.). Oxford: Oxford University Press.

Dehaene, S. (2014). *Consciousness and the brain: Deciphering how the brain codes our thoughts*. New York: Viking.

Dennett, D. (2003). *Freedom evolves*. New York: Penguin.

Dennett, D. C. (2013). *Intuition pumps and other tools for thinking*. New York: W. W. Norton.

Dennett, D. C. (2017). *From bacteria to Bach and back: The evolution of minds*. New York: W. W. Norton.

Devitt, M. (2011). Are unconceived alternatives a problem for scientific realism? *Journal for General Philosophy of Science*, 42(2), 285–293.

Doris, J. M. (Ed.). (2010). *The moral psychology handbook*. Oxford: Oxford University Press.

Douglas, H. E. (2009). *Science, policy, and the value-free ideal*. Pittsburgh: University of Pittsburgh Press.

Dove, G. (2011). On the need for embodied and dis-embodied cognition. *Frontiers in Psychology*, 1(242). doi:10.3389/fpsyg.2010.00242

Doyal, L., & Gough, I. (1991). *A theory of human need*. London: Macmillan.

Dummett, M. (1993). *Origins of analytic philosophy*. Cambridge, MA: Harvard University Press.

Dunn, E. W., Aknin, L. B., & Norton, M. I. (2014). Prosocial spending and happiness: Using money to benefit others pays off. *Current Directions in Psychological Science*, 23(1), 41–47.

Durkheim, E. (1982). *The rules of sociological method* (W. Halls., Trans.). New York: Free Press.

Eco, U. (Ed.). (2004). *History of beauty*. New York: Rizzoli.

Eichenbaum, H. (2014). Time cells in the hippocampus: A new dimension for mapping memories. *Nature Reviews Neuroscience*, 15(11), 732–744.

Eliasmith, C. (2005). Neurosemantics and categories. In H. Cohen & C. Lefebvre (Eds.), *Handbook of categorization in cognitive science* (pp. 1035–1054). Amsterdam: Elsevier.

Eliasmith, C. (2013). *How to build a brain: A neural architecture for biological cognition.* Oxford: Oxford University Press.

Eliasmith, C., Stewart, T. C., Choo, X., Bekolay, T., DeWolf, T., Tang, Y., & Rasmussen, D. (2012). A large-scale model of the functioning brain. *Science, 338,* 1202–1205.

Eliasmith, C., & Thagard, P. (2001). Integrating structure and meaning: A distributed model of analogical mapping. *Cognitive Science, 25,* 245–286.

Elliott, K. C. (2017). *A tapestry of values: An introduction to values in science.* Oxford: Oxford University Press.

Ferrigno, S., Jara-Ettinger, J., Piantadosi, S. T., & Cantlon, J. F. (2017). Universal and uniquely human factors in spontaneous number perception. *Nature Communications, 8.* doi:10.1038/ncomms13968

Findlay, S. D., & Thagard, P. (2012). How parts make up wholes. *Frontiers in Physiology, 3.* doi:10.3389/fphys.2012.00455

Finger, S. (1994). *Origins of neuroscience: A history of explorations into brain function.* New York: Oxford University Press.

Fleck, L. (1979). *Genesis and development of a scientific fact.* Chicago: University of Chicago Press.

Forget, E. L. (2011). The town with no poverty: The health effects of a Canadian guaranteed annual income field experiment. *Canadian Public Policy, 37*(3), 283–305.

Francis, K. B., Howard, C., Howard, I. S., Gummerum, M., Ganis, G., Anderson, G., & Terbeck, S. (2016). Virtual morality: Transitioning from moral judgment to moral action? *PLOS One, 11*(10), e0164374.

Frankland, S. M., & Greene, J. D. (2015). An architecture for encoding sentence meaning in left mid-superior temporal cortex. *Proceedings of the National Academy of Sciences of the United States of America, 112*(37), 11732–11737.

Franklin, J. (2014). *An Aristotelian realist philosophy of mathematics.* Houndmills, UK: Palgrave Macmillan.

French, P. A., & Wettstein, H. K. (Eds.). (2010). *Film and the emotions* (Midwest Studies in Philosophy Vol. xxxiv). Boston: Blackwell.

Gallagher, S., & Zahavi, D. (2012). *The phenomenological mind* (2nd ed.). London: Routledge.

Gazzaniga, M. (2009). *Who's in charge: Free will and the science of the brain.* New York: HarperCollins.

Gilbert, M. (1992). *On social facts.* Princeton, NJ: Princeton University Press.

Glannon, W. (Ed.). (2015). *Free will and the brain: Neuroscientific, philosophical, and legal perspectives.* Cambridge, UK: Cambridge University Press.

Glennan, S. (2017). *The new mechanical philosophy.* Oxford: Oxford University Press.

Goel, A., & Buonomano, D. V. (2014). Timing as an intrinsic property of neural networks: Evidence from in vivo and in vitro experiments. *Philosophical Transactions of the Royal Society B: Biological Sciences, 369*(1637), 20120460.

Goldman, A., & Beddor, B. (2015). Reliabilist epistemology. In E. Zalta (Ed.), *Stanford encyclopedia of philosophy.* https://plato.stanford.edu/entries/reliabilism/

Goldman, A., & Blanchard, T. (2015). Social epistemology. In E. Zalta (Ed.), *Stanford encyclopedia of philosophy.* https://plato.stanford.edu/entries/epistemology-social/

Goldstone, R. L., Marghetis, T., Weitnauer, E., Ottmar, E. R., & Landy, D. (2017). Adapting perception, action, and technology for mathematical reasoning. *Current Directions in Psychological Science, 26*(5), 434–441.

Gopnik, A. (1998). Explanation as orgasm. *Minds and Machines*, 8, 101–118.
Gough, I. (2015). Climate change and sustainable welfare: The centrality of human needs. *Cambridge Journal of Economics*, 39(5), 1191–1214.
Gracyk, T. (2012). *The philosophy of art*. Cambridge, UK: Polity Press.
Greene, J. (2014). *Moral tribes: Emotion, reason, and the gap between us and them*. New York: Penguin.
Griffiths, T. L., Kemp, C., & Tenenbaum, J. B. (2008). Bayesian models of cognition. In R. Sun (Ed.), *The Cambridge handbook of computational psychology* (pp. 59–100). Cambridge, UK: Cambridge University Press.
Guyer, P., & Horstmann, R. (2015). Idealism. In E. Zalta (Ed.). *Stanford encyclopedia of philosophy*. https://plato.stanford.edu/entries/idealism/
Hacking, I. (1999). *The social construction of what?* Cambridge, MA: Harvard University Press.
Hacking, I. (2001). *An introduction to probabiliity and inductive logic*. Cambridge, UK: Cambridge University Press.
Hacking, I. (2014). *Why is there philosophy of mathematics at all?* Cambridge, UK: Cambridge University Press.
Haidt, J. (2003). The moral emotions. In R. Davidson (Ed.), *Handbook of affective sciences* (pp. 852–870). Oxford: Oxford University Press.
Haidt, J. (2012). *The righteous mind: Why good people are divided by politics and religion*. New York: Pantheon.
Hallam, S., Cross, I., & Thaut, M. (2011). *Oxford handbook of music psychology*. Oxford: Oxford University Press.
Hardy, G. H. (1967). *A mathematician's apology*. Cambridge, UK: Cambridge University Press.
Harris, S. (2012). *Free will*. New York: Free Press.
Harris, S., Sheth, S. A., & Cohen, M. S. (2008). Functional neuroimaging of belief, disbelief, and uncertainty. *Annals of Neurology*, 63, 141–147.
Haug, M. C. (Ed.). (2014). *The armchair or the laboratory? Philosophical methodology*. London: Routledge.
Hawking, S. S. W., & Mlodinow, L. (2010). *The grand design*. New York: Bantam.
Hebb, D. O. (1949). *The organization of behavior: A neuropsychological theory*. New York: Wiley.
Hegel, G. (1952). *Hegel's philosophy of right* (T. M. Knox, Trans.). London: Oxford University Press.
Held, V. (2006). *The ethics of care: Personal, political, global*. Oxford: Oxford University Press.
Heller, A. (1976). *The theory of need in Marx*. New York: St. Martin's Press.
Hempel, C. G. (1965). *Aspects of scientific explanation*. New York: Free Press.
Hitchcock, C. (2010). Probalistic causation. In E. Zalta (Ed.), *Stanford encyclopedia of philosophy*. https://plato.stanford.edu/entries/causation-probabilistic/
Holland, J. H., Holyoak, K. J., Nisbett, R. E., & Thagard, P. R. (1986). *Induction: Processes of inference, learning, and discovery*. Cambridge, MA: MIT Press.
Holt, J. (2012). *Why does the world exist? An existential detective story*. New York: W. W. Norton.
Holyoak, K. J., & Powell, D. (2016). Deontological coherence: A framework for commonsense moral reasoning. *Psychological Bulletin*, 142(11), 1179–1203.
Horsten, L. (2012). Philosophy of mathematics. In E. Zalta (Ed.), *Stanford encyclopedia of philosophy*. https://plato.stanford.edu/entries/philosophy-mathematics/
Hrdy, S. B. (2009). *Mothers and others: The evolutionary origins of mutual understanding*. Cambridge, MA: Harvard University Press.
Hull, G., & Peikoff, L. (Eds.). (1999). *The Ayn Rand reader*. New York: Plume.

Huron, D. (2006). *Sweet anticipation: Music and the psychology of expectation*. Cambridge, MA: MIT Press.

Huron, D. (2015). Affect induction through musical sounds: An ethological perspective. *Philosophical Transactions of the Royal Society B: Biological Sciences*, 370(1664). doi:10.1098/rstb.2014.0098

Huron, D., Anderson, N., & Shanahan, D. (2014). "You can't play a sad song on the banjo": Acoustic factors in the judgment of instrument capacity to convey sadness. *Empirical Musicology Review*, 9(1), 29–41.

Huston, J. P., Nadal, M., Mora, F., Agnati, L. F., & Conde, C. J. C. (2015). *Art, aesthetics, and the brain*. Oxford: Oxford University Press.

Hyde, D. C., & Spelke, E. S. (2011). Neural signatures of number processing in human infants: Evidence for two core systems underlying numerical cognition. *Developmental Science*, 14(2), 360–371.

Inzlicht, M., Legault, L., & Teper, R. (2014). Exploring the mechanisms of self-control improvement. *Current Directions in Psychological Science*, 23(4), 302–307.

Irvine, A. D. (Ed.). (2009). *Philosophy of mathematics*. Amsterdam: Elsevier.

Ishizu, T., & Zeki, S. (2011). Toward a brain-based theory of beauty. *PLOS One*, 6(7), e21852.

Jacobs, A. M. (2015). Neurocognitive poetics: Methods and models for investigating the neuronal and cognitive-affective bases of literature reception. *Frontiers in Human Neuroscience*, 9, 186. doi:10.3389/fnhum.2015.00186

James, W. (1948). *Essays in pragmatism*. New York: Hafner.

Jenkins, A. C., Dodell-Feder, D., Saxe, R., & Knobe, J. (2014). The neural bases of directed and spontaneous mental state attributions to group agents. *PLOS One*, 9(8), e105341.

Johnson, M. (2007). *The meaning of the body: Aesthetics of human understanding*. Chicago: University of Chicago Press.

Johnson-Laird, P. N., & Oatley, K. (2016). Emotions in music, literature, and film. In L. F. Barrett, M. Lewis, & J. M. Haviland-Jones (Eds.), *Handbook of emotions* (pp. 82–97). New York: Guilford Press.

Jones, M., & Love, B. C. (2011). Bayesian fundamentalism or enlightenment: On the explanatory status and theoretical contributions of Bayesian models of cognition. *Behavioral and Brain Sciences*, 34, 169–231.

Joyal, C. C., Beaulieu-Plante, J., & de Chantérac, A. (2014). The neuropsychology of sex offenders: A meta-analysis. *Sexual Abuse*, 26(2), 149–177.

Juslin, P. N. (2013). From everyday emotions to aesthetic emotions: Towards a unified theory of musical emotions. *Physics of Life Reviews*, 10(3), 235–266.

Juslin, P. N., & Sloboda, J. (2011). *Handbook of music and emotion: Theory, research, applications*. Oxford: Oxford University Press.

Kahneman, D. (2011). *Thinking, fast and slow*. Toronto: Doubleday.

Kahneman, D., & Tversky, A. (Eds.). (2000). *Choices, values, and frames*. Cambridge, UK: Cambridge University Press.

Kajić, I., Schröder, T., Stewart, T. C., & Thagard, P. (forthcoming). The semantic pointer theory of emotions.

Kania, A. (2012). The psychology of music. In E. Zalta (Ed.), *Stanford encyclopedia of philosophy*. https://plato.stanford.edu/entries/music/

Kashdan, T. B., & McKnight, P. E. (2013). Commitment to a purpose in life: An antidote to the suffering by individuals with social anxiety disorder. *Emotion*, 13(6), 1150.

Kihlstrom, J. F. (1987). The cognitive unconscious. *Science*, 237(4821), 1445–1452.

King, L. A., Heintzelman, S. J., & Ward, S. J. (2016). Beyond the search for meaning: A contemporary science of the experience of meaning in life. *Current Directions in Psychological Science*, 25(4), 211–216.

Kirsch, L. P., Urgesi, C., & Cross, E. S. (2016). Shaping and reshaping the aesthetic brain: Emerging perspectives on the neurobiology of embodied aesthetics. *Neuroscience & Biobehavioral Reviews*, 62, 56–68.

Kitcher, P. (1983). *The nature of mathematical knowledge*. New York: Oxford University Press.

Klein, G. (1999). *Sources of power: How people make decisions*. Cambridge, MA: MIT Press.

Koelsch, S. (2012). *Brain and music*. Chichester, UK: Wiley-Blackwell.

Kuhn, T. S. (1970). *The structure of scientific revolutions* (2nd ed.). Chicago: University of Chicago Press.

Kunda, Z. (1990). The case for motivated reasoning. *Psychological Bulletin*, 108, 480–498.

Lackey, J. (Ed.). (2014). *Essays in collective epistemology*. Oxford: Oxford University Press.

Ladyman, J., & Ross, D. (2007). *Every thing must go: Metaphysics naturalized*. Oxford: Oxford University Press.

Lakoff, G., & Núñez, R., E. (2000). *Where mathematics comes from: How the embodied mind brings mathematics into being*. New York: Basic Books.

Landy, D., Allen, C., & Zednik, C. (2014). A perceptual account of symbolic reasoning. *Frontiers in Psychology*, 5. doi:10.3389/fpsyg.2014.00275

Latour, B. (1987). *Science in action: How to follow scientists and engineers through society*. Cambridge, MA: Harvard University Press.

Latour, B., & Woolgar, S. (1986). *Laboratory life: The construction of scientific facts*. Princeton, NJ: Princeton University Press.

Le Poidevin, R., Andrew, M., Peter, S., & Cameron, R. P. (2009). *The Routledge companion to metaphysics*. New York: Routledge.

Leibovich, T., Katzin, N., Harel, M., & Henik, A. (2017). From "sense of number" to "sense of magnitude"—The role of continuous magnitudes in numerical cognition. *Behavioral and Brain Sciences*, 40. doi.org/10.1017/S0140525X16000960

Levinson, J. (Ed.). (2003). *The Oxford handbook of aesthetics*. Oxford: Oxford University Press.

Levitin, D. (2006). *This is your brain on music*. New York: Dutton.

Lewis, D. (1986). *On the plurality of worlds*. Oxford: Oxford University Press.

Libet, B. (1985). Unconscious cerebral initiative and the role of conscious will in voluntary action. *Behavioral and Brain Sciences*, 8, 529–566.

Lopes, D. M. (2014). *Beyond art*. Oxford: Oxford University Press.

Magnani, L. (2009). *Abductive cognition: The epistemological and eco-cognitive dimensions of hypothetical reasoning*. Berlin: Springer.

Marcus, G. F., & Davis, E. (2013). How robust are probabilistic models of higher-level cognition? *Psychological Science*, 24(12), 2351–2360.

Marmot, M. (2004). *The status syndrome: How social standing affects our health and longevity*. New York: Henry Holt.

Marshall, P. (2010). *Demanding the impossible: A history of anarchism*. Oakland, CA: PM Press.

Maslow, A. H. (1987). *Motivation and personality* (3rd ed.). New York: Harper & Row.

McCauley, R. N. (2007). Reduction: Models of cross-scientific relations and their implications for the psychology-neuroscience interface. In P. Thagard (Ed.), *Philosophy of psychology and cognitive science* (pp. 105–158). Amsterdam: Elsevier.

McCauley, R. N. (2009). Time is of the essence: Explanatory pluralism and accommodating theories about long-term processes. *Philosophical Psychology*, 22, 611–635.

McCauley, R. N., & Bechtel, W. (2001). Explanatory pluralism and the heuristic identity theory. *Theory & Psychology, 11*, 736–760.

McClelland, J. L. (2010). Emergence in cognitive science. *Topics in Cognitive Science, 2*, 751–770.

McDermott, J. H., Schultz, A. F., Undurraga, E. A., & Godoy, R. A. (2016). Indifference to dissonance in native Amazonians reveals cultural variation in music perception. *Nature, 535*(7613), 547–550.

Mele, A. R. (2014). *Free: Why science hasn't disproved free will*. Oxford: Oxford University Press.

Mele, A. R. (2017). *Aspects of agency: Decisions, abilities, explanations, and free will*. Oxford: Oxford University Press.

Mele, A. R. (Ed.). (2015). *Surrounding free will*. Oxford: Oxford University Press.

Menninghaus, W., Wagner, V., Hanich, J., Wassiliwizky, E., Jacobsen, T., & Koelsch, S. (2017). The distancing–embracing model of the enjoyment of negative emotions in art reception. *Behavioral and Brain Sciences, 40*. doi:10.1017/S0140525X17000309

Metz, T. (2015). *Meaning in life*. Oxford: Oxford University Press.

Meyer, L. (2015). Intergenerational justice. In E. Zalta (Ed.), *Stanford encyclopedia of philosophy*. https://plato.stanford.edu/entries/justice-intergenerational/

Miller, S. C. (2013). *The ethics of need: Agency, dignity, and obligation*. New York: Routledge.

Mintzberg, H. (2009). *Managing*. Oakland, CA: Berrett-Koehler.

Morrell, M. E. (2010). *Empathy and democracy: Feeling, thinking, and deliberation*. University Park: Pennsylvania State University Press.

Moser, E. I., Kropff, E., & Moser, M. (2008). Place cells, grid cells, and the brain's spatial representation system. *Annual Review of Neuroscience, 31*, 68–89.

Mullainathan, S., & Shafir, E. (2013). *Scarcity: Why having too little means so much*. New York: Macmillan.

Murphy, G. L. (2002). *The big book of concepts*. Cambridge, MA: MIT Press.

Nagel, T. (2012). *Mind and cosmos*. Oxford: Oxford University Press.

Narveson, J. (2001). *The libertarian idea*. Peterborough, ON: Broadview Press.

Nichols, S. (2011). Experimental philosophy and the problem of free will. *Science, 331*(6023), 1401–1403.

Noë, A. (2009). *Out of our heads*. New York: Hill & Wang.

Noë, A. (2015). *Strange tools: Art and human nature*. New York: Hill & Wang.

Núñez, R., & Cooperrider, K. (2013). The tangle of space and time in human cognition. *Trends in Cognitive Sciences, 17*(5), 220–229.

Nussbaum, M. C. (2000). *Women and human development: The capabilities approach*. Cambridge, UK: Cambridge University Press.

Oatley, K. (2012). *The passionate muse: Exploration of emotion in stories*. Oxford: Oxford University Press.

Oatley, K. (2016). Fiction: Simulation of social worlds. *Trends in Cognitive Sciences, 20*(8), 618–628.

Oishi, S., & Diener, E. (2014). Can and should happiness be a policy goal? *Policy Insights from the Behavioral and Brain Sciences, 1*(1), 195–203.

Oizumi, M., Albantakis, L., & Tononi, G. (2014). From the phenomenology to the mechanisms of consciousness: Integrated Information Theory 3.0. *PLoS Computational Biology, 10*(5), e1003588. doi:10.1371/journal.pcbi.1003588

Olsson, E. (2017). Coherentist theories of epistemic justification. In E. Zalta (Ed.), *Stanford encyclopedia of philosophy*. https://plato.stanford.edu/entries/justep-coherence/

Opfer, J. E., & Siegler, R. S. (2012). Development of quantitative thinking. In K. J. Holyoak & R. G. Morrison (Eds.), *The Oxford handbook of thinking and reasoning* (pp. 585–605). New York: Oxford University Press.

Orend, B. (2002). *Human rights: Concept and context*. Peterborough, UK: Broadview.

Osbeck, L. M., & Held, B. S. (2014). *Rational intuition: Philosophical roots, scientific investigations*. Cambridge, UK: Cambridge University Press.

Paseau, A. (2013). Naturalism in the philosophy of mathematics. In E. Zalta (Ed.), *Stanford encyclopedia of philosophy*. https://plato.stanford.edu/entries/naturalism-mathematics/

Pearl, J. (2000). *Causality: Models, reasoning, and inference*. Cambridge, UK: Cambridge University Press.

Penn, D. C., Holyoak, K. J., & Povinelli, D. J. (2008). Darwin's mistake: Explaining the discontinuity between human and nonhuman minds. *Behavioral and Brain Sciences, 31*, 109–178.

Perlovsky, L. (2014). Aesthetic emotions, what are their cognitive functions? *Frontiers in Psychology, 5*, 98.

Pessoa, L. (2013). *The cognitive-emotional brain: From interactions to integration*. Cambridge, MA: MIT Press.

Peterson, J. B. (2018). *12 rules for life: An antidote to chaos*. Toronto: Random House Canada.

Pezzulo, G., Barsalou, L. W., Cangelosi, A., Fischer, M. H., McRae, K., & Spivey, M. J. (2011). The mechanics of embodiment: A dialog on embodiment and computational modeling. *Frontiers in Psychology, 2*. https://doi.org/10.3389/fpsyg.2011.00005

Pfaff, D. W. (2015). *The altruistic brain*. Oxford: Oxford University Press.

Piantadosi, S. T., Tenenbaum, J. B., & Goodman, N. D. (2016). The logical primitives of thought: Empirical foundations for compositional cognitive models. *Psychological Review, 123*(4), 392–424.

Piccinini, G., & Craver, C. (2011). Integrating psychology and neuroscience: Functional analyses as mechanism sketches. *Synthese, 183*(3), 283–311.

Piff, P. K., Stancato, D. M., Côté, S., Mendoza-Denton, R., & Keltner, D. (2012). Higher social class predicts increased unethical behavior. *Proceedings of the National Academy of Sciences of the United States of America, 109*(11), 4086–4091.

Piketty, T. (2014). *Capital in the twenty-first century*. Cambridge, MA: Harvard University Press.

Plate, T. (2003). *Holographic reduced representations*. Stanford: CSLI.

Plebe, A., & De La Cruz, V. (2016). *Neurosemantics: Neural processes and the construction of linguistic meaning*. Berlin: Springer.

Polya, G. (1957). *How to solve it*. Princeton, NJ: Princeton University Press.

Poston, T. (2014). *Reason and explanation: A defense of explanatory coherentism*. London: Palgrave Macmillan.

Prinz, J. J., & Nichols, S. (2010). Moral emotions. In J. M. Doris (Ed.), *The moral psychology handbook* (pp. 111–146). Oxford: Oxford University Press.

Psillos, S. (1999). *Scientific realism: How science tracks the truth*. London: Routledge.

Pyszczynski, T., Greenberg, J., & Solomon, S. (1999). A dual-process model of defense against conscious and unconscious death-related thoughts: An extension of terror management theory *Psychological Review, 106*, 835–845.

Quine, W. V. O. (1960). *Word and object*. Cambridge, MA: MIT Press.

Radcliff, B. (2013). *The political economy of happiness*. Cambridge, UK: Cambridge University Press.

Ramsey, W. (2013). Eliminative materialism. In E. Zalta (Ed.), *Stanford encyclopedia of philosophy*. https://plato.stanford.edu/entries/materialism-eliminative/

Rasmussen, S. A., Jamieson, D. J., Honein, M. A., & Petersen, L. R. (2016). Zika virus and birth defects—reviewing the evidence for causality. *New England Journal of Medicine*, 374, 1981–1987.

Rawls, J. (1971). *A theory of justice*. Cambridge, MA: Harvard University Press.

Reader, S. (2007). *Needs and moral necessity*. New York: Routledge.

Remmel, R. J., & Glenn, A. L. (2015). Immorality in the adult brain. In J. Decety & T. Wheatley (Eds.), *The moral brain: A multidisciplinary perspective* (pp. 239–251). Cambridge, MA: MIT Press.

Robeyns, I. (2016). The capability approach. In E. Zalta (Ed.), *Stanford encyclopedia of philosophy*. https://plato.stanford.edu/entries/capability-approach/

Robinson, H. (2016). Dualism. In E. Zalta (Ed.), *Stanford encyclopedia of philosophy*. https://plato.stanford.edu/entries/dualism/

Robinson, J. (2005). *Deeper than reason: Emotion and its role in literature, music, and art*. Oxford: Oxford University Press.

Rodman, S. (1961). *Conversations with artists*. New York: Capricorn Books.

Roe, S. (2015). *In Montmartre: Picasso, Matisse and the birth of modernist art*. New York: Penguin.

Rorty, R. (1979). *Philosophy and the mirror of nature*. Princeton, NJ: Princeton University Press.

Ross, D., Ladyman, J., & Kincaid, H. (2013). *Scientific metaphysics*. Oxford: Oxford University Press.

Rumelhart, D. E., & McClelland, J. L. (Eds.). (1986). *Parallel distributed processing: Explorations in the microstructure of cognition*. Cambridge, MA: MIT Press/Bradford Books.

Russo, F., & Williamson, J. (2007). Interpreting causality in the health sciences. *International Studies in the Philosophy of Science*, 21(2), 157–170.

Ryan, R. M., & Deci, E. L. (2017). *Self-determination theory: Basic psychological needs in motivation, development, and wellness*. New York: Guilford Press.

Sachs, M. E., Damasio, A., & Habibi, A. (2015). The pleasures of sad music: A systematic review. *Frontiers in Human Neuroscience*, 9, 404. doi:10.3389/fnhum.2015.00404

Salimpoor, V. N., Zald, D. H., Zatorre, R. J., Dagher, A., & McIntosh, A. R. (2015). Predictions and the brain: How musical sounds become rewarding. *Trends in Cognitive Sciences*, 19(2), 86–91.

Sartwell, C. (2016). Beauty. In E. Zalta (Ed.), *Stanford encyclopedia of philosophy*. https://plato.stanford.edu/entries/beauty/

Schaffer, J. (2016). The metaphysics of causality. In E. Zalta (Ed.), *Stanford encyclopedia of philosophy*. https://plato.stanford.edu/entries/causation-metaphysics/

Schama, S. (2006). *Power of art*. New York: Penguin.

Schiavio, A., Menin, D., & Matyja, J. (2014). Music in the flesh: Embodied simulation in musical understanding. *Psychomusicology: Music, Mind, and Brain*, 24(4), 340.

Schilpp, P. A. (Ed.). (1980). *The philosophy of Bland Blanshard*. Lasalle, IL: Open Court.

Schröder, T., Stewart, T. C., & Thagard, P. (2014). Intention, emotion, and action: A neural theory based on semantic pointers. *Cognitive Science*, 38, 851–880.

Schwartz, S. H. (1992). Universals in the content and structure of values: Theoretical advances and empirical tests in 20 countries. *Advances in Experimental Social Psychology*, 25, 1–65.

Schwitzgebel, E., & Cushman, F. (2015). Philosophers' biased judgments persist despite training, expertise and reflection. *Cognition*, 141, 127–137.

Seager, W., & Allen-Hermanson, S. (2010). Panpsychism. In E. Zalta (Ed.), *Stanford encyclopedia of philosophy*. https://plato.stanford.edu/entries/panpsychism/

Segal, H. D. (2016). Finding a better way: A basic income pilot project for Ontario. https://www.ontario.ca/page/finding-better-way-basic-income-pilot-project-ontario

Sellars, W. (1962). *Science, perception, and reality*. London: Routledge and Kegan Paul.
Sen, A. (1999). *Development as freedom*. New York: Random House.
Sen, A. (2011). *The idea of justice*. Cambridge, MA: Harvard University Press.
Seto, M. C. (2009). Pedophilia. *Annual Review of Clinical Psychology, 5*, 391–407.
Shalvi, S., Gino, F., Barkan, R., & Ayal, S. (2015). Self-serving justifications: Doing wrong and feeling moral. *Current Directions in Psychological Science, 24*(2), 125–130.
Shapiro, L. (Ed.). (2014). *Routledge handbook of embodied cognition*. New York: Routledge.
Shariff, A. F., Greene, J. D., Karremans, J. C., Luguri, J. B., Clark, C. J., Schooler, J. W., . . . Vohs, K. D. (2014). Free will and punishment: A mechanistic view of human nature reduces retribution. *Psychological Science, 25*(8), 1563–1570.
Sheredos, B., Burnston, D., Abrahamsen, A., & Bechtel, W. (2013). Why do biologists use so many diagrams? *Philosophy of Science, 80*(5), 931–944.
Sievers, B., Polansky, L., Casey, M., & Wheatley, T. (2013). Music and movement share a dynamic structure that supports universal expressions of emotion. *Proceedings of the National Academy of Sciences of the United States of America, 110*(1), 70–75.
Simon, D., Stenstrom, D., & Read, S. J. (2015). The coherence effect: Blending cold and hot cognitions. *Journal of Personality and Social Psychology, 109*, 369–394.
Slote, M. (2007). *The ethics of care and empathy*. London: Routledge.
Smee, S. (2016). *The art of rivalry: Four friendships, betrayals, and breakthroughs in modern art*. New York: Random House.
Smith, D. W. (2013). Phenomenology. In E. Zalta (Ed.), *Stanford encyclopedia of philosophy*. https://plato.stanford.edu/entries/phenomenology/
Smolin, L. (2013). *Time reborn: From the crisis in physics to the future of the universe*: New York: Houghton Mifflin Harcourt.
Soares, S., Atallah, B. V., & Paton, J. J. (2016). Midbrain dopamine neurons control judgment of time. *Science, 354*(6317), 1273–1277.
Speaks, J. (2014). Theories of meaning. In E. Zalta (Ed.), *Stanford encyclopedia of philosophy*. https://plato.stanford.edu/entries/meaning/
Stanley, J. (2011). *Know how*. Oxford: Oxford University Press.
Starr, G. G. (2013). *Feeling beauty: The neuroscience of aesthetic experience*. Cambridge, MA: MIT Press.
Steinhardt, P. J., & Turok, N. (2007). *Endless universe: Beyond the Big Bang*. New York: Doubleday.
Steptoe, A., Shankar, A., Demakakos, P., & Wardle, J. (2013). Social isolation, loneliness, and all-cause mortality in older men and women. *Proceedings of the National Academy of Sciences of the United States of America, 110*(15), 5797–5801.
Steup, M. (2005). Epistemology. In E. Zalta (Ed.), *Stanford encyclopedia of philosophy*. https://plato.stanford.edu/entries/epistemology/
Stevenson, B., & Wolfers, J. (2013). Subjective well-being and income: Is there any evidence of satiation? *The American Economic Review, 103*(3), 598–604.
Stiglitz, J. E. (2013). *The price of inequality: How today's divided society endangers our future*. New York: W. W. Norton.
Strecher, V. J. (2016). *Life on purpose: How living for what matters most changes everything*. New York: HarperOne.
Swafford, J. (2013). The most beautiful music in the world. *Slate*. http://www.slate.com/articles/arts/music_box/2013/07/the_most_beautiful_melody_in_the_world_is_it_gershwin_brahms_the_beatles.html

Sytsma, J., & Buckwalter, W. (Eds.). (2016). *A companion to experimental philosophy*. Oxford: Wiley-Blackwell.

Sytsma, J., & Livengood, J. (2016). *The theory and practice of experimental philosophy*. Peterborough, ON: Broadview Press.

Tangney, J. P., Stuewig, J., & Mashek, D. J. (2007). Moral emotions and moral behavior. *Annual Review of Psychology, 58*, 345–372.

Tegmark, M. (2014). *Our mathematical universe*. New York: Knopf.

Thagard, P. (1988). *Computational philosophy of science*. Cambridge, MA: MIT Press.

Thagard, P. (1989). Explanatory coherence. *Behavioral and Brain Sciences, 12*, 435–467.

Thagard, P. (1992). *Conceptual revolutions*. Princeton, NJ: Princeton University Press.

Thagard, P. (1999). *How scientists explain disease*. Princeton, NJ: Princeton University Press.

Thagard, P. (2000). *Coherence in thought and action*. Cambridge, MA: MIT Press.

Thagard, P. (2002). The passionate scientist: Emotion in scientific cognition. In P. Carruthers, S. Stich, & M. Siegal (Eds.), *The cognitive basis of science* (pp. 235–250). Cambridge, UK: Cambridge University Press.

Thagard, P. (2005). Testimony, credibility, and explanatory coherence. *Erkenntnis, 63*, 295–316.

Thagard, P. (2006). *Hot thought: Mechanisms and applications of emotional cognition*. Cambridge, MA: MIT Press.

Thagard, P. (2009). Why cognitive science needs philosophy and vice versa. *Topics in Cognitive Science, 1*, 237–254.

Thagard, P. (2010a). Explaining economic crises: Are there collective representations? *Episteme, 7*, 266–283.

Thagard, P. (2010b). *The brain and the meaning of life*. Princeton, NJ: Princeton University Press.

Thagard, P. (2011). The brain is wider than the sky: Analogy, emotion, and allegory. *Metaphor and Symbol, 26*(2), 131–142.

Thagard, P. (2012a). Cognitive architectures. In K. Frankish & W. Ramsay (Eds.), *The Cambridge handbook of cognitive science* (pp. 50–70). Cambridge, UK: Cambridge University Press.

Thagard, P. (2012b). Coherence: The price is right. *Southern Journal of Philosophy, 50*, 42–49.

Thagard, P. (2012c). *The cognitive science of science: Explanation, discovery, and conceptual change*. Cambridge, MA: MIT Press.

Thagard, P. (2014a). Artistic genius and creative cognition. In D. K. Simonton (Ed.), *Wiley handbook of genius* (pp. 120–138). Oxford: Wiley-Blackwell.

Thagard, P. (2014b). Creative intuition: How EUREKA results from three neural mechanisms. In L. M. Osbeck & B. S. Held (Eds.), *Rational intuition: Philosophical roots, scientific investigations* (pp. 287–306). Cambridge, UK: Cambridge University Press.

Thagard, P. (2014c). Explanatory identities and conceptual change. *Science & Education, 23*, 1531–1548.

Thagard, P. (2014d). The self as a system of multilevel interacting mechanisms. *Philosophical Psychology, 27*, 145–163.

Thagard, P. (2014e). Thought experiments considered harmful. *Perspectives on Science, 22*, 288–305.

Thagard, P. (2018). Social equality: Cognitive modeling based on emotional coherence explains attitude change. *Policy Insights from Behavioral and Brain Sciences, 5*(2), 247–256.

Thagard, P. (2019a). *Brain-mind: From neurons to consciousness and creativity*. Oxford: Oxford University Press.

Thagard, P. (2019b). *Mind-society: From brains to social sciences and professions*. Oxford: Oxford University Press.

Thagard, P., & Aubie, B. (2008). Emotional consciousness: A neural model of how cognitive appraisal and somatic perception interact to produce qualitative experience. *Consciousness and Cognition*, 17, 811–834.

Thagard, P., & Beam, C. (2004). Epistemological metaphors and the nature of philosophy. *Metaphilosophy*, 35, 504–516.

Thagard, P., & Findlay, S. D. (2011). Changing minds about climate change: Belief revision, coherence, and emotion. In E. J. Olsson & S. Enqvist (Eds.), *Belief revision meets philosophy of science* (pp. 329–345). Berlin: Springer.

Thagard, P., & Finn, T. (2011). Conscience: What is moral intuition? In C. Bagnoli (Ed.), *Morality and the emotions* (pp. 150–159). Oxford: Oxford University Press.

Thagard, P., & Schröder, T. (2014). Emotions as semantic pointers: Constructive neural mechanisms. In L. F. Barrett & J. A. Russell (Eds.), *The psychological construction of emotions* (pp. 144–167). New York: Guilford Press.

Thagard, P., & Stewart, T. C. (2011). The AHA! experience: Creativity through emergent binding in neural networks. *Cognitive Science*, 35, 1–33.

Thagard, P., & Stewart, T. C. (2014). Two theories of consciousness: Semantic pointer competition vs. information integration. *Consciousness and Cognition*, 30, 73–90.

Thagard, P., & Verbeurgt, K. (1998). Coherence as constraint satisfaction. *Cognitive Science*, 22, 1–24.

Thagard, P., & Wood, J. V. (2015). Eighty phenomena about the self: Representation, evaluation, regulation, and change. *Frontiers in Psychology*, 6. doi:10.3389/fpsyg.2015.00334

Thompson, E. (2007). *Mind in life: Biology, phenomenology, and the science of mind*. Cambridge, MA: Harvard University Press.

Tolstoy, L. (1995). *What is art?* (R. Pevear & L. Volokhonsky, Trans.). London: Penguin.

Tomasello, M., & Vaish, A. (2013). Origins of human cooperation and morality. *Annual Review of Psychology*, 64, 231–255.

Tomasetti, C., & Vogelstein, B. (2015). Variation in cancer risk among tissues can be explained by the number of stem cell divisions. *Science*, 347(6217), 78–81. doi:10.1126/science.1260825

Tononi, G. (2004). An information integration theory of consciousness. *BMC Neuroscience*, 5. doi:10.1186/1471-2202-5-42

Tononi, G. (2012). *PHI: A voyage from the brain to the soul*. New York: Pantheon.

Tononi, G., Boly, M., Massimini, M., & Koch, C. (2016). Integrated information theory: From consciousness to its physical substrate. *Nature Reviews Neuroscience*, 17(7), 450–461.

Trudeau, P. E. (1998). *The essential Trudeau* (R. Graham, Ed.). Toronto: McClelland & Stewart.

Turri, J. (2017). Compatibilism and incompatibilism in social cognition. *Cognitive Science*, 41(Suppl. 3), 403–424.

Unger, R. M., & Smolin, L. (2014). *The singular universe and the reality of time*. Cambridge, UK: Cambridge University Press.

Valtorta, N. K., Kanaan, M., Gilbody, S., Ronzi, S., & Hanratty, B. (2016). Loneliness and social isolation as risk factors for coronary heart disease and stroke: Systematic review and meta-analysis of longitudinal observational studies. *Heart*, 102(13), 1009–1016.

Van Parijs, P., & Vanderborght, Y. (2017). *Basic income: A radical proposal for a free society and a sane economy*. Cambridge, MA: Harvard University Press.

van Riel, R., & Van Gulick, R. (2014). Scientific reduction. In E. Zalta (Ed.), *Stanford encyclopedia of philosophy*. http://plato.stanford.edu/entries/scientific-reduction/

Vartanian, O., Bristol, A. S., & Kaufman, J. C. (2013). *Neuroscience of creativity*. Cambridge, MA: MIT Press.

Velleman, J. D. (2003). Narrative explanation. *Philosophical Review, 112*, 1–25.

Vertolli, M. O., Kelly, M. A., & Davies, J. (2017). Coherence in the visual imagination. *Cognitive Science, 42*. doi:10.1111/cogs.12569

Villarreal, M. F., Cerquetti, D., Caruso, S., Schwarcz Lopez Aranguren, V., Gerschcovich, E. R., Frega, A. L., & Leiguarda, R. C. (2013). Neural correlates of musical creativity: Differences between high and low creative subjects. *PLoS One, 8*(9), e75427. doi:10.1371/journal.pone.0075427

Wahba, M. A., & Bridwell, L. G. (1976). Maslow reconsidered: A review of research on the need hierarchy theory. *Organizational Behavior and Human Performance, 15*(2), 212–240.

Wegner, D. M. (2003). *The illusion of conscious will*. Cambridge, MA: MIT Press.

Wiggins, D. (1987). *Needs, values, truth*. Oxford: Basil Blackwell.

Wilkinson, R. G., & Pickett, K. (2010). *The spirit level: Why greater equality makes societies stronger*. New York: Penguin.

Williamson, T. (2007). *The philosophy of philosophy*. Malden, MA: Blackwell.

Williamson, T. (2013). *Modal logic as metaphysics*. Oxford: Oxford University Press.

Wilson, D. S. (2015). *Does altruism exist?* New Haven, CT: Yale University Press.

Wilson, E. O. (1998). *Consilience: The unity of knowledge*. New York: Vantage.

Wimsatt, W. C. (2007). *Re-engineering philosophy for limited beings*. Cambridge, MA: Harvard University Press.

Wittgenstein, L. (1968). *Philosophical investigations* (G. E. M. Anscombe, Trans. 2nd ed.). Oxford: Blackwell.

Wolin, R. (2016). *The politics of being: The political thought of Martin Heidegger* (2nd ed.). New York: Columbia University Press.

Woodward, J. (2004). *Making things happen: A theory of causal explanation*. Oxford: Oxford University Press.

Woodward, J. (2014). Scientific explanation. In E. Zalta (Ed.), *Stanford encyclopedia of philosophy*. http://plato.stanford.edu/entries/scientific-explanation/

Zacks, J. M., & Magliano, J. P. (2013). Film, narrative, and cognitive neuroscience. In F. Bacci & D. Melcher (Eds.), *Art and the senses* (pp. 435–455). Oxford: Oxford University Press.

Zaidel, D. W. (2014). Creativity, brain, and art: Biological and neurological considerations. *Frontiers in Human Neuroscience, 8*, 389. doi:10.3389/fnhum.2014.00389

Zajonc, R. B. (1968). Attitudinal effects of mere exposure. *Journal of Personality and Social Psychology, 9*, 1–27.

Zak, P. J. (2012). *The moral molecule: The source of love and prosperity*. New York: Dutton.

Zaki, J., & Cikara, M. (2015). Addressing empathic failures. *Current Directions in Psychological Science, 24*(6), 471–476.

Zeki, S., Romaya, J. P., Benincasa, D. M., & Atiyah, M. F. (2014). The experience of mathematical beauty and its neural correlates. *Frontiers in Human Neuroscience, 8*, 68. doi:10.3389/fnhum.2014.00068

Zunshine, L. (Ed.). (2015). *The Oxford handbook of cognitive literary studies*. Oxford: Oxford University Press.

NAME INDEX

References to figures are denoted by an italic *f* following the page number.

Aaker, J. L., 226
Abrahamsen, A., 117
Abrams, M., 90
Adajian, T., 258
Agnati, L. F., 258
Aknin, L. B., 167, 180
Albantakis, L., 57
Allen, C., 292
Allen-Hermanson, S., 56
Allison, C., 180–81
Anderson, G., 179
Anderson, N., 259
Andrew, M., 116
Aquinas, T., 23
Aristotle, 6, 11, 22, 82, 129, 176, 189, 273–75, 284, 286
Armstrong, D. M., 116
Arneson, R., 205
Aschenbrenner, L., 258–59
Atallah, B. V., 117
Atiyah, M. F., 292
Atkinson, A. B., 205
Aubie, B., 56
Austen, J., 284
Ayal, S., 181

Bacon, F., 11
Baillargeon, R., 146
Banich, M. T., 24
Barbour, J., 108
Barkan, R., 181
Baron-Cohen, S., 166, 175, 180–81
Barsalou, L. W., 56
Bartha, P., 292
Bashour, B., 23
Baumeister, R. F., 226
Bayes, T., 79, 101, 124
Beam, C., 89
Beardsley, M. C., 258–59
Beaulieu-Plante, J., 181
Bechtel, W., 24, 56, 117, 145, 146
Beddor, B., 90
Beethoven, L. van, 32
Bekolay, T., 56
Benatar, D., 226
Benincasa, D. M., 292
Bentham, J., 165
Berger, P. L., 116
Berkeley, G., 27–28
Bernhard, R. M., 179

Bernhardt, B. C., 180
Bethlehem, R. A., 180–81
Blair, J., 181
Blair, K., 181
Blanchard, T., 90
Bloom, P., 169, 180–81
Blouw, P., 23, 56
Boduszek, D., 181
Boly, M., 57
Boomsma, D. I., 226
Boroditsky, L., 117
Bostyn, D. H., 179
Boyd, R., 89
Brattico, E., 259
Brennan, J., 181
Bridwell, L. G., 180
Bristol, A. S., 259
Broadbent, A., 146
Brown, R. C., 292
Brown, S., 259
Buckwalter, W., 24
Bullot, N. J., 258
Bunge, M., 24, 146
Buonomano, D. V., 117
Burnston, D., 117

Cacioppo, J. T., 180, 226
Cacioppo, S., 180, 226
Cameron, R. P., 116
Campbell, D., 89
Camus, A., 254, 259
Cangelosi, A., 56
Cantlon, J. F., 292
Cantor, G., 284
Capitanio, J. P., 180, 226
Caporael, L. R., 89
Cappelen, A. W., 206
Cappelen, H., 23
Carroll, S. B., 259
Caruso, S., 259
Casey, M., 259
Cauchi, A., 181
Cerquetti, D., 259
Cerullo, M. A., 57
Chalmers, D., 42, 56
Chaponis, J., 179
Chatterjee, A., 258
Chekhov, A., 21
Chelnokova, O., 259

Ching, C. M., 180
Choo, X., 56
Church, A. T., 180
Churchill, W., 193
Churchland, P. M., 23, 24, 56
Churchland, P. S., 23, 24, 167, 180
Cikara, M., 180–81
Clark, A., 56
Clark, C. J., 292
Cohen, M. S., 90
Colaço, D., 24
Cole, S. W., 180, 226
Coles, A. I., 180–81
Colyvan, M., 292
Compton, R. J., 24
Conde, C. J. C., 258
Cooperrider, K., 117
Côté, S., 206
Craver, C. F., 24, 56, 145, 146
Crick, F., 69
Crockett, M. J., 180
Cross, E. S., 258
Cross, I., 259
Currie, G., 258
Cushman, F., 23, 181

Dadds, M. R., 181
Dagher, A., 259
Dalton, J., 126, 129
Daly, C., 23
Damasio, A., 259
Dammann, O., 146
Daniels, N., 205
Danks, D., 90
Darby, R. R., 181
Darden, L., 24, 145
Darwin, C., 71, 74, 100, 116, 120–21, 125, 126, 128–29, 131, 271
Davies, J., 90
Davis, E., 90
Dawkins, R., 65–66, 89
Deaton, A., 226
Debowska, A., 181
Decety, J., 179, 180
de Chantérac, A., 181
Deci, E. L., 161, 180
de Fermat, P., 285–86
Dehaene, S., 50, 52–53, 57, 292
De La Cruz, V., 226

Name Index

Demakakos, P., 180
Dennett, D., 2, 3, 23, 24, 45, 56, 57, 89, 292
Descartes, R., 2, 6, 12–13, 27, 42, 71–72, 129, 280
Devitt, M., 117
de Waal, F. B. M., 292–93
Dewey, J., 11
DeWolf, T., 56
Diener, E., 226
Dobson-Stone, C., 181
Dodell-Feder, D., 24
Doris, J. M., 179
Dostoyevsky, F., 214
Douglas, H. E., 89
Dove, G., 56
Doyal, L., 205
Dummett, M., 7–8, 23
Dunn, E. W., 167, 180
Durkheim, E., 114, 117

Ebert, R., 254, 260
Ebstein, R. E., 181
Eco, U., 258–59
Eichele, T., 206
Eichenbaum, H., 117
Eikemo, M., 259
Einstein, A., 71, 77, 108, 109, 129
Eliasmith, C., 3, 12, 23, 25–26, 29–30, 31, 33, 36, 56, 211, 226, 228, 279, 286, 288, 290
Elliott, K. C., 180
Epicurus, 11, 22, 223, 227
Erdos, P., 285–86
Euclid, 278–79, 281, 284

Ferrigno, S., 292
Findlay, S. D., 56, 89
Finn, T., 181
Fischer, M. H., 56
Fleck, L., 104, 117
Fox, M. D., 181
Francis, K. B., 179
Frankland, S. M., 226
Franklin, J., 274–75, 284, 286, 292
Frega, A. L., 259
French, P. A., 259

Galileo, G., 77
Gallagher, P., 179

Gallagher, S., 24
Gandhi, M., 181
Ganis, G., 179
Garbinsky, E. N., 226
Gazzaniga, M., 292
Gendler, T. S., 23
Gerschcovich, E. R., 259
Gettier, E., 61
Gilbert, M., 117
Gilbody, S., 180
Gino, F., 181
Glannon, W., 292
Glenn, A. L., 181
Glennan, S., 145
Gödel, K., 108, 274
Godfrey-Smith, P., 24
Godoy, R. A., 259
Goel, A., 117
Goldman, A., 24, 90
Goldstone, R. L., 292
Goodman, N. D., 90
Goodson, S., 181
Goossens, L., 226
Gopnik, A., 145
Gough, I., 205, 206
Gracyk, T., 258
Greenberg, J., 227
Greene, J., 179–80
Greene, J. D., 179, 226, 292
Griesemer, J. R., 89
Griffiths, T. L., 90
Gummerum, M., 179
Guyer, P., 56, 116

Habibi, A., 259
Hacking, I., 90, 116, 292
Haidt, J., 180
Hallam, S., 259
Hanich, J., 259
Hanratty, B., 180
Hardy, G. H., 292
Harel, M., 292
Harris, S., 90, 292
Haug, M. C., 23
Hausman, D., 24
Hawes, D. J., 181
Hawking, S. S. W., 23
Hawthorne, J., 23
Hebb, D. O., 24, 31, 56

Hegel, G., 27–28, 215–16, 226, 261
Heidegger, M., 9, 24
Heintzelman, S. J., 226
Held, B. S., 23, 90
Held, V., 180–81
Heller, A., 205
Hempel, C. G., 145
Henik, A., 292
Hitchcock, C., 146
Hitler, A., 192–93
Hobbes, T., 12–13, 165
Holland, J. H., 90
Holt, J., 117
Holyoak, K. J., 90, 179–80, 292–93
Honein, M. A., 146
Horn, A., 181
Horsten, L., 292
Horstmann, R., 56, 116
Howard, C., 179
Howard, I. S., 179
Hrdy, S. B., 180
Hugdahl, K., 206
Hull, G., 180
Hume, D., 11
Huron, D., 259
Husserl, E., 9
Huston, J. P., 258
Hyde, D. C., 292
Hyland, P., 181

Inzlicht, M., 292
Irvine, A. D., 292
Ishizu, T., 258–59

Jacobs, A. M., 259
Jacobsen, T., 259
James, W., 38, 116
Jamieson, D. J., 146
Jara-Ettinger, J., 292
Jenkins, A. C., 24
Johnson, M., 259
Johnson-Laird, P. N., 248, 259
Jones, M., 90
Jordania, J., 259
Joyal, C. C., 181
Juslin, P. N., 259

Kahneman, D., 76–77, 90, 292
Kajic, I., 56

Kanaan, M., 180
Kania, A., 259
Kant, I., 6, 27, 165
Karremans, J. C., 292
Kashdan, T. B., 226
Katigbak, M. S., 180
Katzin, N., 292
Kaufman, J. C., 259
Kelly, M. A., 90
Keltner, D., 206
Kemp, C., 90
Keynes, J., 216
Kieran, M., 258
Kierkegaard, S., 214
Kihlstrom, J. F., 24
Kincaid, H., 116
King, L. A., 226
Kirsch, L. P., 258
Kitcher, P., 24, 292
Klein, G., 76–77, 90
Knobe, J., 24
Koch, C., 57
Koelsch, S., 259
Kotovsky, L., 146
Kripke, S., 101
Kropff, E., 117
Kuhn, T., 11, 117
Kunda, Z., 90

Lackey, J., 90
Ladyman, J., 116, 146
Laeng, B., 259
Lakoff, G., 280, 292
Landy, D., 292
Lapicque, L., 31
Latour, B., 90, 116
Lavoisier, A., 120–21
Layton, J., 155, 180
Legault, L., 292
Leibniz, G., 108, 109, 285
Leibovich, T., 292
Leiguarda, R. C., 259
Leknes, S., 259
Lenin, V., 192
Le Poidevin, R., 116
Levinson, J., 258
Levitin, D., 259
Lewis, D., 101, 116
Libet, B., 292

Livengood, J., 24
Locke, J., 11
Locke, K. D., 180
Lopes, D. M., 258
Løseth, G., 259
Love, B. C., 90
Luckmann, T., 116
Lucretius, T., 11
Luguri, J. B., 292

Machery, E., 24
Magliano, J. P., 259
Magnani, L., 89
Mao, Z., 192
Marcus, G. F., 90
Marghetis, T., 292
Marmot, M., 205
Marshall, P., 205
Marx, K., 192, 205
Mashek, D. J., 180
Maslow, A. H., 160, 180
Massimini, M., 57
Matyja, J., 259
Maurud, H., 259
McCauley, R. N., 24, 56, 146
McClelland, J. L., 31, 56, 146
McDermott, J. H., 259
McIntosh, A. R., 259
McKnight, P. E., 226
McRae, K., 56
McTaggart, J., 108
Mele, A. R., 264–65, 267, 268–69, 292
Mendoza-Denton, R., 206
Menin, D., 259
Menninghaus, W., 259
Meskin, A., 258
Metz, T., 226
Meyer, L., 206
Milgram, S., 176
Mill, J. S., 2, 11
Miller, S. C., 180
Mintzberg, H., 24
Mitchell, D. R., 181
Mlodinow, L., 23
Mora, F., 258
Morell, M., 200, 206
Moser, E. I., 117
Moser, M., 117
Moul, C., 181

Moynihan, D., 103
Mullainathan, S., 206
Muller, H. D., 23
Murphy, G. L., 56

Nadal, M., 258
Nagel, T., 56, 290, 293
Narveson, J., 205
Needham, A., 146
Neil, K., 180–81
Newton, I., 80, 108, 120–21, 124, 129, 285
Nichols, S., 180, 292
Nisbett, R. E., 90
Noë, A., 56, 258
Norton, M. I., 167, 180
Núñez, R., 117, 280, 292
Nussbaum, M. C., 189–90, 206

Oatley, K., 248, 253, 259
Oishi, S., 226
Oizumi, M., 57
Olsson, E., 90
Opfer, J. E., 292
Orend, B., 205
Osbeck, L. M., 23, 90
Ottmar, E. R., 292

Papineau, D., 24
Pascal, B., 224
Paseau, A., 292
Pasteur, L., 120–21, 125
Paton, J. J., 117
Peano, G., 279
Pearce, M., 259
Pearl, J., 146
Peikoff, L., 180
Peirce, C., 11, 116, 215
Penn, D. C., 292–93
Perlovsky, L., 259
Pessoa, L., 90
Peter, S., 116
Petersen, L. R., 146
Peterson, J., 291
Peterson, J. B., 293
Pezzulo, G., 56
Pfaff, D. W., 180
Piantadosi, S. T., 90, 292
Piccinini, G., 56
Pickett, K., 205

Piff, P. K., 206
Piketty, T., 206
Planck, M., 109
Plate, T., 56
Plato, 2, 6, 36, 82, 150, 176, 209, 272, 273–74, 275, 282
Plebe, A., 226
Polansky, L., 259
Polya, G., 280, 292
Poston, T., 90, 146
Povinelli, D. J., 292–93
Powell, D., 179–80
Prinz, J. J., 180
Psillos, S., 117
Putnam, H., 36
Pyszczynski, T., 227
Pythagoras, 274, 278, 280–81, 283

Quine, W. V. O., 11, 146

Radcliff, B., 206
Railton, P., 24
Ramón y Cajal, S., 31
Ramsey, W., 56, 145
Rand, A., 156, 167, 180, 181, 200
Ransohoff, K., 179
Rasmussen, D., 56
Rasmussen, S. A., 146
Rawls, J., 165, 183, 186, 205
Read, S. J., 90
Reader, S., 180
Reber, R., 258
Remmel, R. J., 181
Richerson, P. J., 89
Rickard, S., 259
Riegels, J., 259
Rini, R. A., 180
Robeyns, I., 206
Robinson, H., 56
Robinson, J., 259
Robson, J., 258
Rodman, S., 259
Roe, S., 259
Roets, A., 179
Romaya, J. P., 292
Ronzi, S., 180
Rorty, R., 90, 116
Rosenberg, A., 24

Ross, D., 116, 146
Rothko, M., 241, 259
Rumelhart, D. E., 31, 56
Russell, B., 2, 11, 228, 258
Russo, F., 146
Ryan, R. M., 161, 180

Sachs, M. E., 259
Salimpoor, V. N., 259
Sartre, J. P., 9
Sartwell, C., 258–59
Saxe, R., 24
Schaffer, J., 146
Schama, S., 259
Schiavio, A., 259
Schooler, J. W., 292
Schröder, T., 56, 90
Schrödinger, E., 284
Schultz, A. F., 259
Schwarcz Lopez Aranguren, V., 259
Schwartz, S. H., 180
Schwitzgebel, E., 23
Seager, W., 56
Segal, H. D., 206
Sellars, W., 3, 23
Sen, A., 189, 205, 206
Seto, M. C., 181
Sevenhant, S., 179
Shafir, E., 206
Shalvi, S., 181
Shanahan, D., 259
Shankar, A., 180
Shannon, C., 48
Shariff, A. F., 292
Shen, J., 180
Sheredos, B., 117
Sherrington, C., 31
Sheth, S. A., 90
Siburian, R., 179
Siegler, R. S., 292
Sievers, B., 259
Simon, D., 90
Singer, T., 180
Sloboda, J., 259
Slote, M., 180–81
Smee, S., 259
Smith, D. W., 24
Smolin, L., 108, 117

Soares, S., 117
Sober, E., 24
Socrates, 82
Solodkin, E., 23, 56
Solomon, M., 24
Solomon, S., 227
Sørensen, E. Ø., 206
Speaks, J., 226
Specht, K., 206
Spelke, E. S., 292
Spivey, M. J., 56
Stalin, I., 192
Stancato, D. M., 206
Stanley, J., 89
Starr, G. G., 258, 259
Steinhardt, P. J., 108, 117
Stenstrom, D., 90
Steptoe, A., 180
Sterelny, K., 24
Steup, M., 89
Stevenson, B., 205
Stewart, T. C., 50–51, 56, 57, 90, 259
Stich, S., 24
Stiglitz, J. E., 205
Stone, A. A., 226
Strecher, V. J., 226
Stuewig, J., 180
Swafford, J., 259
Sytsma, J., 24

Tang, Y., 56
Tangney, J. P., 180
Tegmark, M., 283, 292
Tenenbaum, J. B., 90
Teper, R., 292
Terbeck, S., 179
Thagard, P., 23, 24, 56, 57, 89–90, 117, 145, 146, 180, 181, 205, 224, 226, 258–59, 292
Thales, M., 11, 22, 24, 99
Thaut, M., 259
Thompson, E., 56
Thomson, W., 116
Tolstoy, L., 241, 259, 284
Tomasello, M., 180
Tomasetti, C., 226
Tononi, G., 57
Trudeau, P. E., 184, 205
Tungodden, B., 206
Turok, N., 108, 117
Turri, J., 24, 292
Tversky, A., 77, 90

Undurraga, E. A., 259
Unger, R. M., 117
Urgesi, C., 258

Vaish, A., 180
Valtorta, N. K., 180
van Andel, E. M., 180–81
Vanderborght, Y., 206
van Gulick, R., 146
Van Parijs, P., 206
van Riel, R., 146
Vartanian, O., 258, 259
Velleman, J. D., 145
Verbeurgt, K., 24, 90
Vertolli, M. O., 90
Villarreal, M. F., 259
Vogelstein, B., 226
Vohs, K. D., 226, 292

Wagner, V., 259
Wahba, M. A., 180
Ward, S. J., 226
Wardle, J., 180
Wassiliwizky, E., 259
Watson, J., 69
Wegner, D. M., 292
Weitnauer, E., 292
Wettstein, H. K., 259
Wheatley, T., 179, 259
Wiggins, D., 205
Wikler, D., 179
Wiles, A., 285–86
Wilkinson, R. G., 205
Williamson, J., 146
Williamson, T., 23, 101, 116
Wilson, D. S., 180
Wilson, E. O., 24
Wimsatt, W. C., 24, 89, 146
Wittgenstein, L., 35, 36, 91, 116
Wolfers, J., 205
Wolin, R., 24
Wood, J. V., 292
Woodward, J., 145, 146
Woolgar, S., 116

Zacks, J. M., 259
Zahavi, D., 24
Zaidel, D. W., 259
Zajonc, R. B., 259
Zak, P. J., 167, 180
Zaki, J., 180–81

Zald, D. H., 259
Zatorre, R. J., 259
Zednik, C., 292
Zeki, S., 258–59, 292
Zhang, H., 180
Zunshine, L., 259

SUBJECT INDEX

References to tables and figures are denoted by an italic *t* or *f* following the page number.

abductive inferences, 68–69
accidents, 214–16
action, in theory of free will, 267–68, 269, 270–71
aesthetics
  beauty in music, 245–47
  beauty in painting, 233–38
  coherence, 229
  creativity in music, 252–53
  creativity in painting, 242–45
  emotional reactions, overview of, 228–29, 255–56, 257
  empathy in literature and film, 253–54
  ethics and, 258
  general discussion, 254–58
  issues and alternatives, 230–32
  other emotions in music, 247–52
  other emotions in painting, 238–41
  overview, 21, 228–30
  social aspects, 230, 232, 245
  three-analysis of art, 230–32
affiliation. *See* relatedness
afterlife, 223–24
agency-based explanations, 137–38

agreement, in science, 98
alternative narratives, 123
ambitious free will, 264–66
analogy
  mathematical, 280
  as mode of empathy, 168, 170–71
  role in generating hypotheses, 68
analytic philosophy, 7–9, 18
anarchism, 185, 194
anger
  in music, 248, 251
  in painting, 239
antimaterialist philosophers, 42–43
antireductionism, 120, 143
antisocial personality, 164, 175
anxiety, in music, 248, 252
approximate correspondence theory of truth, 105–6
a priori ethics, 147
a priori truths, 6, 71
Aristotelian realism, 274–75, 284–85
arithmetic, 278
art, three-analysis of, 230–32. *See also* aesthetics

attention, semantic pointer competition and, 52
auditory representations, 252
automatic decision making, 268
autonomy
    aesthetics, relation to, 255
    basic income as increasing, 201
    constraint satisfaction related to, 218
    freedom indices, 193
    mechanisms for, 162
    needs sufficiency, 187
    overview, 161
    and sense of meaning, 217
axioms, in mathematics, 71, 278–79

basic income, 200–2
Bayes' theorem, 78–81
Beatles, 246–47, 253
beauty
    in mathematics, 281–82
    in music, 245–47
    in painting, 233–38
Beethoven, Ludwig van, 246–47, 253
beliefs. *See also* metaphysics
    generating hypotheses, 68–69
    groups as counterbalance to individual, 85
    just social change, 198
    knowledge, role in, 61–63
    mathematical, 278–79
    probability, 79–80, 81
    religious, explanation of, 100–1
belongingness. *See* relatedness
best explanation, inference to. *See* inference(s)
binding, neural. *See also* Semantic Pointer Architecture
    animal minds, 287–88
    beauty in music, 247
    concepts bound with emotions, 152
    conceptual combination and, 68
    and consciousness, 51–52
    recursive, 51, 167, 234, 287–88
biological analogies of knowledge growth, 65–67
biological needs, 158–59, 162–63, 186–87, 201. *See also* vital needs
bodily perception, emotions as, 38–39, 152, 155, 166–67. *See also* Semantic Pointer Architecture
body, importance to mind, 46

brain. *See also* multilevel materialism; neural mechanisms
    aesthetic judgment in, 236
    caring for others in, 166
    meaning of death, 223
    theories of mind based on, 3, 25–27, 28–29, 42–46
Brain and the Meaning of Life, The (Thagard), 218
Brain-Mind (Thagard), 2, 3
brain revolution, 85–87
broadcasting, neuronal workspace, 49–50

callings, and sense of meaning, 217
capabilities approach to justice, 189–90
careers, and sense of meaning, 217
caring about needs of others, 164–68, 219–20
causal correlations, 75, 134
causality
    free will, 269–70
    general discussion, 144–45
    group, 111
    judgments about cause and effect, 139–40
    mechanistic explanation, 126, 127–28
    in mind, 135–36
    in narrative explanation, 123
    overview, 118–19, 132
    in society, 138
    three-analysis, 132–35
    in world, 136–38
certainty, of mathematics, 280–81
chance, 214–16, 269
change(s)
    existence of, 96–97
    as presupposing existence of time and space, 106
children, meaning related to having, 219–20
chords, musical, 246
Christianity. *See* religion
circularity
    of causality, 137
    of explanatory coherence, 75
coalescence, 86
cognition
    social cognitive-emotional approach, 12–13
    theories of, 3
cognitive-affective maps, 153–54
cognitive appraisal. *See also* Semantic Pointer Architecture
    caring emotions, 166–67

Subject Index

emotions, 38–39, 152, 155
  of music, 247–48
  values as semantic pointers, 152
cognitive role of explanation, 121
cognitive science
  4e cognition, 43–46
  natural philosophy and, 2–3, 17, 21–23
coherence
  aesthetics, 229
  beauty in music, 247
  beauty in painting, 234–36
  emotional, in decision making, 172–73
  ethical, conflicting needs and, 171–74
  hangs together metaphor, 4
  justification of knowledge, 59, 60–61, 73–75, 76, 88
  of materialism, 93–94
  mathematics, 280–81
  natural philosophy, 11, 15–18, 261–62
  philosophical procedure, 4–5
  reality, 116
  semantic pointers, 39
  social cognitivism, 261–62
  truth, theory of, 94, 105–6
coherentism, reliable, 19–20, 61, 76, 179. *See also* epistemology
collaboration, in social epistemology, 83–84
collective mental states, 112–13
combination, conceptual, 67–68, 278, 284
combinations of auditory representations, 252
communication
  of emotion through art, 241
  and meaning in language, 212–13
  multimodal, 212–13
  semantic pointers in, 34–35
  social epistemology, 82–83
  social transmission of knowledge, 70
communism, 185, 192
compassion, 155
competence
  aesthetics, relation to, 255
  basic income as increasing, 201–2
  constraint satisfaction related to, 218
  education and income, relation to, 193
  mechanisms for, 162
  needs sufficiency, 187
  overview, 161
  and sense of meaning, 217
competition, semantic pointer, 50–53
compositional meaning
  of art, 235
  of language, 211, 220
concepts
  generating, 67–68
  mathematical, 275–78
  need for explanation based on, 142
  as semantic pointers, 36–37
  values as associated with, 150–52
conceptual analysis, 7, 8
conceptual change, 85–87
conceptual combination, 67–68, 278, 284
connectionist approach to cognitive science, 43
conscience, 173
consciousness
  free will, 267–68
  information integration, 47–48
  neuronal workspace broadcasting, 49–50
  overview, 47
  in phenomenology, 9–10
  philosophical positions on, 29
  scientific theories of, 99
  semantic pointer theory of, 50–53, 55
  supernatural explanations of, 53–54
consensus, in social epistemology, 84–85
consequentialism. *See also* needs-based consequentialism
  needs sufficiency justice, 188–89
  overview, 149
conservative political parties, 194–95
constraints, in explanatory coherence, 73–74
constraint satisfaction
  aesthetics, 229
  ambitious free will, 265–66
  beauty in music, 247
  beauty in painting, 234–35
  coherence as process of, 15–18
  ethical coherence, 172–74
  life as problem of, 218
contracts, social, 165
convolution, binding by. *See also* Semantic Pointer Architecture
  beauty in music, 247
  concepts bound with emotions, 152
  conceptual combination and, 68
  and meaning in language, 211
  overview, 33
  semantic pointer competition, 51–52
coping, 216
correlations, causal, 75, 134

correspondence theory of truth, 104, 105–6
cosmological natural selection, 108
counterfactuals, 134
countries
　differences in happiness between, 168
　government type and needs in, 193–94
　international justice, 196–97
courts, just, 196
creativity
　machine minds, 288–89
　in music, 252–53
　in painting, 242–45
　procedural, 244, 253
cultural relativism, 147
culture
　judgments of beauty, relation to, 235–36
　memetics, 65–67
　social transmission of knowledge, 69–70
　space and time variations in, 109–10

da Vinci, Leonardo, 244
death, meaning of, 222–24
decision making
　ambitious free will, 265–66
　as coherence problem, 15–16
　conflicting needs and ethical coherence, 171–74
　modest free will, 266–68
　moral responsibility for, 270–71
deductive explanation, 124–25, 128, 140–41
deepity, 45
definitional analysis, 7
degrees of belief, probabilities as, 79–80
de Kooning, Willem, 238–39
deliberate decision making, 268
democracy
　as best overall political system, 192–94
　empathy as contributing to, 200
　liberal, 186
　political parties and ideologies, 194–95
　social, 184
demogrant, 200–2
deontology, 149–50, 151
descriptive information, relevance to normative conclusions, 14–15
desire
　in music, 249
　in painting, 240
determinism, 108, 269

dictatorships, 193–94
differentiation, 86
discrimination, 190–91
disease, 175
disgust
　in music, 251
　in painting, 239
dispositions, 137, 189
disturbing music, 251–52
dogmatism, 15
dualism, 27, 29, 38, 40–41, 54, 93–94
dual process theories, 268

education
　difficulty of, 83
　needs sufficiency and, 187–88
effect, judgments about cause and, 139–40. *See also* causality
elections
　democracy pros and cons, 192–93
　as ethical and emotional, 153–54
eliminative explanation, 128–30, 281
eliminative materialism, 28, 29
embeddedness of minds, 45, 46
embodiment
　of knowledge, 64, 65
　of language, 211
　mathematical concepts, 277–78
　of mind, 43–45, 46
　of sense of space and time, 106–7
emergence. *See also* multilevel emergence
　free will and, 269–70
　in groups, 111
　meaning as, 213, 220–22
　overview, 33
　reduction and, 120, 141–44
emotion(s)
　in analytic versus natural philosophy, 8
　art, impact on, 228–29, 233–41, 245–52, 255–56, 257
　cognitive appraisal, 38–39, 152, 155
　in collaboration, 84
　in decision making, 172–73
　epistemic, 69
　in evaluation of art, 243–44
　explanation, aspects of, 121, 130–31
　in mathematics, 281–82
　meaning of life, relation to, 225
　moral, 154–56, 170

in music, 245–52
need for explanation based on, 142
nested, 241
in painting, 233–41
as part of knowledge, 59, 69
rational, 156–57, 222 (*see also* needs-based consequentialism)
semantic pointer competition and, 52
semantic pointer theory of, 38–39
social cognitivism, 12–13, 261–62
theories of, 3
value maps, 153–54
values as associated with, 150–52

empathy
analogy mode, 168, 170–71
general discussion, 167–71, 178
just social change and, 199–200
in literature and film, 253–54
mirror neurons, 168–69, 170, 199–200
moral emotions, 155
reverse, in art, 241
simulation mode, 169, 170, 254
towards future generations, 191

empiricism, 72, 274, 283
enactive, minds as, 45, 46
entities, theoretical, 97, 111
envy, 187–88
epistemic emotions, 69
epistemology. *See also* explanation
conceptual change and brain revolution, 85–87
defining knowledge, 59, 61–65
ethics and, 179
foundationalism, 60, 71–72
free will, 268–69
general discussion, 88–89
growth of knowledge, 59–60, 65–70
issues and alternatives, 59–61
justification, 60, 70–78
metaphysics and, 118–19
overview, 19–20
probability, 78–81
reliable coherentism, 19–20, 61, 76, 179
skepticism, 59, 60, 72, 214
social, 82–85
social transmission of knowledge, 60, 69–70
theory of mind and, 58–59

equality, 184, 187–88, 196–97. *See also* justice

erotic paintings, 240
ethical intuitions, 173
ethics. *See also* justice; meaning
aesthetics and, 258
conflicting needs and ethical coherence, 171–74
empathy, 168–71
evil, explaining, 174–76
free will, 270–71
general discussion, 176–79
issues and alternatives, 148–49
moral emotions, 154–56
needs-based consequentialism, 148, 157–68
objective values and rational emotions, 156
overview, 20, 147–48
reflective equilibrium, 183
values, 150–54

events
existence of, 96–97
as presupposing existence of time and space, 106

everything happens for a reason concept, 214–16

evidence, in coherence-based justification, 75–76

evil, explaining, 174–76, 178

evolution
cultural, 65–67
explanatory coherence, 74

excitatory links, 172

exemplars, in three-analysis
of art, 230–31
of causality, 132
of explanation, 120–21
of injustice, 204
of justice, 182, 204
of knowledge, 64
of number concept, 276
of philosophy, 3–4
of semantic pointers, 34
of values, 151
of vital needs, 158

existence
changes, processes, and mechanisms, 96–97
general discussion, 114–15
group minds, 112–13
groups, 110–11
idealism, 102–3
issues and alternatives, 93–94

existence (Cont.)
  objects, properties, and relations, 94–96
  overview, 92, 94
  possible worlds, 101
  procedure for determining, 115
  scientific realism, 97–99
  social constructivism, 103–4
  social facts, 113–14
  supernatural entities, 99–101
experience(s)
  in phenomenology, 9–10
  semantic pointer theory of consciousness, 53
experimental philosophy, 18–19
experimentation, in scientific realism, 98
explanation
  causality, 132–40
  cognitive role of, 121
  deductive, 124–25, 128, 140–41
  eliminative, 128–30, 281
  emotional and social aspects, 130–31
  general discussion, 144–45
  issues and alternatives, 119–20
  justification of knowledge, 71
  mechanistic, 125–28, 140–41
  narrative, 122–23, 128, 138
  overview, 20, 118–19
  reduction and emergence, 140–44
  styles of, 120–30
  three-analysis of, 120–22
explanations, in three-analysis
  of art, 231
  of causality, 132
  of explanation, 120
  of injustice, 204
  of justice, 204
  of knowledge, 64
  of number concept, 276, 277
  of philosophy, 3–4
  of semantic pointers, 34
  of values, 151
  of vital needs, 158
explanatory coherence, 73–76, 105
explanatory hypotheses, generating, 68–69
explanatory identities, 30, 42–43
expressivism, neural, 254–55
extended mind hypothesis, 45–46

face-to-face meetings, role in consensus, 84–85

facts
  as dispensable, 96
  as interconnected with values, 14–15
  social, existence of, 110
false cause, fallacy of, 139
fear
  of death, 222–24
  inference driven by, 100
  in music, 252
  in painting, 239
fictionalism, 274, 283, 285
Fifth Symphony (Beethoven), 246–47
films
  empathy in, 171, 253–54
  multimodal explanations of, 256
firing patterns, neuronal, 32, 33. *See also* Semantic Pointer Architecture
First Nations, discrimination against, 190–91
first-person experience, role in knowledge, 65
flourishing, 189
formalism, 274, 283
foundationalism, 60, 71–72
4e cognition, 43
Fraser Institute Freedom Index, 193
Freedom Index, Fraser Institute, 193
freeish will, 267–68
free will
  knowledge, 268–69
  mind, 265–68
  morality and meaning, 270–71
  overview, 178, 263–65
  reality, 269–70
frequencies, probabilities as, 80–81
functionalism, 28–29, 41–42
fundamentally mathematical, world as, 283
future generations
  just social change and, 198
  from needs-sufficiency perspective, 191

general beliefs, formation of, 68
generality of philosophy, 2–3
generation of cultural ideas, 66
genetics, memetics as analogy to, 65–67
geometry, 278–79, 280–81
Gini index, 196–97
giving relationships, 220
goals
  of art, 230
  cognitive appraisal theory of emotion, 38
  conflicting, and ethical coherence, 171–74

meaningful pursuit of, 218–19
role in generating hypotheses, 69
gods, existence of, 99–101
governments, just
  alternative views, 184–86
  basic income, 200–2
  general form of government, 192–94
  international justice, 196–97
  legal justice, 196
  overview, 192
  particular parties and ideologies, 194–95
Goya, Francisco, 239
Green parties, 195–96
grief
  in music, 248, 250–51
  in painting, 239–40
group minds, 110, 112–13
group processes, in just social change, 199
groups
  emergent properties of, 142
  of neurons, 32, 33
  reality of, 110–14
growth of knowledge, 59–60, 65–70
guaranteed annual income, 200–2
gut reactions, 170

hangs together metaphor for philosophy, 4, 17, 261–62
happiness
  caring for others, relation to, 167–68
  government policies, relation to, 195
  versus meaning, 220
  as measure for morality and justice, 163
  in music, 248–49
  in painting, 240
  political system, relation to, 193
hierarchy of needs, 160
historical approach to philosophy, 6
hobbies, and sense of meaning, 217
horror
  in music, 252
  in painting, 239
Human Development Index, 193
humanities, natural philosophy and, 21–22
human rights, relation to vital needs, 187
humor
  in music, 249
  in painting, 240–41
hypotheses, generating, 68–69

idealism
  consciousness, 29, 54
  existence, 93–94, 95, 102–3
  implausibility of, 41
  melding Platonism with, 282–83
  primacy of mind, 27–28
  truth, 94
ideas, spread of according to memetics, 65–67
ideologies, political, 194–95
IIT (information integration theory) of consciousness, 47–48
imagery, 243
imagination, 7, 42–43, 243
immorality, explaining, 174–76, 178
implicit knowledge, 62
income, basic, 200–2
indigenous people, discrimination against, 190–91
individualism, 156
inequality, 187–88, 190, 196–97. See also justice
infants, conception of causality in, 133
inference(s)
  in analytic versus natural philosophy, 8
  conflicting needs and ethical coherence, 172–74
  generating hypotheses, 68
  judgments about cause and effect, 139–40
  motivated, 100, 176
  to multilevel materialism, 40–42
  narrative explanation, improving, 123
  role in religious belief, 100
  semantic pointers, 39
information integration theory (IIT) of consciousness, 47–48
injustice, three-analysis of, 203, 204. See also justice
innateness of mathematical concepts, 275
instrumental needs, 157, 159, 162–63
intentions, 267–68
interdependent mechanistic systems, 13
international justice, 196–97
intersubjectivity of evidence in science, 75
introspection, 9–10, 18
intuition(s)
  in analytic versus natural philosophy, 8
  diversity in philosophical, 18
  ethical, 173
  justification of knowledge, 76–77
  pure-reason approach to philosophy, 7
  reflective equilibrium, 183

Inuit, discrimination against, 190–91
isolation, social, 161–62

jobs, and sense of meaning, 217
judgments, ethical. *See* ethics
judgments about cause and effect, 139–40
justice
    basic income, 200–2
    general discussion, 202–5
    governments, 192–97
    international, 196–97
    issues and alternatives, 184–86
    legal, 196
    needs sufficiency, 186–91
    overview, 20, 182–83
    social change, 197–200
    three-analysis of, 182, 203, 204
justification of knowledge. *See also* metaphysics
    coherence-based, 59, 60–61, 73–75, 76, 88
    definitions of knowledge, 61, 63
    evidence, 75–76
    foundationalism, 71–72
    general discussion, 88
    intuition, 76–77
    overview, 60, 70–71
    probability, 78–81
    truth and, 63

knowledge. *See also* explanation; justification of knowledge; mathematics
    beliefs, role in, 61–63
    biological analogies, 65–67
    coherence, 59, 73–75
    concepts, generating, 67–68
    conceptual change and brain revolution, 85–87
    defining, 59, 61–65
    embodiment of, 64, 65
    emotions as part of, 59
    ethics and, 179
    foundationalism, 60, 71–72
    free will, 268–69
    general discussion, 88–89
    growth of, 59–60, 65–70
    hypotheses, generating, 68–69
    interconnections of reality and, 118–19
    issues and alternatives, 59–61
    nonverbal, 59, 62–63
    overview, 19–20
    probability, 78–81
    reliable coherentism, 19–20, 61, 76, 179
    skepticism related to, 59, 60, 72, 214
    social epistemology, 82–85
    social transmission, 60, 69–70, 88–89
    theory of mind and, 58–59
    three-analysis of, 63–65
    transbodiment of, 64–65
    truth, role in, 61, 63
knowledge-how, 62, 88
knowledge-of, 62, 65, 88
knowledge-that, 62, 65, 88

language
    in analytic philosophy, 7–8
    mathematical knowledge and, 275
    meaning in, 20–21, 207–8, 210–13, 220, 221–22, 224
    and sense of space and time, 106–7
legal justice, 196, 270–71
levels, reduction of, 143
liberal democracy, 186
liberal political parties, 194–95
libertarianism, 185, 188–89
life, meaning of
    constraint satisfaction, 218
    everything happens for a reason concept, 214–16
    free will and, 271
    general discussion, 225–26
    issues and alternatives, 209
    as multilevel emergence, 220–22
    needs, relation to, 216–17
    needs of others, 219–20
    nihilism, 214
    overview, 20–21, 207–8, 213–14
literature
    empathy in, 171, 253–54
    multimodal explanations of, 256
logic
    in analytic versus natural philosophy, 9
    modal, 101
logicism, 274, 283
loneliness, 161–62
love, in music, 249
lust
    in music, 249
    in painting, 240

Subject Index

manipulation account of causality, 133–34
maps, value, 153–54
Maslow's hierarchy of needs, 160
materialism. *See also* multilevel materialism
   existence, 93, 94–95, 97–99
   truth, 94
   versions of, 28–29, 41
mathematics
   axioms and theorems, 71, 278–79
   coherence and necessity, 280–81
   emotions, 281–82
   information integration theory of consciousness, 48
   issues and alternatives, 273–74
   metaphors and analogies, 280
   in mind, 275–78
   overview, 263–64, 272–73
   in society, 285–86
   in world, 282–85
mattering, 150
McCartney, Paul, 247
meaning
   of death, 222–24
   free will, 270–71
   general discussion, 224–26
   issues and alternatives, 209
   in language, 209–13, 220, 221–22
   of life, 213–22
   as multilevel emergence, 213
   in natural philosophy, 13
   overview, 20–21, 207–8
   semantic pointers and, 34–35
mechanisms. *See also* molecular mechanisms; neural mechanisms; Semantic Pointer Architecture; social mechanisms
   for biological needs, 159
   causality and, 134, 136–38
   defining, 30
   existence of, 96–97
   identifying in narrative explanation, 123
   multilevel, 26–27, 29, 62–63, 142
   probability, 80
   for psychological needs, 161–63
   role in meaning, 221–22
   social cognitive-emotional approach, 12–13
   truth, 106
mechanistic explanation, 125–28, 140–41
melodies, 246, 247
memes/memetics, 65–67

mental illnesses, 162, 175
mental processes. *See* mind(s)
mental representations. *See also* Semantic Pointer Architecture
   causality in world, 136–37
   meaning in language, 210–13
   semantic pointers, 31–36, 37
mental states, collective, 112–13
mere exposure effect, 237
meta-classification, 86–87
metaphors, mathematical, 280
metaphysics. *See also* explanation
   epistemology and, 118–19
   ethics and, 179
   existence, 94–104
   free will, 269–70
   general discussion, 114–16
   groups and society, 110–14
   issues and alternatives, 93–94
   mathematics, 283–85
   multilevel materialism, 40–42
   overview, 20, 92–93
   space and time, 106–10
   truth, 104–6
methodological individualism, 110, 113
mincome, 200–2
Mind and Cosmos (Nagel), 290
mind-body dualism. *See* dualism
mind-brain identity theory, 28–29, 41
mind(s). *See also* multilevel materialism; Semantic Pointer Architecture
   animal, 286–88
   brain-based theories, 3, 25–27, 28–29, 42–46
   causality in, 135–36
   as coherence engine, 17
   consciousness, 47–54
   ethics and approach to, 178
   free will, 265–68
   general discussion, 54–55
   group, existence of, 112–13
   in idealism, 102
   issues and alternatives, 27–29
   knowledge and, 58–59
   language and, 210
   machine, 288–89
   mathematics in, 275–78
   meaning of death, 223
   mechanistic explanation in, 127
   in natural philosophy, 13, 19

mind(s) (Cont.)
  neural mechanisms, 29–31
  overview, 25–27, 261
  reality as dependent on, 98–99
  semantic pointers, 31–39
  social facts, 114
  space and time in, 106–7
  theories of, 3
Mind-Society (Thagard), 2, 3, 113, 128, 212
mirror neurons, 168–69, 170, 199–200
modal logic, 101
modal retention, semantic pointers, 51–52
modest free will, 264–65, 266–68
modus ponens, 279
molecular mechanisms. *See also* Semantic Pointer Architecture
  aesthetic judgment, 236–37
  for care, 166
  relevant to mind, 26–27, 29, 30
  role in meaning, 221–22
monism, neutral, 28, 29
moral emotions, 154–56, 170
morality. *See also* justice; meaning
  aesthetics and, 258
  conflicting needs and ethical coherence, 171–74
  empathy, 168–71
  evil, explaining, 174–76
  free will, 270–71
  general discussion, 176–79
  issues and alternatives, 148–49
  moral emotions, 154–56
  naturalistic explanations of, 100
  in natural philosophy, 13
  needs-based consequentialism, 148, 157–68
  objective values and rational emotions, 156–57
  overview, 20, 147–48
  reflective equilibrium, 183
  values, 150–54
mortality, meaning of, 222–24
motivated inference, 100, 176
movies
  empathy in, 171, 253–54
  multimodal explanations of, 256
multilevel emergence
  creativity in painting as, 245
  and emotional impact of art, 236–38, 241–42
  of evil, 175–76
  free will and, 269–70
  meaning as, 213, 220–22
  reductionism and, 142, 143
multilevel materialism. *See also* Semantic Pointer Architecture
  versus alternative explanations of mind, 29
  consciousness, 47–54
  general discussion, 54–55
  inference to best explanation to, 40–42
  overview, 19, 25–27
  philosophical objections to, 42–46
  reality, 115
  semantic pointers, 31–39
multilevel mechanisms, 26–27, 29, 62–63, 142
multimodal communication, 212–13
multimodal neural processes, semantic pointers as, 32–34. *See also* Semantic Pointer Architecture
multimodal rules
  axioms as, 279
  built out of semantic pointers, 62–63
  in conception of causality, 133, 135–36
  overview, 45
  procedural creativity, 244
  simulation mode of empathy, 169, 170, 254
music
  beauty in, 245–47
  creativity in, 252–53
  other emotions in music, 247–52

narrative explanation, 122–23, 128, 138
natural philosophy. *See also* multilevel materialism; *specific areas of philosophy*
  versus alternatives, 262, 263
  versus analytic philosophy, 8–9
  book overview, 19–21
  central principles of, 11–12
  coherence, 15–18, 261–62
  elements of, 12–19
  experimental philosophy, 18–19
  free will, 264–71
  future of, 289–90
  general discussion, 21–23
  improved theories, need for, 271
  mathematics, 272–86
  nonhumans, 286–89
  normative procedure, 13–15
  overview, 2, 11, 261–62

questions remaining unresolved in, 263–64, 289
rules for philosophical life, 291
social cognitivism, 12–13, 261–62
natural selection, cosmological, 108
necessary truths, 6, 8–9, 101, 281
needs-based consequentialism
conflicting needs and ethical coherence, 171–74
empathy, 168–71
evil, explaining, 174–76
free will and, 270–71
general discussion, 177–79
needs of others, 163–68
overview, 148, 157
psychological needs, 159–63
three-analysis of vital needs, 158–59
vital versus instrumental needs, 157
needs-sufficiency theory of justice
alternative views, 188–90
applications of, 190–91
basic income, 200–2
general discussion, 203–5
just governments, 192–97
just social change, 197–200
needs and justice, 186–88
overview, 186
negative constraints
in coherence, 15–16
in ethical coherence, 172
in explanatory coherence, 73–74
negative values, in value maps, 153–54
nested emotions, 241
neural binding. *See* binding, neural
neural expressivism, 254–55
neural materialism. *See* multilevel materialism
neural mechanisms. *See also* Semantic Pointer Architecture
for care, 165–67
in contemporary neuroscience, 28
for empathy, 168–70
for psychological needs, 161–63
relevant to mind, 26–27, 29–31
role in meaning, 221–22
neural representations, 50–52. *See also* Semantic Pointer Architecture
neural synchrony, 51
neuroaesthetics, 236, 257

neuronal workspace broadcasting, 49–50
neurons
breakthroughs concerning, 31
semantic pointers, 31–36
neuropsychological theory of values, 151–52
neutral monism, 28, 29
nihilism, 214
nonhumans
animal minds, 286–88
machine minds, 288–89
overview, 264, 286
nonverbal knowledge, 59, 62–63
normative procedure, 9, 13–15, 230, 262–63
normativity, of philosophy, 2–3. *See also* aesthetics; epistemology; justice; meaning; morality
norms, social, 199
notes, musical, 246
numbers
defined, 276
need for better account of, 285
three-analysis of, 276–77

objective probabilities, 80–81
objective values, 156–57. *See also* needs-based consequentialism
objectivism, 156
objectivity
of explanation, 131–32
in judgments of beauty, 235–36
of value judgments, 152
objects
existence of, 94–96
in information integration theory of consciousness, 48
as part of thinking, 45
O'Keeffe, Georgia, 244
ontology, 20
openness of decisions, 265–66
others, needs of, 163–68, 219–20
Our Mathematical Universe (Tegmark), 283

painting
beauty in, 233–38
creativity in, 242–45
other emotions in, 238–41
panpsychism, 28, 29, 41
parallel constraint satisfaction
aesthetics and, 229

Subject Index

parallel constraint satisfaction (*Cont.*)
  ambitious free will, 265–66
  coherence as process of, 15–18
  ethical coherence, 172–74
  semantic pointers, 39–40
parenting, meaning related to, 219–20
parodies, in music, 249
parties, political, 194–95
Passionate Muse, The (Oatley), 253
patterns
  in explanation, 120, 121
  of neuron firing, 32, 33
pedophilia, 175
people, as theoretical entities, 111
perception, bodily, emotions as, 38–39, 152, 155, 166–67. *See also* Semantic Pointer Architecture
perceptual learning, 67
phenomenology, 9–10, 18
philosophy. *See also* natural philosophy; *specific areas of philosophy*
  analytic, 7–9, 18
  approaches to, 5–12, 262, 263
  book overview, 19–21
  brain-based theories, objections to, 42–46
  coherence, 15–18, 261–62
  experimental, 18–19
  free will, 264–71
  future of, 289–90
  general discussion, 21–23
  hangs together metaphor, 4, 17, 261–62
  historical approach, 6
  improved theories, need for, 271
  mathematics, 272–86
  nonhumans, 286–89
  normative procedure, 13–15
  overview, 1–3, 261–62
  phenomenology, 9–10
  philosophical procedure, 4–5
  postmodernism, 10
  pure reason, 6–7
  questions remaining unresolved in, 263–64, 289
  religion, 5
  rules for philosophical life, 291
  social cognitivism, 12–13, 261–62
  three-analysis of, 3–5
physics, Theory of Everything in, 283
physiological perception. *See also* Semantic Pointer Architecture
  emotions as, 38–39, 152, 155, 166–67
  of music, 247–48
pitch, 246
plants, attributing minds to, 286–87
Platonism, 273–74, 275, 282–83
play, and sense of meaning, 217
pleasure, as measure for morality and justice, 163
political justice. *See* governments, just; justice
political parties, 194–95
positive constraints, in coherence, 15–16, 73–74, 172
positive values, in value maps, 153–54
possible worlds, 101
postmodernism, 10, 156, 184–85
poverty, basic income programs for, 200–2. *See also* justice
practical usefulness, in justification of knowledge, 71
pragmatics, 39
prediction, through deductive explanation, 124
prejudice, 190–91
premature elimination, 129–30
principle of sufficient reason, 108
probability
  causality and, 134
  in knowledge justification, 78–81
  weakness of deductive explanation, 125
procedural creativity, 244, 253
processes, existence of, 96–97
proof, mathematical, 279
properties, existence of, 94–96
psychological effects of explanation, 121
psychological mechanisms, role in meaning, 221–22
psychological needs. *See also* needs-based consequentialism; needs-sufficiency theory of justice; *specific needs*
  basic income, 201–2
  constraint satisfaction related to, 218
  Maslow's hierarchy of needs, 160
  mechanisms for, 161–63
  needs sufficiency, 187
  overview, 159–61
  and sense of meaning, 216–18
psychological theory of decision making, 270–71
psychopathy, 164, 175
punishment, justification of, 271
pure reason, 6–7, 182–83

pure reflection, 9–10
purpose
  of art, 232
  sense of, 207 (see also meaning)
Pythagorean theorem, 278, 279, 280–81

qualitative experiences, 53
quantum gravity, 283–84
quantum theory, 215

racial discrimination, 190–91
rational emotions, 156–57, 222. See also needs-based consequentialism
rationalist foundationalism, 71–72
rationality
  emotion and explanation, relation to, 131
  and explanation as social process, 131–32
  real is rational and rational is real concept, 215–16
  of value judgments, 152
realism
  Aristotelian, 274–75, 284–85
  scientific, 20, 93, 94–95, 97–99, 115
reality. See also explanation
  ethics and, 179
  existence, 94–104
  free will, 269–70
  general discussion, 114–16
  groups and society, 110–14
  issues and alternatives, 93–94
  knowledge and, 118–19
  mathematics and, 283–85
  overview, 20, 92–93
  real is rational and rational is real concept, 215–16
  space and time, 106–10
  truth, 104–6
reason
  in analytic versus natural philosophy, 8
  everything happens for a reason concept, 214–16
  pure, 6–7, 182–83
reclassification, 86
recreation, and sense of meaning, 217
recursive binding, 45, 51, 167, 234, 287–88. See also Semantic Pointer Architecture
reduction/reductionism, 29, 41, 120, 130, 140–44
reflection, in phenomenology, 9–10

reflective equilibrium, 183
relatedness
  aesthetics, relation to, 255
  basic income as increasing, 202
  constraint satisfaction related to, 218
  happiness and, 193
  needs sufficiency, 187
  overview, 160, 161
  and sense of meaning, 216–17, 219
relations
  existence of, 94–96
  and psychological vital needs, 161
  and reality of space and time, 109
relativism, 147, 156, 184–85
relativity theory, 108, 109
reliability of evidence, 75
reliable coherentism, 19–20, 61, 76, 179. See also epistemology
religion
  evil, explaining, 174–75
  explanation of belief in, 100–1
  free will, 270
  meaning of death, 223–24
  meaning of life, 214
  morality and, 147
  philosophy and, 5
  purpose of art, 232
  supernatural free will, 264–65
repeatability, as feature of evidence, 75
representations. See also Semantic Pointer Architecture
  causality in world, 136–37
  neural, and consciousness, 50–52
  semantic pointers, 31–36, 37
resilience, 216
responsibility for actions, 270–71
reverse empathy, in art, 241
revision, and deductive account of reduction, 141
right. See morality
rights, relation to vital needs, 187
robustness, as feature of evidence, 75
Roots of Empathy, 199–200
Rothko, Mark, 241
rules for philosophical life, 291

sadness
  in music, 248, 250–51
  in painting, 239–40
Saturn Devouring his Son (Goya), 239

science
- causality, 134–35
- deductive explanations in, 124
- eliminative explanation, 128–30
- mechanistic explanation, 126, 127
- narrative explanation in, 122
- natural philosophy and, 2–3, 17, 21–23
- postmodernism and, 10
- reality of time in, 109
- reductionism, 140–44

scientific realism
- existence, 93, 94–95, 97–99
- overview, 20, 115

selection of ideas, cultural, 66

self, multilevel mechanism theory of, 62–63, 142

self-evident truths, 48

self-interest, 165

Semantic Pointer Architecture (SPA)
- aesthetics, 228, 256–57
- ambitious free will, 265–66
- animal minds, 287–88
- caring about needs of others, 166–67
- collaboration, 83–84
- communication, 212–13
- compared to alternatives, 40–42
- concepts, 36–37
- conceptual change, 87
- conceptual combination, 68
- consciousness, 50–53, 55
- consensus, 84–85
- creativity in music, 252–53
- creativity in painting, 242–45
- emotions, 38–39, 155
- ethics, 178
- explanation as social process, 131
- group minds, 112
- inference and coherence, 39
- intuitions, 78
- knowledge, 62–63, 64, 88–89
- mathematics, 275–78, 279–80
- meaning in language, 208, 211, 212–13, 224
- mind, theory of, 31–39, 54–55
- music, emotions in, 245–52
- painting, emotions in, 233–35, 238–41
- philosophical objections to, 42–46
- semantic pointers, defining, 31–36
- simulation mode of empathy, 169
- social cognitivism, 12

social epistemology, 82
social transmission of knowledge, 70
truth, 105
values, 151–52, 177
vital needs, 162

sensory experience
- concept learning based on, 67
- in empiricist justification, 72

sensory-motor character of causality, 133, 135–36

sensory-motor information
- mathematical knowledge and, 275, 277, 279
- in semantic pointers, 35
- and sense of space and time, 106–7

sentences, meaning of, 211, 224

sex
- in music, 249
- in painting, 240

shocking art, 240, 249
signatures of consciousness, 49–50
simulation mode of empathy, 169, 170, 254
simulations, probability, 81
skepticism, 59, 60, 72, 214

social aspects
- aesthetics, 230, 232, 237–38, 245
- of explanation, 130–31
- of meaning in language, 212–13
- of meaning of life, 219–20

social causes of evil, 176
social change, just, 197–200
social cognitivism, 12–13, 261–62. *See also* natural philosophy; *specific areas of philosophy*

social constructivism
- as challenge to scientific realism, 98–99
- existence, 94, 95, 103–4
- mathematics, 274
- postmodernism, 10
- space and time, 109–10
- truth, 94, 105

social contracts, 165
social democracy, 184, 194–95
social dependence, history of in humans, 164–65

social epistemology
- collaboration, 83–84
- communication, 82–83
- consensus, 84–85
- overview, 82

social facts, existence of, 110, 113–14
social interaction
  group minds, 112
  theories of, 3
socialism, 185, 195–96
social isolation, 161–62
social justice. *See* justice
social mechanisms. *See also* Semantic Pointer Architecture
  aesthetics, 237–38, 245
  mathematics, 285
  in natural philosophy, 262–63
  relevant to mind, 26–27, 29, 46
  role in meaning, 221–22
  social change and, 127
social norms, 199
social process, meaning as, 35
social sciences
  deductive explanations, 124
  eliminative explanation, 129–30
  mechanistic explanation, 127
  narrative explanation in, 122
  natural philosophy and, 2–3, 17, 21–23
  reductionism, 140–44
social transmission of knowledge, 60, 69–70, 88–89, 101
society
  causality in, 138
  mathematics in, 285–86
  meaning of worlds as connected to, 212–13
  reality of, 110–14
  space and time in, 109–10
sociopathy, 164, 175
sorrow
  in music, 248, 250–51
  in painting, 239–40
souls
  existence of, 99–101
  supernatural free will, 264–65
SPA. *See* Semantic Pointer Architecture
space, reality of, 106–10
spirits, existence of, 99–101
standard examples. *See* exemplars, in three-analysis
Starry Night, The (van Gogh), 233–34, 235, 242
statistical explanations, 125
stories, explanation through, 122–23
strong emergence, 144
structuralism, 274

subjective probabilities, 79–81
subjectivism, 15
sufficient reason, principle of, 108
supernatural
  existence of, 99–101
  in explanations of consciousness, 53–54
  in explanations of meaning, 209
  meaning of life, 214
supernatural free will, 264–65
supervenience, reduction of, 143–44
surprise
  in music, 249
  in painting, 240
surveys, in experimental philosophy, 18
symbolic approach to cognitive science, 43, 44
sympathy, 155
synchrony, neural, 51
syntax, 39

techniques, creativity in artistic, 244
technology
  machine minds, 288–89
  scientific realism and, 99
teleology, 290
theorems, in mathematics, 278–79
theoretical entities, 97, 111
theories, development of, 18, 19
Theory of Everything, 283
Thinking, Fast and Slow (Kahneman), 76–77
thought experiments
  in analytic versus natural philosophy, 8
  in ethics, 148
  in experimental philosophy, 18
  in justification of knowledge, 71–72, 77
  related to justice, 186
thoughts, in analytic versus natural philosophy, 9
three-analysis
  of art, 230–32
  of causality, 111, 132–35
  of concepts, 37
  conceptual analysis, 7
  of conceptual change, 87
  of explanation, 120–22
  of injustice, 203, 204
  of justice, 182, 203, 204
  of knowledge, 63–65
  of number concept, 276–77
  of philosophy, 3–5

three-analysis (Cont.)
  of semantic pointers, 34
  of values, 150–51
  of vital needs, 158–59
time, reality of, 106–10
transbodiment
  of knowledge, 64–65
  of language, 211
  mathematical concepts, 277–78
  overview, 45
  recursive binding, 45, 51, 167, 234, 287–88
transmission
  of cultural ideas, 66–67
  social, of knowledge, 60, 69–70
trust, in collaboration, 84
truth(s)
  a priori, 6, 71
  in analytic versus natural philosophy, 8–9
  approximate correspondence theory, 105–6
  causal explanations and, 137–38
  coherence theory of, 105–6
  correspondence theory of, 104, 105–6
  in definitions of knowledge, 61, 63
  justification of knowledge, 63, 71–72
  mathematical, 280–81
  in metaphysics, 94, 104–6
  necessary, 6, 8–9, 101, 281
  pure-reason approach, 6–7
  self-evident, 48
tychism, 215
typical features, in three-analysis
  of art, 231
  of causality, 111, 132, 133
  of explanation, 120, 121
  of injustice, 204
  of justice, 204
  of knowledge, 64
  of number concept, 276, 277
  of philosophy, 3–4
  of semantic pointers, 34
  of values, 151
  of vital needs, 158–59

ugliness
  in mathematics, 282
  in music, 251–52
  objectivity in judgments of, 236
  in painting, 238–39
uncertainty, coherence and, 17

United Nations Human Development Index, 193
usefulness, in justification of knowledge, 71
utilitarianism, 149, 151, 171, 188

value maps, 153–54
values
  conflicting, and ethical coherence, 171–74
  as interconnected with facts, 14–15
  just social change, 198
  in natural philosophy, 262
  objective, 156–57
  overview, 150
  as semantic pointers, 151–52, 177
  social groups as counterbalance to individual, 85
  social transmission of knowledge, 69–70
  three-analysis, 150–51
  value maps, 153–54
van Gogh, Vincent, 233–34, 235, 242, 244
verbal representations, in music, 252
verbal rules, in conception of causality, 133, 136
vital needs. *See also* needs-based consequentialism; needs-sufficiency theory of justice
  aesthetics, relation to, 255
  animal minds, 288
  defined, 157
  versus instrumental needs, 157, 162–63
  meaning of life, relation to, 214, 216–22, 225
  as measure for morality and justice, 162–63
  psychological needs as, 159–61
  relation to rights, 187
  three-analysis of, 158–59
voting
  conflicting needs and ethical coherence, 171, 172, 174
  democracy pros and cons, 192–93
  as ethical and emotional, 153–54

wants, versus vital needs, 157, 159, 162–63
weak emergence, 144
well-being
  relation to caring for others, 167–68
  relation to government policies, 195
will. *See* free will
Woman III (de Kooning), 238–39
women, discrimination against, 191

words, meaning of, 210–13, 224
word-to-word meaning, 211, 212
word-to-world meaning, 211, 212, 224–25, 284
work, and sense of meaning, 217
workspace broadcasting, neuronal, 49–50
world. *See also* metaphysics
    causality in, 136–38
    as independent of mind and society, 102
    mathematics in, 282–85
    meaning of words as connected to, 210–11
    mechanisms in, 127
    social constructivism, 103–4
    space and time in, 108–9
World Happiness Report, 193
wrong. *See* morality

Yankovic, Weird Al, 249
"Yesterday" (Beatles), 246–47